T0192107

# Communications in Computer and Information Science 2059

## Rationale

The CCIS series is devoted to the publication of proceedings of computer science conferences. Its aim is to efficiently disseminate original research results in informatics in printed and electronic form. While the focus is on publication of peer-reviewed full papers presenting mature work, inclusion of reviewed short papers reporting on work in progress is welcome, too. Besides globally relevant meetings with internationally representative program committees guaranteeing a strict peer-reviewing and paper selection process, conferences run by societies or of high regional or national relevance are also considered for publication.

## Topics

The topical scope of CCIS spans the entire spectrum of informatics ranging from foundational topics in the theory of computing to information and communications science and technology and a broad variety of interdisciplinary application fields.

## Information for Volume Editors and Authors

Publication in CCIS is free of charge. No royalties are paid, however, we offer registered conference participants temporary free access to the online version of the conference proceedings on SpringerLink (http://link.springer.com) by means of an http referrer from the conference website and/or a number of complimentary printed copies, as specified in the official acceptance email of the event.

CCIS proceedings can be published in time for distribution at conferences or as postproceedings, and delivered in the form of printed books and/or electronically as USBs and/or e-content licenses for accessing proceedings at SpringerLink. Furthermore, CCIS proceedings are included in the CCIS electronic book series hosted in the SpringerLink digital library at http://link.springer.com/bookseries/7899. Conferences publishing in CCIS are allowed to use Online Conference Service (OCS) for managing the whole proceedings lifecycle (from submission and reviewing to preparing for publication) free of charge.

## Publication process

The language of publication is exclusively English. Authors publishing in CCIS have to sign the Springer CCIS copyright transfer form, however, they are free to use their material published in CCIS for substantially changed, more elaborate subsequent publications elsewhere. For the preparation of the camera-ready papers/files, authors have to strictly adhere to the Springer CCIS Authors' Instructions and are strongly encouraged to use the CCIS LaTeX style files or templates.

## Abstracting/Indexing

CCIS is abstracted/indexed in DBLP, Google Scholar, EI-Compendex, Mathematical Reviews, SCImago, Scopus. CCIS volumes are also submitted for the inclusion in ISI Proceedings.

## How to start

To start the evaluation of your proposal for inclusion in the CCIS series, please send an e-mail to ccis@springer.com.

Hai Jin · Yi Pan · Jianfeng Lu
Editors

# Data Science and Information Security

First International Artificial Intelligence Conference, IAIC 2023
Nanjing, China, November 25–27, 2023
Revised Selected Papers, Part II

 Springer

*Editors*
Hai Jin
Huazhong University of Science
and Technology
Wuhan, Hubei, China

Yi Pan
Chinese Academy of Science
Shenzhen, China

Jianfeng Lu 🆔
Nanjing University of Science
and Technology
Nanjing, China

ISSN 1865-0929          ISSN 1865-0937 (electronic)
Communications in Computer and Information Science
ISBN 978-981-97-1279-3          ISBN 978-981-97-1280-9 (eBook)
https://doi.org/10.1007/978-981-97-1280-9

This Springer imprint is published by the registered company Springer Nature Singapore Pte Ltd.
The registered company address is: 152 Beach Road, #21-01/04 Gateway East, Singapore 189721, Singapore

Paper in this product is recyclable.

# Preface

These conference proceedings are a collection of the papers accepted by IAIC 2023 – the 2023 International Artificial Intelligence Conference, held on November 25–27, 2023 in Nanjing, China.

The conference was organized by Nanjing University of Science & Technology, and Tech Science Press. IAIC 2023 aimed to provide a platform for the exchange of ideas and the discussion of recent developments in artificial intelligence. The conference showcased a diverse range of topics, including machine learning, natural language processing, computer vision, robotics, and ethical considerations in AI, among others.

The reviewing process for IAIC 2023 was meticulous and thorough. We received an impressive number of qualified submissions, reflecting the growing interest and engagement in the field of artificial intelligence. The number of the final accepted papers for publication is 86. The high standard set for acceptance resulted in a competitive selection, with a commendable acceptance rate that attests to the caliber of the contributions presented at the conference.

We extend our gratitude to the authors for their outstanding contributions and dedication, as well as to the reviewers for ensuring the selection of high-quality papers, which made these conference proceedings possible.

We also would like to thank the organizers and sponsors whose generous support made IAIC 2023 possible. Their commitment to advancing the field of artificial intelligence is commendable, and we acknowledge their contributions with sincere appreciation. The logos of our esteemed sponsors can be found on the following pages.

We hope this volume serves as a valuable resource for researchers, academics, and practitioners, contributing to the ongoing dialogue that propels the field forward.

December 2023                                   IAIC 2023 Organizing Committee

# Organization

## General Chairs

Hai Jin      Huazhong University of Science and Technology, China

Yi Pan      Shenzhen Institute of Advanced Technology, Chinese Academy of Sciences, China

Jianfeng Lu      Nanjing University of Science and Technology, China

## Technical Program Chairs

Yingtao Jiang      University of Nevada Las Vegas, USA

Q. M. Jonathan Wu      University of Windsor, Canada

## Technical Program Committee Members

Yudong Zhang      University of Leicester, UK

Shuwen Chen      Jiangsu Second Normal University, China

Xiaoyan Zhao      Nanjing Institute of Technology, China

Wentao Li      Southwest University, China

Chao Zhang      Shanxi University, China

Huiyan Zhang      Chongqing Technology and Business University, China

Tao Zhan      Southwest University, China

Muhammad Attique Khan      HITEC University, Pakistan

Tallha Akram      COMSATS University Islamabad, Pakistan

Zhewei Liang      Mayo Clinic, USA

Yi Ding      University of Electronic Science and Technology of China, China

Xianhua Niu      Xihua University, China

Yingjie Zhou      Sichuan University, China

Dajiang Chen      University of Electronic Science and Technology of China, China

Fang Liu      Hunan University, China

| | |
|---|---|
| Zhiping Cai | National University of Defense Technology, China |
| Zongshuai Zhang | Chinese Academy of Sciences, China |
| Daniel Xiapu Luo | Hong Kong Polytechnic University, China |
| Jieren Cheng | Hainan University, China |
| Xinwang Liu | National University of Defense Technology, China |
| Qiang Liu | National University of Defense Technology, China |
| Xiangyang (Alex X.) Liu | Michigan State University, USA |
| Wei Fang | Nanjing University of Information Science and Technology, China |
| Victor S. Sheng | Texas Tech University, USA |
| Jinwei Wang | Nanjing University of Information Science and Technology, China |
| Leiming Yan | Nanjing University of Information Science and Technology, China |
| Jian Su | Nanjing University of Information Science and Technology, China |
| Zheng-guo Sheng | University of Sussex, UK |
| Si-guang Chen | Nanjing University of Posts and Telecommunications, China |
| Yanchao Zhao | Nanjing University of Aeronautics and Astronautics, China |
| Hao Han | Nanjing University of Aeronautics and Astronautics, China |
| Hao Wang | Ratidar Technologies LLC, China |

## Publication Chair

| | |
|---|---|
| Zhihua Xia | Jinan University, China |

## Publicity Chairs

| | |
|---|---|
| Lei Chen | Shandong University, China |
| Yuan Tian | Nanjing Institute of Technology, China |

# Organization Committee Members

| | |
|---|---|
| Laith Abualigah | Al Al-Bayt University, Jordan |
| Muhammad Azeem Akbar | LUT University, Finland |
| Farman Ali | Sejong University, South Korea |
| Shuwen Chen | Jiangsu Second Normal University, China |
| Chien-Ming Chen | Nanjing University of Information Science and Technology, China |
| Dajiang Chen | University of Electronic Science and Technology of China, China |
| Ting Chen | University of Electronic Science and Technology of China, China |
| Ke Feng | National University of Singapore, Singapore |
| Honghao Gao | Shanghai University, China |
| Xiaozhi Gao | University of Eastern Finland, Finland |
| Ke Gu | Changsha University of Science and Technology, China |
| Mohammad Kamrul Hasan | Universiti Kebangsaan Malaysia, Malaysia |
| Celestine Iwendi | University of Bolton, UK |
| Heming Jia | Sanming University, China |
| Deming Lei | Wuhan University of Technology, China |
| Peng Li | Nanjing University of Aeronautics and Astronautics, China |
| Huchang Liao | Sichuan University, China |
| Mingwei Lin | Fujian Normal University, China |
| Anfeng Liu | Central South University, China |
| Xiaodong Liu | Edinburgh Napier University, UK |
| Niancheng Long | Shanghai Jiao Tong University, China |
| Jeng-Shyang Pan | Shandong University of Science and Technology, China |
| Danilo Pelusi | University of Teramo, Italy |
| Kewei Sha | University of Houston, USA |
| Shigen Shen | Huzhou University, China |
| Xiangbo Shu | Nanjing University of Science and Technology, China |
| Adam Slowik | Koszalin University of Technology, Poland |
| Jin Wang | Changsha University of Science and Technology, China |
| Kun Wang | Fudan University, China |
| Changyan Yi | Nanjing University of Aeronautics and Astronautics, China |
| Yudong Zhang | University of Leicester, UK |
| Chengwen Zhong | Northwestern Polytechnic University, China |

Junlong Zhou                    Nanjing University of Science and Technology,
                                   China
Xiaobo Zhou                     Tianjin University, China
Fa Zhu                          Nanjing Forestry University, China

# Contents – Part II

# Knowledge Graph Reasoning with Bidirectional Relation-Guided Graph Attention Network

Rui Wang[✉] ⓘ and Yongli Wang ⓘ

NanJing University of Science and Technology, NanJing 210000, China
rui@njust.edu.cn

**Abstract.** Graph convolutional neural networks (GCN) have demonstrated superior performance in graph data modeling and have been widely used in knowledge inference research in recent years. However, knowledge graph is a heterogeneous multi-relational connected graph with complex interactions between neighboring nodes, and most existing GCN-based methods aggregate neighborhood information with the same importance, which leads to the loss of important semantics in the context. In addition, most GAT-based methods consider the neighborhood as a whole and ignore the direction information of the relationship. To this end, we propose a bidirectional relation-guided graph attention network (BR-GAT), which utilizes a bidirectional self-attention mechanism to compute the importance of neighboring nodes, computes the importance of the neighborhood on the representation of relations through a relation-specific mechanism, and finally fuses the joint propagation of neighboring information to update the representations of entities and relations. We conduct link prediction experiments on three standard datasets, and the results demonstrate that BR-GAT outperforms several state-of-the-art models.

**Keywords:** Knowledge graph · Attention mechanism · Graph convolutional network

## 1 Introduction

Knowledge graphs represent entities and relationships in the real world through graph structures, realizing the combination of artificial intelligence and human knowledge, and providing new ideas for understanding and processing complex information for research on artificial intelligence applications such as knowledge quiz [1, 2], information retrieval [3, 4] and recommendation [5–7]. However, knowledge graphs are often incomplete and may also contain some inaccurate information. Real-world information data contains multifaceted information, which determines the heterogeneity and semantic richness of knowledge graphs. Therefore, how to use the heterogeneity and rich semantic information of the knowledge graph (KG), and complete the complementation of the KG through reasoning methods is a very important and difficult task.

Currently, the graph neural network approach is the mainstream research method, because graph neural networks can consider both semantic and structural information, with better interpretability and stronger reasoning ability, and the based-GNN models have achieved superior performance in knowledge reasoning tasks. However, KGs are highly diverse and include a large amount of semantic information, designing GNNs for heterogeneous graphs is extremely difficult. (1) Neglecting the heterogeneous multi-relational connectivity property of knowledge graphs. As shown in Fig. 1, relational connections in knowledge graphs can be roughly categorized into three kinds: (i) the same relations connecting different entities; (ii) different relations connecting different entities; and (iii) the same entities connected by different relations. The regional data cannot be adequately grabbed by previous GCN-based models, which aggregate neighborhood information through a static attention mechanism. (2) The direction of relations also has some impact on message aggregation. The graph attention-based approaches [8–12] introduces the attention mechanism into GCNs to selectively aggregate messages, and although the performance is greatly improved, it ignores the influence of relationship information in different directions, and few graph convolutional neural network-based research works have taken the direction information of the relationship into account.

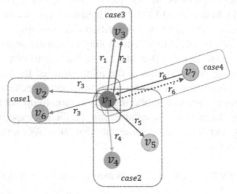

**Fig. 1.** Examples of heterogeneous multi-relationship graph connections, where $v_1 \sim v_7$ represent different entities and $r_1 \sim r_6$ are different relationships. Different colors represent different relationships, and the figure above shows several common relational connections of kg. The inward relationship $r_6$ of the central entity $v_1$ in the blue dashed box (case 4), $r_6$ inversely transforms into the inverse relation $r_6'$. (Color figure online)

Therefore, we propose a bidirectional relation-guided GAT (BR-GAT) for knowledge reasoning, which can selectively aggregate key neighborhood information based on the interactions between nodes and the direction of the relations, when it comes to intricate, diverse, multi-relational relationships. In particular, we postulate that the neighborhood and the relationship are intimately connected to the attention that the central node pays to it. As shown in Fig. 2. Our specific work includes:

- Our new framework, BR-GAT, is a GAT-based encoder that can adapt to various heterogeneous multi-relationship connectivity scenarios. It integrates a bidirectional self-attention mechanism into the entity representation learning process by combining relationship directions to determine the relative importance of distinct neighboring nodes.

- Designing a relation-specific attention mechanism to assess the importance of neighbors for relation representation learning encapsulates neighborhood details into the relational embedded representation.
- Our experiments on popular datasets demonstrate that BR-GAT achieves superior performance compared to other models in link prediction tasks.

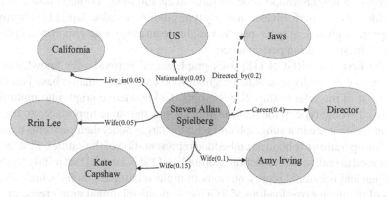

**Fig. 2.** Example of BR-GAT used for KG. The solid relations in the figure represent the original relations, the dashed relations are the reverse relations, and the numbers represent the attention of the central node to the neighboring nodes.

## 2 Related Work

### 2.1 Neural Network Based Approach

The neural network-based approach utilizes multiple nonlinear representation layers combined to extract knowledge graphs deep features of the spectrum for representation and then carry out knowledge inference. ConvE [13] is a model based on convolutional neural networks. It aims to mine local relationships between entries of different dimensions in each entity embedding or relationship embedding, however, ignores global relationships between entries of the same dimension in ternary embeddings. InteractE [14] extends ConvE by focusing on feature ordering of reshaped embeddings of head entities and relations.

### 2.2 Methods Based on Graph Neural Networks

With the increasing scale of knowledge graph and deepening of application scenarios, traditional representation learning models are inefficient and cannot meet the demands of large-scale KG inference tasks. Traditional neural network methods mainly focus on semantic information acquisition, ignoring the KG's structural details, and have low interpretability. Given that GNNs are able to modify graph structures and extract contextual data, which has attracted extensive attention from the academic community in recent years. HGNN-AC [15] first obtains the topological embedding of nodes using the exist-ing HIN-Embedding method, and uses the topological relationships between the nodes as a guide to complete the attributes of unattributed nodes by weighted aggregation of nodes with attributes. Zhao et al. [16] suggested a multi-relationship graph attention network model, which captures the neighborhood information by learning a semantic subgraph composed of three of a heterogeneous relational graph to obtain embeddings and aggregating neighborhood node features to capture neighborhood information, the model achieved good performance in entity type prediction tasks. AttEt [17] improves the entity type through adding type-specific weights to an entity's neighborhood knowledge to improve the speculation performance.

For the first time, RGCN [18] uses graph neural networks for knowledge graph embedding to weightedly aggregate neighborhood entity information based on the type and direction of the relationship. SACN splits the knowledge graph into multiple sub-graphs based on the type of the relationship, assigns the same importance to the neigh-boring entities that are in a subgraph and with the target node, then performs a weighted aggregation operation to obtain an embedded representation of the entities. Finally, Conv-TransE is used to realize the prediction. CompGCN [19] is a model that jointly learns node embeddings and relational representations in multi-relational graphs, which solves the problem of parameter overloading of RGCNs in multi-relational graph representations.

To further accurately aggregate neighborhood information, graph attention networks are also introduced into knowledge graph embedding. DisenKGAT [20] decouples entity different thematic semantics for decoupling, predicting the corresponding thematicized semantic representations according to a given query, realizing dynamic representations in different scenarios, and effectively solving the problem of complex multi-semantic knowledge graph complementation. DRR-GAT [21] learns different embeddings from different relations by introducing graph attention networks, dynamically forming local embeddings of relations and combining them with global embeddings of relations to form the final embedding of the relationship, develops a new message passing mech-anism to capture the embedding of entities, and the model obtains good performance. D-AEN [22] takes direction information into account for measuring the importance of neighboring entities, and also fully considers the influence of neighboring entities on the representation of the relationship, and demonstrates that the entity embedding and the relationship embedding can be optimized for each other.

## 3   Methodology

The model of this paper is the encoder-decoder framework, the structural design of the encoder is shown in Fig. 3. The encoder is used to generate an embedded of entities and relations containing the neighborhood topology and global structure, and can be used to aggregate n-hop neighborhood information by stacking multiple encoders. MRGAT23 proves that the assumption that different weights cannot be assigned to neighboring entities is not reasonable. This is because different semantic surroundings in the KG have different impacts on the central entity. In addition, this paper argues that the direction information of relationships also has an impact on the aggregation of information of central entities. Therefore, it considers the inverse relationship in the KG., and firstly, the neighboring entities of the central entity are divided into two groups (original and reverse relationship) according to the two directions of the edges, and then the attention scores of the central entity and its neighboring entities are calculated separately, and the neighboring information is aggregated according to the attention scores to learn the embedding of the relationship and the entity together. Finally, the generated embedded representations of relations and nodes are input to the decoder.

Define a heterogeneous graph $G = (V, R)$ made up of a set of node sets V ($v \in V$) and a set of edge sets R ($r \in R$). Referring to the idea of Vashishth, Sanyal, Nitin and Talukdar [14], the incoming relation of the central entity is converted into an inverse relation. This is shown in Case 4 in Fig. 1. Specifically, for the neighboring entities of $v_i$, the edges to the central entity are reversed. Therefore, the reverse relation is defined as $R_{rev}$, $R_{rev} = \{r_k^{-1} | r_k \in R\}$, and the reverse triad is $T_{rev}$, $T_{rev} = \{(v_j, r_k^{-1}, v_i) | (v_i, r_k, v_j) \in T\}$, which will be added to the representation learning. Then, the relations connected to the center entity are classified into two types: primitive type and reverse type. In addition, define $R = R_{out} \cup R_{rev}$ and $T = T_{out} \cup T_{rev}$, with $R_{out}$ out and $T_{out}$ denoting the outgoing edges of the central entity and the triples from the central entity to the neighboring entities, respectively. For entity $v_i$, define the set of neighbor entities as $N_{v_i}$.

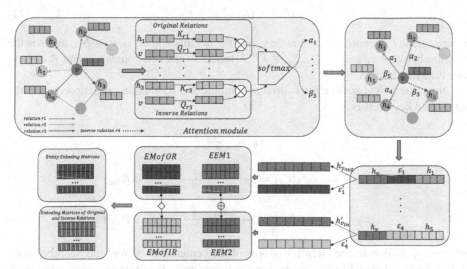

**Fig. 3.** Overview of the model encoder. Different colored edges in the figure represent different types of relations, solid line represents the original relation and dashed line represents the reversed relation. $K_{r1}$ and $Q_{r1}$ denote the query matrix and key matrix associated with the original relation $r_1$, and $K_{r3}$ and $Q_{r3}$ denote the query matrix and key matrix associated with the reversed relation $r_3$, respectively. Firstly, the attention module is used to calculate the attention scores $\alpha_1$, $\alpha_2$, $\beta_3$, $\alpha_4$, $\beta_5$ of the central entity to the neighboring entities $h_1 \sim h_5$ according to the pointing types of the relations, respectively. $h_v$, $\varepsilon_1$, $h_1$ denote the initial information of the central entity $v$, the relation $r_1$, and the neighboring entity $v_1$, respectively, and the neighboring information is aggregated by using the multiple attention mechanism, $EEM1$, $EEM2$, $EMogOR$, $EMofIR$ denote the embedded matrices in the two cases of original and inverse relations, respectively, and finally the learning of the common representation of entities and relations is obtained by $\diamond$ joining and $\oplus$ summing.

## 3.1 Encoder

### Self-Attention Layer

The central node in the KG gives distinct consideration to its neighbor nodes, which is related to the neighbor nodes themselves as well as the edges between them. Based on node-to-node interactions, we compute neighbors' attention scores using the self-attention mechanism. In the first place, for every relationship, there are two defined matrices: a query matrix that projects the center node's vector $v_i$ as a query vector, and a key matrix that maps the center node's neighbor nodes ($v_j \in N_{vi}$) to the key space. The dot product between the query and key vectors is then computed according to the two edge pointing types (original or inverse), respectively, to obtain the attention scores between pairs of nodes. Since the KG is diversified, a single projection of transfromer cannot cope with the situation of dealing with multiple nodes and multiple edges in a heterogeneous graph.

Define $H \in R^{N \times D}$ to denote the initial embedding of the node, and $H' \in R^{N \times D}$ to denote the node representation at the output, where $N$ is the number of nodes, and $D$ indicates the input vector's dimension. The node pair representations are first projected into the query and key vectors as follows:

$$q_i^r = h_i Q_r, \, k_j^r = h_j K_r \tag{1}$$

where $h_i$ and $h_j$ denote the core node $v_i$ and the neighbor node $v_j$, $v_j \in N_{vi}$, and $r$ is the relation between $v_i$ and $v_j$. $Q_r, K_r \in R^{D \times P}$ based on the relationship $r$, and $P$ is these two matrices' dimension.

Next, the SoftMax function is used to compute and normalize the dot product of the query and key vectors:

$$e_{ij} = \frac{q_i^r \cdot k_j^r}{\sqrt{P}} \tag{2}$$

$$\alpha_{ij} = softmax\left(e_{ij}\right) = \frac{\exp(e_{ij})}{\sum_{(r_r, v_k) \in N_{vi}} \exp(e_{ik})}, \, (r \in R_{out}) \tag{3}$$

$$\beta_{ij} = softmax\left(e_{ij}\right) = \frac{\exp(e_{ij})}{\sum_{(r_r, v_k) \in N_{vi}} \exp(e_{ik})}, \, (r \in R_{rev}) \tag{4}$$

where $\alpha_{ij}$, $\beta_{ij}$ denote the attention scores of the center node $v_i$ and the neighbor node $v_j$ in the two relationship directions. For the center node, the attention score varies with different connections of the heterogeneous multi-relationship and different relationship directions.

### Learning Representation of Entities

We learn the representation of entities by aggregating the neighborhood-related triples of central entity, and encapsulate the neighborhood multi-relationship type information and direction information to measure the importance of neighborhoods using a two-way self-attention mechanism. According to Nathani et al.10, the representation of ternary $t_{ij}^k = \left(v_i, r_k, v_j\right) \in T$ is learned as follows:

$$tr_{ijk} = W_1 \cdot Concat\left(h_i, h_j, \varepsilon_k\right) \tag{5}$$

where $h_i, h_j \in R^D$, $\varepsilon_k \in R^D$ are the head, tail entities' and the relation' $(v_i, v_j, r_k)$ original embedding vectors, $W_1 \in R^{D \times 3D}$ is a linear transformation matrix.

The portrayal of entities $v_i$ is reloaded by aggregating the representations of the related ternary groups, according to the above setup as shown below:

$$h_{i_{out}} = f\left(W_O \sum_{(r_k,v_j)\in N_{vi}} \alpha_{ij} tr_{ijk}\right), r_k \in R_{out} \tag{6}$$

$$h_{i_{in}} = f\left(W_I \sum_{(r_k,v_j)\in N_{vi}} \beta_{ij} tr_{ijk}\right), r_k \in R_{rev} \tag{7}$$

$$h'_i = h_{i_{out}} + h_{i_{in}} + h_i W_e \tag{8}$$

where $h_{i\_out}$ and $h_{i\_in}$ denote the total of the inverse and original relations' ternary representations. $f$ is a nonlinear activation function. $W_O \in R^{D' \times D}$, $W_I \in R^{D' \times D}$ and $W_e \in R^{D' \times D}$ are learnable transformation matrices. To ensure that the original data from the core node is not lost, the transformed initial representation is merged into the final representation.

During training, multi-head attention stabilizes the convergence of parameters, and can focus on different subspaces to obtain richer feature information. Therefore, m independent attention heads are used to learn entity representations, and finally they are stitched together to generate the final representation:

$$h'_i = Concat\left(h'_{i1}, h'_{i2}, \ldots, h'_{im}\right) W_m \tag{9}$$

where $W_m \in R^{mD' \times D'}$ is converting the spliced vectors to their original dimensions.

**Learning Representations of Relationships**
Similar to the relational learning process of 22, for a relation $r_k \in R$, the embedding of the relation is revised by selectively combining the linked triples' representations. The attention score of the ternary $t^k_{ij} = \left(v_i, r_k, v_j\right) \in T$ for relation $r_k$ is computed as follows:

$$\gamma_k = softmax(e_{ij}) = \frac{\exp(e_{ij})}{\sum_{(v_m,v_n)\in N_{r_k}} \exp(e_{mn})} \tag{10}$$

where $N_{r_k} = \{(v_i, v_j)|(v_i, r_k, v_j) \in T\}$ denotes the associated triad of relation $r_k$.
The new relation is represented as follows:

$$\varepsilon_k = f_r\left(W_R \sum_{(v_i,v_j)\in N_{r_k}} \gamma_k tr_{ijk}\right) \tag{11}$$

where, $f_r$ denotes the nonlinear activation function of the relation and $W_R \in R^{D' \times D}$ is a graph convolution kernel for the relation.

In addition, the original representation of the relation is also included in the final representation of the relation. Additionally, to be able to capture more information throughout the representation learning process of the relation, a multi-head attention technique is used.

$$\varepsilon'_k = \varepsilon'_k + W_r \varepsilon_k \tag{12}$$

$$\varepsilon'_k = Concat\left(\varepsilon'_{i1}, \varepsilon'_{i2}, \ldots, \varepsilon'_{im}\right)W_m \tag{13}$$

## 3.2 Decoder

The decoder predicts the validity of a given triad by using a score system, which can be any of the embedded models TransE, DistMult, ConvE, etc., In this paper, ConvE is used as the decoder. Its definition of the ternary scoring function is as follows:

$$f\left(v_i, r_k, v_j\right) = g\left(vec\left(g\left(\left[\overline{h_i}; \overline{\varepsilon_k}\right]*\omega\right)\right)W\right)h_j \tag{14}$$

where $g$ is the mechanism of activation, $\overline{h_i}, \overline{\varepsilon_k} \in R^{D_1 D_2}$ denotes the 2D reshaping of the encoder model outputs of $h_i, \varepsilon_k \in R^{D'}, D' = D_1 D_2$. A 2D convolution process and a collection of convolution kernels are indicated by the symbols $*$ and $\omega$, respectively. A vector can be created from a tensor using the vectorization operation $vec(\cdot)$.

## 4 Experiment

### 4.1 Datasets and Baselines

We chose datasets WN18RR, FB15K-237, and YAGO-10 for our link prediction experiments. Detailed information about these three datasets is shown in Table 1.

**Table 1.** Detailed information of the datasets

| Datasets | Entities | Relations | Edges | | |
|---|---|---|---|---|---|
| | | | Train | Valid | Test |
| MN18RR | 40943 | 11 | 86835 | 3034 | 3134 |
| FB15k-237 | 14541 | 237 | 272115 | 17535 | 20466 |
| YAGO3-10 | 123182 | 37 | 1079040 | 5000 | 5000 |

In our tests, we contrast BR-GAT with other currently used techniques, which we divide into two groups based on whether they are GNN-based methods or not: GNN-based methods and other methods. GNN-based methods are R-GCN, CompGCN, and MRGAT; other methods include TransE, ComplEx, and ConvE.

### 4.2 Evaluation Protocol

According to most of the baseline methods, a linked prediction task is performed for each ternary in the test set to evaluate the model's performance. Firstly, the ternary is rebuilt utilizing the candidate entities in the KG to swap out the head or tail entities. Then the reconstructed ternary is scored by the proposed model and sorted in descending rank based on the results, and finally the rank of the original ternary in the reconstructed ternary is calculated. However, considering that there are a large number of one-to-many, many-to-one, and many-to-many relationships in the KG, the corrupted triples will lead to the degradation of the model's performance, therefore, the corrupted triples should be filtered out of the knowledge graph before sorting, and we adopt the "Filter" setting strategy in TransE. Based on this, we used the mean-reverse rank (MRR) and Hits@1, Hits@3, Hits@10 to evaluate our proposed model. Because the evaluation process of the head and tail entities is consistent, the final results are determined by averaging the ratings of the head and tail entities.

### 4.3 Experiment Setup

We jointly trained encoder and decoder models for the experiments. The experimental hyperparameter settings were chosen as: learning rate [0.01,0.005,0.0003], entity and relation embedding dimensions [100,200,300,400,500], batch size [128,256,512], label smoothing [0.1,0.2,0.3], number of attention headers [1–4], convolution kernel size [3, 5, 7], number of number of filters [100,200,300], and the value of Dropout varies from 0.0 to 0.5.

### 4.4 Results and Analysis

We compare and analyze BR-GAT with other existing knowledge graph embedding models in terms of both the overall performance of the model and the performance results for different relation types, the overall performance comparison of BR-GAT is shown in Tables 2, 3 and 4, and Tables 5 and 6 exhibit a comparison of the performance outcomes for various relation kinds.

**General Results**
The overall results on the datasets WN18RR, FB15k-237, and YAGO3-10 are shown in Table 2, 3 and 4. We have analyzed Table 2, 3 and 4 in depth, and the results are as follows: firstly, when compared to the approaches given in this paper, BR-GAT demonstrates relative advantage in all evaluation criteria on all three datasets. The findings generally demonstrate the efficacy of the encoder suggested in this work and demonstrate knowing the KG's local domain and structural information may significantly enhance link prediction task performance. Secondly, observations reveal that BR-GAT executes significantly superior to other models on the YAGO3-10 dataset. This demonstrates the better generalization ability of our model on large-scale datasets. Finally, compared with ConvE, a knowledge embedding model that also uses ConvE as a decoder, our model shows significant improvement in all metrics. This proves that our encoder is more advanced than ConvE.

In conclusion, our approach clearly outperforms the knowledge embedding models utilized in this paper in every assessment criteria. In contrast to conventional knowledge embedding models: TransE, ConvE, ComplEx, etc., BR-GAT is able to grab the topological structure details in the KG. Moreover, BR-GAT also shows greater performance compared to GNN-based approaches: R-GCN, CompGCN, MRGAT, etc., since the encoder we designed is able to focus more on critical neighborhood characteristics in several situations with intricate and varied linkages.

**Table 2.** Link prediction outcomes of BR-GAT on dataset MN18RR evaluated by MRR and Hits@N

|  | MN18RR | | | |
|--|--------|--|--|--|
|  | Hits@1 | Hits@3 | Hits@10 | MRR |
| TransE | – | – | 0.501 | 0.226 |
| ComplEx | 0.158 | 0.275 | 0.428 | 0.247 |
| ConvE | 0.400 | 0.440 | 0.520 | 0.430 |
| R-GCN | – | – | – | – |
| CompGCN | 0.443 | 0.494 | 0.546 | 0.479 |
| MRGAT | 0.443 | 0.501 | 0.568 | 0.481 |
| **BR-GAT** | **0.450** | 0.499 | **0.581** | **0.490** |

**Table 3.** Link prediction outcomes of BR-GAT on dataset FB15k-237 evaluated by MRR and Hits@N

|  | FB15k-237 | | | |
|--|-----------|--|--|--|
|  | Hits@1 | Hits@3 | Hits@10 | MRR |
| TransE | – | – | 0.465 | 0.294 |
| ComplEx | 0.158 | 0.275 | 0.428 | 0.247 |
| ConvE | 0.237 | 0.356 | 0.501 | 0.325 |
| R-GCN | 0.151 | 0.264 | 0.417 | 0.249 |
| CompGCN | 0.264 | 0.390 | 0.535 | 0.355 |
| MRGAT | 0.266 | 0.386 | 0.542 | 0.358 |
| **BR-GAT** | **0.351** | **0.480** | **0.620** | **0.443** |

## Different Relationship Categories

In order to evaluate BR-GAT's ability to model different types of relationships in heterogeneous graphs, we also conducted comparative experiments. Connections based on relationships in KG, they are categorized into four types: one-to-one, one-to-many, many-to-one, and many-to-many relations. Based on these four categories, the relationships in the dataset FB15k-237 are divided (the dataset contains 7.2% one-to-one relationships, 11.0% one-to-many relationships, 34.2% many-to-one relationships, and 47.6% many-to-many relationships) and link prediction experiments are done, and the results are shown in Table 5, 6 and 7. We took the experimental results from Hits@10 for analysis, in addition, the average results of the head and tail entity prediction scores on Hits@10 were also used for comparison. The results are obtained from Li, Wang et al.9 and MRGAT23.As can be observed from Tables 5 and 6, our model presents better results on all four types of relations, which proves the power of our model in modeling multi-relational heterogeneous graphs. From Table 7, observations reveal that

**Table 4.** Link prediction outcomes of BR-GAT on dataset YAGO3-10 evaluated by MRR and Hits@N.

|  | YAGO3-10 | | | |
| --- | --- | --- | --- | --- |
|  | Hits@1 | Hits@3 | Hits@10 | MRR |
| TransE | – | – | – | – |
| ComplEx | 0.260 | 0.400 | 0.550 | 0.360 |
| ConvE | 0.350 | – | 0.620 | 0.440 |
| R-GCN | – | – | – | – |
| CompGCN | **0.448** | 0.556 | 0.679 | 0.535 |
| MRGAT | 0.439 | 0.561 | 0.698 | 0.552 |
| **BR-GAT** | 0.440 | **0.566** | **0.712** | **0.566** |

BR-GAT performs better in aggregating complex relations, although its performance is not as prominent as TransE in aggregating one-to-one simple relations, which just shows that BR-GAT is more suitable for modeling knowledge graphs with complex structures.

**Table 5.** Link prediction results of BR-GAT on the FB15k-237 dataset by relation categories

| Models /Hits@10 | 1–1 | | | 1-N | | |
| --- | --- | --- | --- | --- | --- | --- |
|  | Head | Tail | Avg | Head | Tail | Avg |
| TransE | 0.537 | 0.521 | 0.529 | 0.573 | 0.052 | 0.312 |
| ConvE | 0.250 | 0.258 | 0.254 | 0.603 | 0.132 | 0.368 |
| BR-GAT | 0.568 | 0.601 | **0.585** | 0.632 | 0.268 | **0.450** |

**Table 6.** Link prediction results of BR-GAT on the FB15k-237 dataset by relation categories

| Models /Hits@10 | N-1 | | | N-N | | |
| --- | --- | --- | --- | --- | --- | --- |
|  | Head | Tail | Avg | Head | Tail | Avg |
| TransE | 0.070 | 0.833 | 0.452 | 0.347 | 0.508 | 0.428 |
| ConvE | 0.147 | 0.865 | 0.506 | 0.426 | 0.581 | 0.504 |
| BR-GAT | 0.570 | 0.882 | **0.726** | 0.563 | 0.654 | **0.609** |

**Table 7.** Hits@N results on the FB15k-237 dataset by relation categories.

|  | 1-1 | 1-N | N-1 | N-N |
|---|---|---|---|---|
|  | Hits@10 | Hits@10 | Hits@10 | Hits@10 |
| TransE | **0.407** | 0.399 | 0.381 | 0.51 |
| ConvE | 0.401 | 0.41 | 0.397 | 0.531 |
| BR-GAT | 0.401 | **0.427** | **0.450** | **0.570** |

**Table 8.** Ablation study results

|  | WN18RR | | | |
|---|---|---|---|---|
|  | Hits@1 | Hits@3 | Hits@10 | MRR |
| RemoveMB-GAT | 0.436 | 0.483 | 0.542 | 0.470 |
| RemoveMR-GAT | 0.443 | 0.500 | 0.561 | 0.484 |
| RemoveBR-GAT | 0.443 | 0.501 | 0.568 | 0.481 |
| BR-GAT | **0.450** | **0.499** | **0.581** | **0.490** |

## 4.5 Hyperparametric Analysis

**Embedding Dimensions**
In order to study the effect of embedded dimensions, we set the dimensions as 100, 200, 300, 400 and 500, and conduct experiments on the datasets MN18RR, FB15k-237. The MRR scores are shown in Fig. 4. From the figure, it can be observed that on both datasets, when the embedding dimension rises, the MRR scores first rise and subsequently fall. Because when the embedding dimension is small, the embedding representation does not adequately preserve the structural details in the KG, and when the number of embedding dimensions exceeds a definite limit, it degrades the model's consistency performance, and the redundant embedding dimensions make the model overfitting the training data, which diminishes the model's ability to generalize.

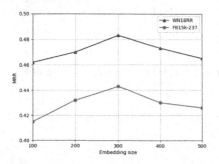

**Fig. 4.** Different embedding dimensions

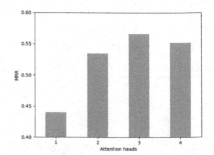

**Fig. 5.** Number of attention heads

**Number of Attention Heads.**
To explore the impact of attention heads on the performance of BR-GAT, we performed experiments on the dataset YAGO3-10. The quantity of focus heads was set to [1–4]. In the Fig. 5, the highest MRR score is achieved when the number of attentional heads is 3. This illustrates that an appropriate number of attentional heads facilitates the aggregation of more neighborhood information to learn the embedding of entities and relations, thus enhancing the robustness of the model, but too many attentional heads can lead to overfitting of the training set, which leads to the model being overly sensitive to the noisy and redundant information in the training data.

## 4.6 Ablation Studie

To confirm the model's validity even further, we introduced three variants of BR-GAT and conducted experiments on the dataset MN18RR. The variants arc as follows: (1) RemoveMB-GAT: removes the self-attention layer from BR-GAT and aggregates neighborhood information with the same importance by considering the whole domain as a whole; (2) RemoveMR-GAT: removes the self-attention layer from BR-GAT without calculating the attention scores between nodes, and the neighborhood information is of the same importance for the learning of entities; (3) RemoveBR-GAT: learn the representation of entities and relations without considering the importance of the direction of relations, and consider the whole neighborhood as a whole. The results are shown in Table 8.

According to Table 8, first of all, the performance of all indicators of the variant RemoveMB-GAT is not as good as the performance of all indicators of the other two variants, which indicates that the mechanism of bi-directional self-attention has a bigger influence on the outcomes of the experiment. It is shown that it is necessary to take into account the influence of interactions between nodes and the influence of the direction of the relationship when evaluating neighborhood information's significance for entity-relationship embedding representation. Second, compared with RemoveMR-GAT and RemoveBR-GAT, the performance of BR-GAT is significantly improved in all the metrics, which demonstrates how significantly learning entity-relationship representations is facilitated by combining neighborhood data with varying levels of attention. The above observations also demonstrate the effectiveness of the individual components of BR-GAT.

## 5 Conclusion

In this paper, we propose a new GNN-based encoder, BR-GAT, that learns the entity and relationship embedding simultaneously by aggregating the neighborhood details and is capable of adapting to various scenarios of heterogeneous multi-relationship connectivity. It does this by computing the attention scores between neighboring nodes through a bidirectional self-attention layer. To improve the model's interpretability, a bi-directional self-attention mechanism is specifically created to monitor the core node's level of attention to its nearby nodes. We conduct a thorough assessment of BR-GAT on a benchmark dataset, proving its superiority over comparable state-of-the-art models.

# References

1. Sorokin, D., Gurevych, I.: Modeling semantics with gated graph neural networks for knowledge base question answering. In: Proceedings of the 27th International Conference on Computational Linguistics, COLING 2018, Santa Fe, New Mexico, USA, 20–26 August 2018, pp. 3306–3317. Association for Computational Linguistics (2018)
2. Huang, X., Zhang, J., Li, D., Li, P.: Knowledge graph embedding based question answering. In: Proceedings of the Twelfth ACM International Conference on Web Search and Data Mining, pp. 105–113 (2019)
3. Chen, L., Tu, D., Lv, M., Chen, G.: A knowledge-based semisupervised hierarchical online topic detection framework. IEEE Trans. Cybern. **49**(9), 3307–3321 (2018)
4. Li, F., Li, Y., Shang, C., Shen, Q.: Fuzzy knowledge-based prediction through weighted rule interpolation. IEEE Trans. Cybern. **50**(10), 4508–4517 (2019)
5. Rosa, R.L., Schwartz, G.M., Ruggiero, W.V., Rodríguez, D.Z.: A knowledge-based recommendation system that includes sentiment analysis and deep learning. IEEE Trans. Industr. Inf. **15**(4), 2124–2135 (2018)
6. Shao, B., Li, X., Bian, G.: A survey of research hotspots and frontier trends of recommendation systems from the perspective of knowledge graph. Exp. Syst. Appl. **165**, 113764 (2021)
7. Wang, H., Zhang, F., Wang, J., Zhao, M., Li, W., Xie, X., et al.: Exploring high-order user preference on the knowledge graph for recommender systems. ACM Trans. Inf. Syst. (TOIS) **37**(3), 1–26 (2019)
8. Li, Z., Liu, H., Zhang, Z., Liu, T., Xiong, N.N.: Learning knowledge graph embedding with heterogeneous relation attention networks. IEEE Trans. Neural Netw. Learn. Syst. **33**(8), 3961–3973 (2021)
9. Li, Q., Wang, D., Feng, S., Niu, C., Zhang, Y.: Global graph attention embedding network for relation prediction in knowledge graphs. IEEE Trans. Neural Netw. Learn. Syst. **33**(11), 6712–6725 (2021)
10. Nathani, D., Chauhan, J., Sharma, C., Kaul, M.: Learning attention-based embeddings for relation prediction in knowledge graphs. In: Proceedings of the 57th annual meeting of the association for computational linguistics, pp. 4710–4723 (2019)
11. Zhang, Z., Zhuang, F., Zhu, H., Shi, Z., Xiong, H., He, Q.: Relational graph neural network with hierarchical attention for knowledge graph completion. In: Proceedings of the AAAI conference on artificial intelligence, vol. 34, pp. 9612–9619 (2020)
12. Zhao, Y., Zhou, H., Xie, R., Zhuang, F., Li, Q., Liu, J.: Incorporating global information in local attention for knowledge representation learning. In Findings of the association for computational linguistics: ACL-IJCNLP 2021, pp. 1341–1351 (2021)
13. Dettmers, T., Minervini, P., Stenetorp, P., Riedel, S.: Convolutional 2d knowledge graph embeddings. In Thirty-Second AAAI Conference on Artificial Intelligence (2018)
14. Vashishth, S., Sanyal, S., Nitin, V., Agrawal, N., Talukdar, P.: Interacte: Improving convolution-based knowledge graph embeddings by increasing feature interactions. In: Proceedings of the AAAI Conference on Artificial Intelligence, vol. 34, pp. 3009–3016 (2020)
15. Jin, D., Huo, C., Liang, C., Yang, L.: Heterogeneous graph neural network via attribute completion. In: Proceedings of the Web Conference 2021, pp. 391–400 (2021)
16. Zhao, Y., Zhou, H., Zhang, A., Xie, R., Li, Q., Zhuang, F.: Connecting embeddings based on multiplex relational graph attention networks for knowledge graph entity typing. IEEE Trans. Knowl. Data Eng. **35**(5), 4608–4620 (2022)
17. Zhuo, J., Zhu, Q., Yue, Y., Zhao, Y., Han, W.: A neighborhood-attention fine-grained entity typing for knowledge graph completion. In: Proceedings of the Fifteenth ACM International Conference on Web Search and Data Mining, pp. 1525–1533 (2022)

18. Schlichtkrull, M., Kipf, T.N., Bloem, P., van den Berg, R., Titov, I., Welling, M.: Modeling relational data with graph convolutional networks. In: Gangemi, A., et al. (ed.) The Semantic Web, ESWC 2018, LNCS, vol. 10843, pp. 593–607. Springer, Cham (2018). https://doi.org/10.1007/978-3-319-93417-4_38
19. Vashishth, S., Sanyal, S., Nitin, V., Talukdar, P.: Composition-based multi-relational graph convolutional networks. In: International Conference on Learning Representations (2020)
20. Wu, J., Shi, W., Cao, X., Chen, J., Lei, W., Zhang, F., et al.: DisenKGAT: knowledge graph embedding with disentangled graph attention network. In: Proceedings of the 30th ACM International Conference on Information & Knowledge Management, pp. 2140–2149 (2021)
21. Zhang, X, Zhang, C, Guo, J, Peng, C, Niu, Z, Wu, X , et al.: Graph attention network with dynamic representation of relations for knowledge graph completion. Expert Syst. Appl. **219**, 119616 (2023)
22. Fang, H, Wang, Y, Tian, Z, Ye, Y.: Learning knowledge graph embedding with a dual-attention embedding network. Expert Syst. Appl. **212**, 118806 (2023)
23. Li, Z, Zhao, Y, Zhang, Y, Zhang, Z.: Multi-relational graph attention networks for knowledge graph completion. Knowl.-Based Syst. **251**, 109262 (2022)

# Research on PCB Defect Detection Using 2D and 3D Segmentation

Lin Hua[1], Kuiyu Li[2], Lunxin Cheng[2], Yifan Chen[2], Dongfu Yin[1], and Fei Richard Yu[1(✉)]

[1] Guangdong Laboratory of Artificial Intelligence and Digital Economy (SZ), Shenzhen, China
{hualin,yindongfu,yufei}@gml.ac.cn
[2] China GridCom Co., Ltd., Shenzhen, China
{likuiyu,chenglunxin,chenyifan}@sgchip.sgcc.com.cn

**Abstract.** The detection of defects in PCB components has always been a popular field, with the main challenge being the presence of defects that cannot be recognized by 2D machine vision. This paper investigates PCB component defect detection using a combination of 2D and 3D techniques. We propose some image enhancement and contour extraction methods, and employ Contrast Limited Adaptive Histogram Equalization (CLAHE) to correct the color differences caused by inter-board illumination. By incorporating morphological adaptive anisotropic diffusion filtering, we have improved the performance of PCB board and component contour extraction. And we find that contour extraction based on polygon approximation outperforms diagonal circular regression. In the 2D defect detection method, masked template matching has been used to reduce the calculation of similarity involving non-component pixels, thereby improving recognition accuracy. To address defects that cannot be identified by 2D techniques, we have generated the original point cloud by fusing RGB-D data of PCB components. By utilizing RANSAC and DBSCAN algorithms, the point cloud is segmented into reference planes of component, upper planes of component and point clouds of pins. Based on the geometric description operators for calculating depth of point clouds, features have be made such as difference of depth between reference and upper planes and cross-sectional area to identify defects. Experimental results demonstrate that the combination of 2D and 3D machine vision techniques can accurately address various types of PCB component defects.

**Keywords:** Defect detection · Image enhancement · Masked template matching · Point cloud segmentation

## 1 Introduction

### 1.1 Background

The AOI (Automated Optical Inspection) of PCBs (Printed Circuit Boards) is one of the commonly used techniques for defect detection. With the widespread

H. Jin et al. (Eds.): IAIC 2023, CCIS 2059, pp. 16–28, 2024.
https://doi.org/10.1007/978-981-97-1280-9_2

adoption of machine vision-based automated inspection systems, these systems are capable of rapidly and accurately detecting various defects on PCBs, such as soldering issues, shorts, opens, and component defects. However, due to the complex three-dimensional structure of PCBs and the high density of components, it is challenging to comprehensively capture PCB information solely from two-dimensional images. Additionally, the side-view imaging is prone to occlusion, making it difficult to differentiate small defects from normal samples in 2D images. This limitation can result in inaccurate detection and localization of certain defects, such as component lifting on one end or false soldering of pin connections. To address this issue and detect finer defects, an increasing number of AOI systems have adopted three-dimensional inspection techniques. Three-dimensional inspection provides additional surface and depth information, enabling better analysis and identification of various defects, including lifted pins and component irregularities (floating height). In summary, the adoption of three-dimensional inspection allows for more comprehensive and accurate detection of PCB defects, thereby enhancing product quality and production efficiency. With the advancements in 3D imaging technology and cost reduction, the application of three-dimensional inspection in the PCB manufacturing industry is becoming increasingly widespread.

## 1.2 Related Works

The 2D defect detection of PCB components can be categorized into traditional feature extraction methods and deep learning-based methods, with different techniques employed for each category. Some traditional methods include RF-based EWMA sliding window method [7], AdaBoost [1], $\beta$-VAE [17], wavelet transform [11], Gabor filters [9], and normalized cross-correlation template matching method [6]. On the other hand, recent advancement shave introduced deep learning-based methods such as DAGMM [18], DeepSVDD [15], skip-connected convolutional autoencoders [10] YOLOV4-MN3 [13], and Mask R-CNN with geometric attention-guided mask branches [12]. However, it is crucial to consider the unique challenges and factors present in the industrial domain when applying these methods.

In contrast, three-dimensional defect detection of PCB components is less developed compared to 2D-AOI technology. Some studies have explored the use of 3D deep learning methods combined with spatial erasing and MRI volume for anomaly segmentation [2]. Others have focused on 2D imagenet-based feature extraction combined with FPFH [16] for 3D feature extraction and the use of PatchCore [14] for improved results [5]. Bergmann, P et al. utilized a 2D imagenet-based feature extraction method to extract RGB features and FPFH for 3D feature extraction. They demonstrated that combining the features using PatchCore helps improve the P-ROC [3]. Additionally, methods have been proposed for unsupervised detection of geometric anomalies in high-resolution 3D point clouds by constructing expressive teacher networks. These advancements highlight the need for better 3D methods in detecting certain types of anomalies that may be challenging for 2D-only approaches [4].

However, due to the minuteness of component defects and environmental conditions such as uneven lighting, noise, and color variations, defects that lack color information cannot be detected solely relying on 3D point clouds and their topological information. Furthermore, the characteristics of 3D structured light cameras limit their output to high-resolution grayscale and depth images.

To address small-sample defects on PCBs, this paper proposes two defect detection algorithms: a 2D traditional defect algorithm based on template matching and a 3D defect detection algorithm that utilizes depth information. The experimental results on 2D and 3D instance image datasets demonstrate the effectiveness of these two defect detection algorithms in detecting various types of 2D and 3D defects, including component defects, misalignment, floating height, and lifted pins.

## 2    Methodology

### 2.1    Image Preprocessing

To begin with, we are focus on extracting the maximum outer contour of each PCB, it is important to eliminate or minimize the texture information in the PCB background. Dirt or indentations on the PCB base can hinder the extraction of PCB edges, thus requiring an initial image cropping step.

**Background Removal.** Initially, we have subjected the original image in Fig. 1(a) to threshold transformation, emphasizing the differentiation between the green and the black background of PCB by extracting and thresholding the colors of the PCB substrate. Following the transformation, we have applied the Canny operator for dual-threshold gradient segmentation, resulting in a binary image that preserves true edges. Subsequently, morphological erosion has been performed to eliminate small noise and unwanted details from the edge image. Finally, the Hough transform algorithm have been used to select the four outer contours of the PCB. This process not only removes burrs and excessive background from the image but also retains crucial edge context information.

By using the filtered long edges, we have calculated the four intersecting vertices, $E$, $F$, $G$, and $H$, of the PCB as shown in Fig. 1(b). We then crop the image by expanding around these vertices, which is done to obtain local contextual information of the PCB edges and reduce background edges as illustrated in Fig. 1(c).

**PCB Maximum Outer Contour Segmentation.** After the background removal process, we have converted the processed PCB image to grayscale and has undergone adaptive thresholding, median filtering, morphological erosion, and Canny edge detection to enhance edge extraction. The largest outer contour is extracted based on its area, and then a polygonal approximation is performed for accurate fitting. We also have tried other methods, such as minimum bounding rectangle and Harris corner detection,which are result in gaps between the

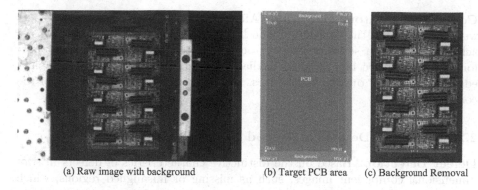

(a) Raw image with background       (b) Target PCB area     (c) Background Removal

**Fig. 1.** Data: (a) Raw image, (b) target area, and (c) the cropped image with background removed.

contour and PCB boundary and not perform well. To address this, we have applied a perspective transformation to rectify the segmented foreground of the PCB to its original position in the image.

**Image Enhancement.** To address the issue of uneven illumination and improve the recognition rate of component contours, we have employed Contrast Limited Adaptive Histogram Equalization (CLAHE) for contrast enhancement. CLAHE redistributes the pixel values of the $V$ channel in the HSV color space, preserving image details. Comparative analysis shows that CLAHE outperforms other traffic image enhancement methods [7]. However, CLAHE has a limitation in locally allocating pixel values, which will be demonstrated in the experimental section. After removing the background, data enhancement is performed on the PCB foreground to suppress noise.

$$S(i) = 255 * \frac{\text{CDF(i)}}{N-1} \tag{1}$$

$$C = \frac{M}{L} \left[ 1 + \frac{\alpha}{100} \left( \max\left(CDF'(i)\right) - 1 \right) \right] \tag{2}$$

where $M$ is the number of local sub-blocks in the histogram, $\alpha$ is the clipping factor, and L is the number of gray levels.

We observe that CLAHE method has improved the overall brightness of the image but it also has intensified the reflection effects of black components, making their contours more difficult to recognize. However, CLAHE enhances local feature information of solder joints, improving the distinguishability of components with diffuse reflective materials and pin textures. In the experimental section, we will make a comparison between PCB images with CLAHE applied and the component contours obtained from preprocessed original images in 2D defect detection experiments.

**Component Contour Detection.** After obtaining the complete PCB image, we have extracted the contours of the black components on the PCB. This is achieved through a series of image processing steps including grayscale transformation, adaptive binarization, median filtering, morphological erosion, and edge detection. By outlining the edges, we obtain the detected contours of the components.

## 2.2   2D Defect Detection Method

**Defect Detection.** Based on the distinctive characteristics of defects in two-dimensional component images, such as missing or misaligned regions, which exhibit discontinuities in RGB channel feature values, gradient and edge calculations are well-suited for their detection. However, due to limited availability of defect samples, training deep learning models for this small-sample object detection problem is challenging. Therefore, template matching is employed as an alternative approach. Template matching offers the advantage of fixed component positions on the PCB, facilitating accurate component localization. Moreover, this paper utilizes template matching with a mask for defect identification. By directly calculating the pixel values within the component domain, it achieves more precise results compared to using pixel points of ground truth boxes as input for CNN networks.

**Masked Templates Creating.** We use obtained component contours for localizing and filtering the required detection contours based on their aspect ratio, area, and approximate position. To improve defect detection accuracy, we have made a mask during template creation, converting the template into a binary image with pixel values of 0 and 1. The mask ensures that certain areas, indicated by 0 pixels, which are ignored during template matching, preventing the inclusion of non-component pixels in the scoring process.

**Masked Template Matching.** We apply masked template matching method to incorporate a mask image into the template matching process, specifying the regions eligible for matching. This means that only pixels in the mask image with non-zero values are considered for matching calculations. In this method, the following mathematical formula can be used:

$$R(x,y) = \frac{\sum_{x',y'} \left( (T(x',y') * I(x+x',y+y')) * M(x',y') \right)^2}{\sqrt{\sum_{x',y'} (T(x',y') * M(x',y'))^2 * \sum_{x',y'} (I(x+x',y+y')) * M(x',y') \right)^2}}$$

(3)

where $R(x,y)$ represents the value at the pixel position in the output image, indicating the normalized similarity score of the match, $T(x',y')$ is the pixel value in the template image, $M(x',y')$ is the pixel value in the mask image, and $I(x+x',y+y')$ is the pixel value at the corresponding position in the target image.

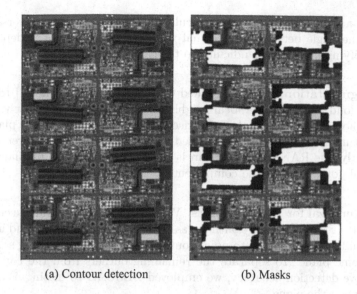

(a) Contour detection                    (b) Masks

**Fig. 2.** Mask creation based on contour detection.

Unlike previous template matching methods, the masked template matching method includes a multiplication term $M(x', y')$ with the mask image. It implies that only the pixel values at corresponding positions in the template and target images multiplied with non-zero pixel values in the mask image are considered for calculating the match degree. The corresponding results are shown in Fig. 2.

### 2.3  3D Defect Detection Method

**Detection Strategy.** In 2D images of PCB components, various defects such as missing components and misalignment can be observed. However, certain defects may not be easily identifiable solely in 2D, necessitating the use of depth information from PCB RGB-D data for detection. For example, floating defects exhibit uneven component heights with one end raised, and their detection involves assessing the height difference within a threshold. On the other hand, lifted pin defects are characterized by raised component pins that are not properly soldered, requiring measurement of pin heights relative to the depth of component's reference plane. When the height difference exceeds a threshold, component is classified as a defect.

**Filtering Inlier Points by Normal Vectors.** The algorithm for fitting the reference plane from point clouds involves several steps. We start with RGB-D data turning to Point Clouds and calculating the gradients of the normal vectors, followed by segmentation to enhance accuracy and remove interference. Then, the normal vectors are computed, some parallel and some perpendicular to the camera. A kd-tree is constructed, and the dot product of the normal vector and the Z-axis vector (0, 0, 1) is calculated. By applying the inverse cosine function, the angle between the normal vector and the camera is determined, then identifying points perpendicular to the camera.

**Plane Segmentation.** We has applied the RANSAC algorithm [8] to cluster the point cloud and identified local neighborhoods of points. A larger value of k has been chosen to minimize the influence of fluctuating points. The plane with the highest number of points is segmented and considered as the reference plane. Subsequently, the RANSAC algorithm is iteratively applied to fit planes using the original point cloud of the components.

**Outlier Removal for Planar Points.** We found the original Ransac clustering segmentation result is in many outliers, as the segmented planes extend infinitely as shown in Fig. 3. This leads to points outside the components being segmented into the same plane. To separate the remaining outliers from the components and improve detection accuracy, we employed DBSCAN to distinguish outliers from points on the plane.

**Fig. 3.** 3D segmentation using Ransac algorithm.

We apply DBSCAN algorithm to perform clustering by identifying core points and density reachability. It assigns points within a certain threshold of density in the neighborhood to the same cluster, while marking points in sparse regions as outliers.

**Geometric Descriptors of Defects.** In the real PCB point cloud, we filtered the vectors perpendicular to the 3D camera by computing the normal vectors and generate a collection of vectors. The DBSCAN method is used to select the pins and plane regions on the components that need to be detected. The depth deviation between these point cloud collections and the reference plane is then calculated. The depth deviation between the reference plane $D_{\text{base-plane}}$

and the pin $D_{\text{pin}}$, as well as the depth deviation between the reference plane and the component plane $D_{\text{component}}$, are compared with a standard threshold to determine if they exceed the threshold. The standard depth deviation is shown in Fig. 4.

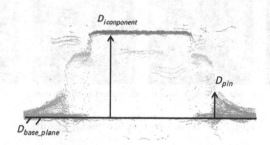

**Fig. 4.** Geometric descriptors of defects.

# 3   Experiment Results

## 3.1   2D Defect Detection Results

We conducted experiments on a two-dimensional image dataset which is consists of 200 PCB images captured by a high-resolution industrial camera. The dataset contains a total of 1325 defects. Experimental tests were conducted on the proposed two-dimensional defect detection algorithm presented in this paper, and the results were evaluated.

To verify the general performance of our proposed image processing methods. This paper presents a 2D defect detection algorithm using masked normalized cross-correlation template matching. The template creation process involves image preprocessing, specifically cropping to eliminate background edge noise. Histogram statistics are computed for the G channel, with the black background area comprising approximately 35% of the image. Pixel points corresponding to the top 65% of the histogram are extracted, and a threshold of 255 is set for all three channels (Fig. 5).

To verify the effect of image enhancement on component contour extraction, we compared the results of contour extraction using CLAHE with those using unenhanced images.

In Fig. 6, the enhanced image using CLAHE shows that the small components with pin legs have been almost accurately identified based on their black main contours as shown in Fig. 6(b). However, the largest component with black, due to intensified reflections, exhibits less effective contour extraction compared to the original image. The brightness enhancement improves the edge extraction results for tightly bonded components.

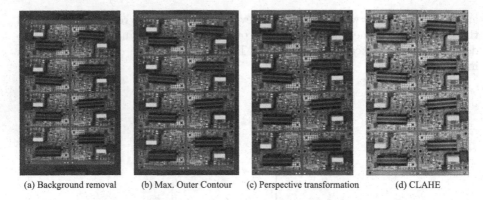

(a) Background removal    (b) Max. Outer Contour    (c) Perspective transformation    (d) CLAHE

**Fig. 5.** PCB image preporocessing results.

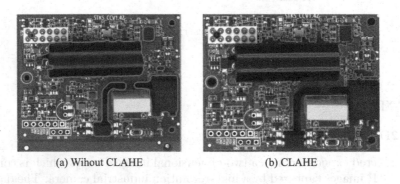

(a) Wihout CLAHE                    (b) CLAHE

**Fig. 6.** Mask creation based on contour detection.

After extracting the contours of the components, we create a mask. And we input the image with component contours and calculate the minimum bounding rectangle of the contours as the template for matching. The pixels within the component contour inside the rectangular box are set to a threshold value of 255, while the pixels outside the contour are set to 0.

From the mask being detected, we evaluate the detection results by comparing the similarity of the foreground with the instance and the standard template. We use recall, precision, and F-score to assess the final detection performance. In Table 1, we compare the methods of image enhancement and mask template creation for the same component contours with the original. The results show that the CLAHE+Mask Template method performs the best.

**Table 1.** 2D Defect Detection Results

| Method | Recall | Precision | F-Score |
|---|---|---|---|
| Seg | 0.82 | 0.75 | 0.78 |
| Seg.+Mask | 0.92 | 0.93 | 0.92 |
| Seg.+CLAHE | 0.85 | 0.79 | 0.82 |
| Seg.+CLAHE+Mask | 0.94 | 0.96 | 0.95 |

## 4   3D Defect Detection Results

We created a three-dimensional image dataset consisting of 200 RGB-D data of PCBs, which are similar to the 2D images. We conducted experiments on our proposed two-dimensional defect detection algorithm in order to use 2D contours to localize components in RGB-D images. Then we evaluated the experimental results.

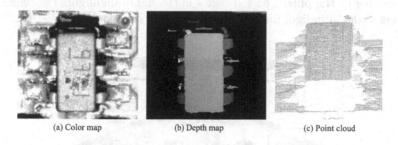

(a) Color map             (b) Depth map             (c) Point cloud

**Fig. 7.** 3D point cloud processing.

In the collected RGB-D images of the PCBs, we matched the maximum outer contour of the RGB-D PCB, obtained through extraction, scaling, and transformation, with the 2D PCB contour. We used the contour detection of the 2D image to locate the components. By segmenting the component contours, we extracted the RGB-D images for defect detection and performed point cloud fusion to obtain three-dimensional data samples of the components.

Figure 7 demonstrates RGB-D images turning into point cloud of the component, Fig. 7(c) shows us an entire 3D pcs of component, the point cloud of the reference plane with green points pin leg clouds with yellow, and segmented upper plane with orange, respectively. It can be observed from the figure that the fitted planes are divided into upper and lower planes, but still with some outliers.

By employing DBSCAN density clustering, we can identify clusters of point clouds corresponding to the planes and the pins. Figure 8 demonstrates that the outliers of point cloud is filtered by DBSCAN and planes is segmented.

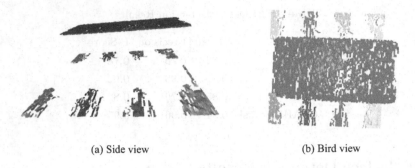

(a) Side view                    (b) Bird view

**Fig. 8.** Component plane segmentation in 3D.

After plane segmentation, we further apply DBSCAN clustering to the remaining points to segment the pins as shown in Fig. 9. Unlike plane segmentation, pins exhibit curved surfaces in the point cloud data. By employing DBSCAN density clustering, we can successfully identify clusters of point clouds corresponding to the pins. The red boxes in the figure highlights the segmented component pins in indigo, indicating the slightly raised portions.

**Fig. 9.** Component pin segmentation in 3D.

**Table 2.** Component plane misalignment detection results.

| Method | Recall | Precision | F-Score |
|---|---|---|---|
| Seg | 0.64 | 0.69 | 0.66 |
| Seg.+DBSCAN | 0.88 | 0.92 | 0.90 |

The planes and pins detection results from segmentation are shown in Tables 2 and 3.

**Table 3.** Component pin lifting detection results.

| Method | Recall | Precision | F-Score |
|---|---|---|---|
| Seg | 0.71 | 0.78 | 0.74 |
| Seg.+DBSCAN+ | 0.92 | 0.94 | 0.93 |

# 5   Conclusion

In this paper, we propose a 2D and 3D PCB defect detection method to identify various types of defects in components. For small sample sizes, we employ traditional methods for 2D detection, taking into account edge extraction, image enhancement, and use mask-matching computations to improve the detection results. However, a limitation is the inability to effectively address component reflection issues. For defects that cannot be identified in 2D, we utilize 3D techniques to analyze and identify commonly occurring defect features. This method has been successfully applied in engineering practices.

# References

1. Alelaumi, S.M., Wang, H., Lu, H., Yoon, S.W.: A predictive abnormality detection model using ensemble learning in stencil printing process. IEEE Trans. Compon. Packag. Manuf. Technol. **10**, 1560–1568 (2020)
2. Bengs, M., Behrendt, F., Krüger, J., Opfer, R., Schlaefer, A.: Three-dimensional deep learning with spatial erasing for unsupervised anomaly segmentation in brain MRI. Int. J. Comput. Assist. Radiol. Surg. **16**, 1413–1423 (2021)
3. Bergmann, P., Fauser, M., Sattlegger, D., Steger, C.: MVTec AD - a comprehensive real-world dataset for unsupervised anomaly detection. In: 2019 IEEE/CVF Conference on Computer Vision and Pattern Recognition (CVPR), pp. 9584–9592 (2019)
4. Bergmann, P., Jin, X., Sattlegger, D., Steger, C.: The MVTec 3D-AD dataset for unsupervised 3D anomaly detection and localization. arXiv:2112.09045 (2021)
5. Bergmann, P., Sattlegger, D.: Anomaly detection in 3D point clouds using deep geometric descriptors. In: 2023 IEEE/CVF Winter Conference on Applications of Computer Vision (WACV), pp. 2612–2622 (2022)
6. Crispin, A.J., Rankov, V.: Automated inspection of PCB components using a genetic algorithm template-matching approach. Int. J. Adv. Manuf. Technol. **35**, 293–300 (2007)
7. Ding, C., Zhang, Z., Li, F., Zhang, J.: Traffic image dehazing based on HSV color space. In: 2021 33rd Chinese Control and Decision Conference (CCDC), pp. 5442–5447 (2021)
8. Fischler, M.A., Bolles, R.C.: Random sample consensus: a paradigm for model fitting with applications to image analysis and automated cartography. Commun. ACM **24**, 381–395 (1981)
9. Jing, J., Jing, J., Zhang, H., Li, P.: Improved Gabor filters for textile defect detection. Procedia Eng. **15**, 5010–5014 (2011)

10. Kim, J., Ko, J., Choi, H., Kim, H.: Printed circuit board defect detection using deep learning via a skip-connected convolutional autoencoder. Sensors (Basel) **21**, 4968 (2021)
11. Li, W.C., Tsai, D.M.: Wavelet-based defect detection in solar wafer images with inhomogeneous texture. Pattern Recognit. **45**, 742–756 (2012)
12. Lian, J., Wang, L., Liu, T., Ding, X., Yu, Z.: Automatic visual inspection for printed circuit board via novel mask R-CNN in smart city applications. Sustain. Energy Technol. Assess. **44**, 101032 (2021)
13. Liao, X., Lv, S., Li, D., Luo, Y., Zhu, Z., Jiang, C.: YOLOv4-MN3 for PCB surface defect detection. Appl. Sci. **11**, 11701 (2021)
14. Roth, K., Pemula, L., Zepeda, J., Scholkopf, B., Brox, T., Gehler, P.: Towards total recall in industrial anomaly detection. In: 2022 IEEE/CVF Conference on Computer Vision and Pattern Recognition (CVPR), pp. 14298–14308 (2021)
15. Ruff, L., et al.: Deep one-class classification. In: International Conference on Machine Learning (2018)
16. Rusu, R.B., Blodow, N., Beetz, M.: Fast point feature histograms (FPFH) for 3D registration. In: 2009 IEEE International Conference on Robotics and Automation, pp. 3212–3217 (2009)
17. Ulger, F., Yuksel, S.E., Yilmaz, A.: Anomaly detection for solder joints using $\beta$-VAE. IEEE Trans. Compon. Packag. Manuf. Technol. **11**, 2214–2221 (2021)
18. Zong, B., et al.: Deep autoencoding gaussian mixture model for unsupervised anomaly detection. In: International Conference on Learning Representations (2018)

# A Secure Circulation Mechanism of Personal Data Based on Blockchains

Tianqi Cai[1]([✉])(iD), Yuhan Dong[1], Zitao Xuan[1], Lei Wang[1], Kun Huang[2], Zhide Li[1], and Hengjin Cai[2]

[1] Tsinghua Shenzhen International Graduate School, Tsinghua University, Guangdong 518055, China
`cai.tianqi@sz.tsinghua.edu.cn`
[2] School of Computer Science, Wuhan University, Hubei 430072, China

**Abstract.** This paper discusses the privacy protection and security application of personal data from the perspective of building private chains. Individuals have control over their private chains, and their data can be transferred based on the owner's active hashed interactions, which is a mechanism allowing the step of reaching consensus to be separated from the data ontology, keeping a balance among efficiency, security and privacy. Confronting with heterogeneous and multi-modal personal data, it is able to trace, verify and promote transactions of tokenized data with personal trustworthy AI agents. The system is conducive to attracting high-quality data into transactions by making it easier for superior data to be circulated, and the circulation records further increasing endorsement and adding values for the high-quality data, so as to break through "the Market for Lemons".

**Keywords:** Data Sharing · Data Transaction · Active Hash Interaction Network

## 1 Introduction

The digital economy has emerged as a pivotal driver of global economic progress, encompassing economic activities rooted in digital technology. This domain encompasses research and development, digital production, trading, and the management of digital assets [1]. Within the sphere of the digital economy, the healthcare sector is actively pursuing avenues for digital transformation. This initiative aims to augment the quality and efficiency of medical services through the integration of digital technologies, ultimately bolstering the competitiveness and innovation capacity of the healthcare industry. The circulation of personal health data pertains to the sharing and exchange of health-related information among healthcare institutions, between such institutions and individuals, as well as among individuals themselves. Leveraging digital technology to securely facilitate the circulation of personal health data holds the potential to elevate the quality and efficiency of healthcare services, consequently fostering innovation and progress within the medical domain.

The conventional methods of storing and exchanging health data predominantly revolve around a limited number of healthcare institutions or data centers, giving rise to

H. Jin et al. (Eds.): IAIC 2023, CCIS 2059, pp. 29–44, 2024.
https://doi.org/10.1007/978-981-97-1280-9_3

concerns of data centralization and centralized control. Conversely, a blockchain-based mechanism for the secure circulation of personal health data can achieve decentralized data storage and exchange, mitigating issues related to data centralization and centralized control, thus enhancing the security and reliability of the data. Personal health data involves privacy and sensitive information. Conventional methods of data exchange can lead to potential breaches of individual privacy. In contrast, a blockchain-based mechanism for the secure circulation of personal health data can implement security measures such as encryption and anonymity, safeguarding individual privacy. Moreover, this mechanism can leverage smart contracts, allowing for data ownership and authorization to be established through automated agreements, which can enhance the credibility and reliability of the data.

In the field of personal health, the distinctive characteristics of personal information pose significant challenges for designing effective regulatory solutions. Some researchers propose that, in order to strike a balance between information protection in the era of big data and the demands of industry development, relevant laws and regulations should explicitly state that personal medical information lacks property attributes and that the ownership of personal information is an inalienable personality right. Special protections should be provided for personal medical information, limited solely to medical and scientific research purposes, and any use beyond these purposes must undergo strict anonymization procedures. Additionally, it is suggested to establish a dedicated administrative oversight body for personal information, providing comprehensive review and supervision services for the protection of personal medical information [2]. Research teams also provide recommendations from a technological perspective, emphasizing the governance of personal health and medical data through information technology and big data. When hospitals face the challenge of incompatible business management systems, they need to integrate various business platforms, eliminate the so-called information silos, and achieve interoperability and integrated applications of data within the hospital. This lays the foundation for establishing high-quality big data assets, which serve as the basis for operational data analysis and the realization of comprehensive intelligent operational management in hospitals [3–5].

It is evident that there is still room for development and optimization in the implementation and technological aspects of the security, ownership, and application of personal health data. With this foundation, personal health data can be collected, circulated, and utilized in a rational manner, unlocking higher value to advance the progress of medical research.

The selection of data security technologies each possesses distinctive characteristics and limitations, making it challenging to strike a balance between security, efficiency, and flexible scalability. For instance, in the domain of system access control, on one hand, there is a need for a type of access control technology that supports fine-grained, dynamically manageable permissions. On the other hand, it also requires a certain level of learning capability to extract and acquire relevant knowledge from vast amounts of data, which can be used to enhance the security of access control. Regarding data ownership and governance, these are primarily established through a top-down mechanism.

However, in reality, due to the complexity of individual data usage scenarios, it is difficult to formulate comprehensive and universally applicable mechanism rules to constrain different data rights.

In the realm of privacy protection, researchers have proposed various methods such as differential privacy, homomorphic encryption, and secure multi-party computation. These methods allow for model training and inference while safeguarding data privacy. Privacy-preserving computing techniques have the potential to give rise to the market for lemons [6]. In situations of information asymmetry, buyers may only be willing to pay an average price. Sellers holding high-quality goods are unwilling to sell at a low price and exit the market with their premium products. Buyers holding low-quality goods may intentionally withhold complete information and profit by selling at the average price. As a result, market product quality gradually declines, and the average price decreases accordingly. High-quality products are increasingly driven out, leading to the phenomenon of low-quality driving out high-quality in the market for lemons.

In scenarios involving upstream and downstream participants, as well as collaborative cooperation, when high-quality data is not provided or is provided incompletely, one's own party stands to benefit significantly, while the other parties may not easily discern this. Solely relying on privacy-preserving computation cannot address issues related to the trustworthiness of data sources.

Exploring how to balance the conflicts among the stakeholders of data arising from the circulations of personal data is of paramount importance in driving the development of the digital economy.

## 2    Related Work

Blockchain technology is a distributed technology capable of implementing data sharing in a decentralized and transactional manner. It holds significant potentials in the protection and utilization of privacy in medical data. In addition to features such as decentralization and traceability, blockchain technology's smart contracts can automatically execute corresponding operations when specific conditions are met. This allows for a more automated data management and access control. When combined with cryptographic techniques, precise authorization for medical data can be implemented, ensuring that only authorized users can access the specific data. Healthcare institutions and researchers can securely share data through blockchain platforms, thereby promoting medical research and collaboration.

In practical applications, blockchain technology has already been implemented in areas such as medicine traceability and electronic health records. Employing blockchain technology to track the production, transportation, and sales process of pharmaceuticals ensures the authenticity of medicines, thereby preventing counterfeit medicines from entering the market and enhancing patient medication safety. Storing patients' electronic health records on the blockchain ensures that only authorized healthcare professionals can access them. This increases patients' trust in the privacy of their health records, while reducing the risk of data tampering or leakage.

Blockchain technology can be leveraged in the healthcare sector to achieve a delicate balance between the privacy and accessibility of personal medical data. Building upon the characteristics of blockchain, Dagher et al. [7] introduced the Ancile framework for personal electronic health records, which employs smart contracts and advanced encryption methods to effectively control access and ensure privacy protection. The framework also explores the data interaction relationships between patients, hospitals, and third-party agents, offering a potential solution to the longstanding privacy issues in the healthcare industry.

In addition to privacy and security concerns, standardization of data poses a common challenge in this domain. As medical data originates from different hospitals and institutions, they may adhere to varying formats and standards, potentially leading to issues of inconsistency and interoperability. To address this, Reichert et al. [8] propose integrating data from diverse healthcare institutions and sources into a centralized data warehouse. This involves preprocessing, data model design, and data warehouse construction steps to standardize and enable interoperability of medical data. Through this framework, different healthcare institutions can share data across different information systems, ultimately enhancing patient care.

The decentralized nature and consensus mechanism of blockchain itself may introduce performance issues, particularly when dealing with large volumes of medical data that require real-time accuracy. This presents a challenge to system efficiency. In response to these performance concerns, some papers propose the use of technologies such as sidechains, sharding, and consensus mechanisms to enhance performance. One paper introduces a medical data management system called HealthChain [9], which employs a sidechain-based solution. It segments the medical information exchange system into multiple sidechains, with each sidechain responsible for handling a specific type of medical data or transaction. This approach improves data storage and processing speed. Additionally, the authors suggest a sharding-based solution, dividing the main chain into multiple segments, each tasked with processing a portion of transactions, thereby accelerating transaction processing.

Trustworthy Artificial Intelligence (AI) can be understood as the inherent quality of AI technology itself [10]. Developing trustworthy AI has become a global consensus. The practical application areas of trustworthy AI include finance, healthcare, transportation, energy, and agriculture. Taking healthcare as an example, trustworthy AI can be used to assist doctors in disease diagnosis and treatment plan selection. This technology typically employs feature importance analysis and visualization techniques to help doctors better understand the decision-making process of the model, thereby enhancing diagnostic accuracy and treatment effectiveness. However, some AI systems have issues of unpredictability and explanation, which raises concerns about trust and security. In order to address these problems, trustworthy AI has been proposed as a new paradigm and has gradually gained widespread attention. Kumar [11] proposes that with the advent of the big data era, AI algorithms can use collected large amounts of data to improve the level of service, but at the same time, completely opening private data to machines has caused a crisis of trust in machines for some people. There are concerns about whether machines will leak user data or use it for illegal activities. The core of trustworthy AI lies in reliability, transparency, and interpretability. Mauricio [12] analyzes the interpretability of AI in clinical radiology and summarizes methods to improve algorithm interpretability such as guiding backpropagation and concept vector regression. Marco

[13] introduces the emerging technology LIME for explaining any classifier. In addition, AI interpretability can be enhanced through methods like model explanation, local sensitivity analysis, feature importance analysis, and visualization.

There are some practical implementations. The Korean project MediBloc [14] utilizes blockchain technology to establish a decentralized medical data exchange platform. Patients have autonomous control over their medical data and can choose whether to share it with healthcare institutions or researchers. Apple's ResearchKit [15], an open-source medical research framework, along with an application for recruiting research participants, enables medical researchers to collect user health data while safeguarding user privacy. Privacy measures include allowing users to choose to participate in specific studies and to withdraw at any appropriate time. Apple emphasizes the users' controls over their data, which are only shared with specific research projects. Google Health [16] introduced an anonymized data sharing program that allows researchers to use anonymized Google health data for medical research. Google claims to ensure the anonymity of user data, with all personal identifying information removed, and data can only be provided to qualified research institutions under appropriate privacy protection measures.

Compared to data from enterprises and organizations, which can be regulated and managed from a top-down approach, personal data exhibits distributed, heterogeneous, and multimodal features. While blockchain technology offers many advantages, its practical application also requires consideration of issues related to performance, cost, regulatory compliance, and so on. Health and medical data involve personal privacy, and in principle, data ownership should belong to the individual. However, in actual practice, from the perspective of data rights, there are many stakeholders involved in personal health data. Most health data is collected, analyzed, and stored by machines in hospitals or institutions. From the standpoint of data application, the value of an individual health data sample is limited. However, collectively categorized data based on specific indicators holds higher research value. Due to the characteristics of personal data, questions about how to protect the security and privacy of personal data, how to obtain user permission or authorization, how to collect a sufficient and valid data set as the foundation for research applications, and how to reasonably and compliantly allocate the value generated from data circulation are all extensions of the security, trustworthiness, and availability issues associated with personal data.

## 3 The Mechanism for Secure Circulation of Personal Data Based on AHIN

We further decompose the mechanism for privacy protection and secure circulation of personal data into three levels, including the trusted proof of existence at the personal data ontology layer, the trusted traceability at the node interaction layer, and the trusted circulation at the application layer.

### 3.1 Trusted Proof of Existence at the Personal Data Ontology Layer

Firstly, from the perspective of security, we initiate the construction of an individual data chain at the ontology layer. In response to the user's ownership demands over their own data, the established personal data chain manages data in the forms of tokens, ensuring that the data unequivocally belongs to the user. This safeguards personal data privacy and lays the foundation for interactions and trustworthy traceability among nodes.

**Personal-owned Private Chain.** Considering that personal data generation and utilization primarily occur on personal hardware devices, such as mobile devices and personal computers, a personal data chain should be deployable and operable on such hardware. In designing the framework for the personal data chain, the distinctive features of personal hardware environments need to be considered and therefore, lightweight applications are provided, along with plenty room for scalability. Starting from the user's rights to their data, a mainstream blockchain technology is adopted for building the private chain. This allows users to be at the center of their personal data chain, granting them autonomy over their data. Additionally, mainstream blockchain API interfaces are provided to establish channels for subsequent inter-chain interactions.

**Permission for Third Parties to Provide Data.** Depending on the specific application scenario, users may authorize qualified third parties to provide data, thereby enhancing data trustworthiness. For example, in the context of personal health and medical scenarios, a majority of an individual's health examination data is provided by institutions such as hospitals and physical examination centers. With user permission, personal health data can be transmitted from the institution to the user's private chain, thereby bolstering the credibility of professional data.

**Tokenized Data Management.** The ontology of the personal data chain is determined by the user, meaning they independently decide which data to upload to the chain. Then the data can be tokenized, which clarifies data ownership, allowing for a clear and transparent flow of data in the form of tokens. Different tokens can be matched with default personal privacy protection settings based on user operations and data classification. For instance, data intended for full sharing can default to plaintext, while directionally shared data can be encrypted. Access keys or access times can be set, and data owners can also manually adjust encryption methods.

In the context of personal health data management, as shown in Fig. 1, users initially deploy their private chains on personal devices such as smartphones or computers. Subsequently, they can record their personal information on their private blockchains. Personal health data can be either user-generated or imported from trusted third parties like healthcare institutions. The personal data is stored on the private chain in the forms of tokens to ensure immutability once it is recorded on the blockchain, thereby laying the foundation for data exchange and circulation.

### 3.2 Trusted Traceability of the Node Interaction Layer

**Active Hash Interaction Network (AHIN)** [17]. The concept of Active Hashed Interaction Network involves proactive hash interactions between individual private data chains. Within this network: each chain or personal system is privately owned by the user; users autonomously decide to interact with other chains, which includes initiating associations and accepting associations.

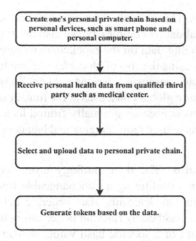

Fig. 1. The primary process for constructing a personal data ontology.

In this context, when Alice's Chain actively associates with Bob's Chain, it means that Alice's Chain shares the hash value of its block summary with Bob's Chain. On the other hand, when Bob's Chain accepts the association with Alice's Chain, it means that Bob's Chain incorporates the hash value shared by Alice's Chain into its next block (see Fig. 2).

```
input received_hashValue, my_presentBlock
if received_hashValue.sender is in my_whitelist
then add received_hashValue to my_presentBlock
else post received_hashValue.sender
    get myResponse
    if myResponse equals acceptance
    then add received_hashValue to my_presentBlock
end
```

Fig. 2. The pseudocode of accepting associations.

Moreover, a node can also choose to hash and transmit semantically meaningful data. These interactions contain semantic information digests. Even if a particular chain is forged, the attacker would need to match the history records of interactions and the relevant data on the interacting data chain. This significantly increases the cost of a successful attack, ensuring data reliability while greatly reducing storage requirements.

**The Consensus Mechanism Separated from the Data Ontology Layer.** Since the inception of blockchain technology, reaching consensus has been based on the data ontology itself, emphasizing the immutability of data. In theory, once data is stored on the blockchain, it becomes immutable. However, actual validation is quite limited. Validating blockchain timestamped data, especially for blocks with a long-time interval, poses significant challenges. This not only raises questions about the verifiability of data or the acceptable level of validation complexity, but also limits the application scenarios of blockchain. In short, placing data on the blockchain requires consensus on the data ontology, significantly increasing the cost of blockchain technology. Through active hash interactions, we can separate consensus from the data ontology layer. It means that the data ontology does not require global consensus at the time of storing new data on one's own chain. Instead, it forms consensus gradually from a local level as nodes interact. This approach satisfies diverse user requirements and balances efficiency, security, and privacy.

**Traceable Interaction Paths.** The data ontology layer maintains the data that is uploaded to the blockchain, constituting an unchangeable foundational layer of data. How to interpret local data and determine data ordering between multiple chains is achieved through consensus reached by relevant nodes through hashed interactions. By tracing the transmission path of a specific hash value, we can trace back to the initial node that emitted it and find the relevant interaction trajectory. An important aspect of blockchain technology is its intrinsic ability to provide an inherent sense of time in the digital world, such as using block numbers to establish the chronological order of different blocks. In a blockchain network with multiple coexisting chains, it is challenging to establish a unified system for quick sequencing. This mechanism emphasizes the traceability and verifiability of data, enabling comprehensive sequencing of blocks on different chains through relationship clues between multiple chains.

Active hash interaction is built upon a certain foundation of trust, relying not only on historical diachronic relationships but also anticipating future connections. This trust foundation stems from our real-world associations, manifested in our continuous maintenance of personal data within the blockchain network and the hash interactions with other nodes. For instance, in the scenario depicted in Fig. 3, Alice performs a hash operation on a specific data block and subsequently sends the generated hash value to both Bob and Emily. If Bob is unfamiliar with Alice, he may choose not to further propagate the hash value. Emily, who has a closer connection with Alice, may include the hash value in her current data block, generating a new hash value and transmitting it to others, thereby forming a potential propagation path.

## 3.3  Trusted Circulation at the Application Level

**Personal Trustworthy AI Agents.** The traditional agency model is prone to agency problems. This is because the interests of the agent and principal are not completely aligned. In situations where the principal is in an information disadvantage and cannot fully supervise the agent, the agent has a motivation to act in their own interest, potentially to the detriment of the principal's interests. This phenomenon, resulting in harm to the principal's interests, is known as the agency problem. In our mechanism, on the one hand, trustworthy AI agents are AI data processing assistants managed autonomously by users, rather than being entirely represented by third-party natural persons. On the other hand, a trustworthy AI agent serves only one user, eliminating the need to make choices among the interests of multiple principals. This greatly mitigates agency problems.

The initial state of the AI agent is consistent, equipped with basic natural language processing capabilities. When a user creates a new AI agent, they automatically become the sole owner of that AI agent, responsible for management tasks such as activation or deactivation. The AI agent cannot be transferred to others, and the user's association with their AI agent can be verified through the use of public and private keys, ensuring that an AI agent has only one user. The AI agent's access to user data, learning, and utilization is entirely authorized and confirmed by the user, and it operates within a specified device environment.

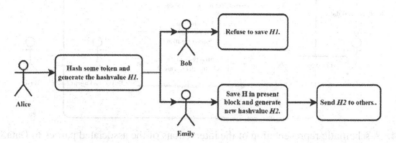

**Fig. 3.** The Diagram illustrating node interaction paths.

**Demand Matching Based on AI Agents.** Data trading parties engage in demand matching through their own AI agents on the secure platform DataStill, as shown in Fig. 4.

The responsibilities of a seller's AI agent encompass: acquiring authorization for personal data access; categorizing and packaging diverse datasets, furnishing verifiable data descriptions; associating the data with the respective buyer agent, aiding the buyer agent in data verification, establishing mutual agreement on data pricing, and facilitating equitable benefit-sharing with relevant data owners.

The responsibilities of a buyer's AI agent encompass: acquiring the purchase demand and agent authorization from the buyer; comprehending the requirements, identifying potentially suitable data targets, and confirming pricing with the seller's agent; furnishing feedback information to the buyer to facilitate the completion of the transaction.

**Data Validation Based on the AI Agents.** The verifiability of data is an important premise for data to reveal its value, which means that relevant parties can test the characteristics of data to ensure the credibility and availability of data. In view of the credible problem of the data source, it is mainly solved through the way of personal data tokenization. A token clearly records the corresponding circulation record, which is convenient for quick verification. In view of the verification of data characteristics, we match the requirements and data characteristics through personal AI agents, and abstract data description of the results of data verification, to ensure that the data details are not disclosed during the verification.

An advantage of bottom-up data management mode is that the data rights are clear from the beginning, which can solve the problem of data rights and compliance use. The corresponding limitation is the different cognitive levels and expertise of different users, which cannot guarantee that all users can operate personal chain in an appropriate way. In view of this problem, third-party services are allowed to provide users with personalized suggestions for data management on the premise of privacy protection, so that the data can be safely circulated and create further values.

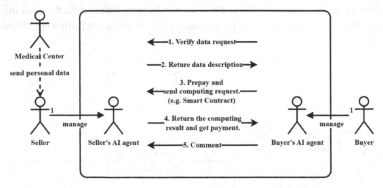

**Fig. 4.** A schematic representation of the interactions of the associated parties in DataStill.

## 4   Data Description and Demo

The security platform demo is named as DataStill, where trustworthy AI agents can engage in demand matching and verification of data characteristics. The trustworthy agent of the seller is capable of summarizing data description without disclosing specific data details. Through this data description, the trustworthy agents of both the buyer and seller can collaboratively generate validation smart contracts for confirming if the original data complies with the data description, as illustrated in Fig. 5.

The key technologies of DataStill can be primarily divided into three parts.

**Verification Contract Mechanism.** The trustworthy AI agent of the seller provides an initial draft of the verification contract. Both the buyer and seller execute a one-time verification contract to ensure the consistency of data source and format, and obtain the verification results. This process ensures that the trading data is genuine, trustworthy, valid, and compliant with the buyer's requirements. It also guarantees the privacy rights of the seller, eliminating information asymmetry and distrust issues in the data matching process.

**Universal Token Model.** Applicants use tokenized data, meaning that each user's data is treated as a unique non-fungible token (NFT). In the transaction process, NFTs serve as unique identifiers, ensuring the identity and ownership of the data. This also enables the traceability of transaction history and ensures the ownership of the data.

**Encryption Mechanism.** During data interactions, data is desensitized and encrypted based on the results of data classification to prevent the leakage of user's private data. This achieves comprehensive data privacy protection throughout the entire transaction process within a trusted execution environment. In addition, the buyer's agent confirms semi-structured data requirements and encrypts the requirement file with their own key. The seller's agent decides whether to encrypt using their own key based on the encrypted requirement, further ensuring the security and verifiability of the data.

DataStill can be divided into three subsystems, including Data Presentation, Data On-chain, and Demand Matching. These subsystems collaborate through services between different implementation layers, as shown in Fig. 6.

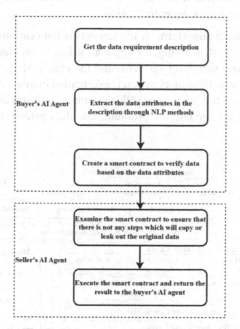

**Fig. 5.** The process of data verification.

The comprehensive diagram detailing the processes and interactions between DataStill's subsystems is presented in Fig. 7. When new data arrives in DataStill, it initially enters the Data On-chain Subsystem. It can then be accessed through the Data Presentation Subsystem and may be activated, either automatically or manually, to undergo data matching and trading within the Demand Matching Subsystem via a trustworthy AI agent. Following this process, the data is once again made accessible through the Data Presentation Subsystem. The specific workflow and components of these three subsystems are elaborated upon below.

**The Data Presentation Subsystem.** It implements collaborations through the system presentation layer's data dashboard, the application layer's on-chain and off-chain data interaction services, and the chaincode layer's token services. This subsystem desensitizes data leaving the blockchain network to prevent off-chain leakage of user privacy, ensuring comprehensive data privacy protection throughout the entire transaction process.

**The Data On-chain Subsystem.** It implements collaborations through the system presentation layer's upload records, on-chain data management, the application layer's on-chain and off-chain data interaction services, and the chaincode layer's token services. This subsystem demonstrates the complete process of users uploading data onto the blockchain and converting it into tokens, implementing digital ownership rights. Through collaborative filtering of uploaded semi-structured data, abnormal data is alerted, and based on user operational attributes, a combination grading is established. The side chains are built based on different attribute combinations, with the data ontology as the main chain, thus realizing data decentralization.

**The Demand Matching Subsystem.** It implements collaborations through the system presentation layer's demand matching, the application layer's on-chain and off-chain data interaction services, data matching services, and the chaincode layer's comprehensive services. The demand matching mechanism implemented in this subsystem matches the trustworthy AI agents of the buyer and seller to obtain high-quality data, avoiding the occurrence of the market for lemons problem, and data pricing can be implemented through negotiation.

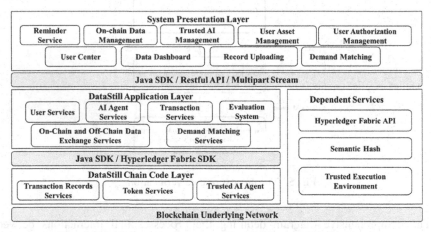

**Fig. 6.** The Schematic diagram of DataStill's system architecture.

A demo is conducted using a de-identified dataset provided by a hospital system as the basis. The dataset includes 9,649 records of patients with basic conditions such as diabetes and hypertension, containing data such as blood pressure, fasting blood sugar, height, weight, waist circumference, and so on. Several users are simulated as some nodes based on this dataset, and each user has one's own dedicated trustworthy AI agent. Users can choose to hash their data, then send the hash in a directed manner to a target user. The receiver can choose to include the hash value in his/her current block and continue passing it on, gradually forming a propagation chain. Afterward, it is possible to trace whether and how the hash value has been disseminated.

User data is stored in a private chain on their own device, giving them control over their data. When a user wants to share a specific piece of data (as shown in Fig. 8), one first hashed the selected data record, generating a corresponding hash value (as shown in Fig. 9). Then the user can select the recipient and propagate the hash value in a directed manner. If the recipient chooses to continue propagating the received hash value, they include it in their current data block, generating a new hash value (as shown in Fig. 10), and pass it on to others (as shown in Fig. 11). The initiator can verify whether the recipient has disseminated the hash value through a search process, as illustrated in Fig. 12.

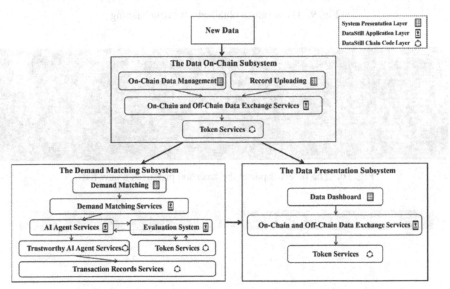

**Fig. 7.** The Comprehensive diagram of processes and interactions between DataStill's subsystems.

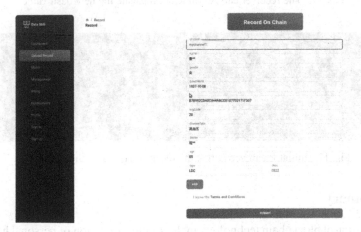

**Fig. 8.** The data owner hashes the selected data.

**Fig. 9.** Hash values obtained after data hashing.

**Fig. 10.** The receiver packs the hash and produces a new hash.

**Fig. 11.** The receiver can continue to propagate the new hash value.

**Fig. 12.** Initiator can verify that the receiver has stored the hash value.

## 5 Summary

The utilization of blockchain technology in the secure circulation of personal health data represents a specific application of artificial intelligence. Access permissions for data can be controlled through protocols or encryption methods. Challenges in dealing with large-scale data processing and varying data source structures across different institutions have been addressed by some proposed frameworks, which have shown improvements in processing efficiency and standardizing data formats to a certain extent. However, practical implementation still confronts with challenges such as technical barriers, cost requirements, and establishing user trust on centralized platforms. There remains room for many solutions for further improvements.

Firstly, blockchain technology entails a certain level of complexity, requiring a degree of technical understanding and expertise. This necessitates the provision of user-friendly tools and platforms to simplify and streamline the user experience. Secondly, establishing and maintaining a blockchain system may require significant financial cost. Therefore, choosing the appropriate blockchain solution based on the application scenario is crucial. Public blockchains or configurable private chains may be more cost-effective options compared to fully customized private solutions. The token-based mechanism may also lower maintenance costs while ensuring security [18]. Leveraging cloud services and collaborative models can also help reduce overall costs. Thirdly, instilling user trust in centralized platforms to protect personal data and comply with privacy regulations presents challenges. Many solutions emphasize the importance of privacy protection, adopting techniques like zero-knowledge proofs, multi-party computation, and other privacy-preserving technologies to ensure that user data is accessed only when necessary. Alternatively, independent third-party audits of the platform can be allowed to verify data security and privacy protection measures. Sponsored token technology, backed by qualified initiators, can also boost participant confidence [19].

By emphasizing the secure circulation of personal health data, we leverage blockchain technology in the establishment of individual private chains. This is carried out in a user-driven, bottom-up approach to ensure data security, consensus reaching, and secure circulation. Beginning with the storage end, user ownership of personal data is clearly defined, forming the foundation of the blockchain network. Through users' active hash interactions, data exchange and incremental consensus among nodes are established. Users manage their dedicated trustworthy AI agents, which handle the tasks of demand matching and quality verification of data. The final approval is made by the owner of the trustworthy AI agent, avoiding direct contact between data trading parties, reducing manual operations. This approach enables data being organized and authorized while safeguarding the privacy of personal data, providing a foundation for the secure protection and compliant application of personal health data. This paper substantiates the viability of the proposed mechanism through the examination of anonymized personal health data.

# References

1. Skare, M., Maria, D. L. M. D. O., Ribeiro-Navarrete, S.: Digital transformation and European small and medium enterprises (SMES): a comparative study using digital economy and society index data. International journal of information management (2023)
2. Xu, H.: Legal protection of personal medical data in the context of big data. Library **11**, 38–45 (2019)
3. Zhao, Y.: The governance of big data promotes lean operations in public hospitals. China Health Human Resources **10**, 10–14 (2021)

4. Gu, R., Wu, Z., Shi, H.: New approach for graded and classified cloud data access control for publish security based on TFR model. Comput. Sci. **47**(z1), 400–403 (2020)
5. Zhao, Q.: Construction and application of the special disease database based on the scientific research platform of big data. China Digit. Med. **15**(12), 89–92 (2020)
6. Akerlof, G.A.: The market for "lemons": Quality uncertainty and the market mechanism. Uncertainty in economics. Academic Press, Cambridge, pp. 235–251 (1978)
7. Dagher, G.G., Mohler, J., Milojkovic, M., et al.: Ancile: privacy-preserving framework for access control and interoperability of electronic health records using blockchain technology. Sustain. Cities Soc. **39**, 283–297 (2018)
8. Lenz, R., Reichert, M., Weber, B.: A framework for integrating process-oriented information systems and data-driven decision support systems. Decis. Support. Syst. **56**, 219–235 (2013)
9. Chenthara, S., Ahmed, K., Wang, H., et al.: Healthchain: a novel framework on privacy preservation of electronic health records using blockchain technology. PLoS ONE **15**(12), e0243043 (2020)
10. He, J.: Secure and trustworthy artificial intelligence. Inf. Secur. Commun. Priv. **310**(10), 7–10 (2019)
11. Kumar, A., Braud, T., Tarkoma, S., Hui, P.: Trustworthy AI in the age of pervasive computing and big data. In: 2020 IEEE International Conference on Pervasive Computing and Communications Workshops (PerCom Workshops), pp. 1–6. IEEE (2020)
12. Reyes, M., Meier, R., Pereira, S., et al.: On the interpretability of artificial intelligence in radiology: challenges and opportunities. Radiol. Artif. Intell. **2**(3), e190043 (2020)
13. Ribeiro, M.T., Singh, S., Guestrin, C.: "Why should I trust you?": Explaining the predictions of any classifier, pp. 1135–1144. ACM (2016)
14. Singh, C., Chauhan, D., Deshmukh, S.A., et al.: Medi-Block record: Secure data sharing using block chain technology. Inf. Med. Unlocked **24**(9), 100624 (2021)
15. Jardine, J., Fisher, J., Carrick, B.: Apple's ResearchKit: smart data collection for the smartphone era? J. R. Soc. Med. **108**(8), 294–296 (2015)
16. Brandt, R., Rice, R.: Building a better PHR paradigm: lessons from the discontinuation of Google health. Health Policy Technol. **3**(3), 200–207 (2014)
17. Cai, H., Cai, T.: An architecture for web 3.0 and the emergence of spontaneous time order. https://arxiv.org/abs/2202.10619. Accessed 30 Aug 2023
18. Cai, T., Cai, H., Wang, H., et al.: Analysis of blockchain system with token-based bookkeeping method. IEEE Access **7**, 50823–50832 (2019)
19. Cai, T., Cai, H., Lee, D., et al.: ITO: the sponsored token technology. J. Br. Blockchain Assoc. **4**(1), 1–12 (2020)

# Road Traffic Waterlogging Detection Based on YOLOv5

Jianqiang Liu, Yujie Shang, Xingyao Li, Huizhen Hao[✉], and Peng Geng

School of Information and Communication Engineering, Nanjing Institute of Technology, Nanjing 211167, China
haohuizhen@njit.edu.cn

**Abstract.** In view of the frequent occurrence of waterlogging in urban areas and the problems that traditional waterlogging monitoring methods consume a lot of human and material resources with high cost and low timeliness, an improved YOLOv5 waterlogging detection method for road traffic is proposed, which enhances the feature extraction of road traffic waterlogging information by feature extraction of waterlogging in urban waterlogging scenarios, and adds the CBAM attention mechanism in the backbone network; and adds a CIoU loss function to optimize the model in the prediction layer to improve the identification accuracy of road traffic waterlogging so as to construct a road traffic waterlogging detection model. In the prediction layer, a CIoU loss function is added to optimize the model and improve the detection accuracy of road water, thus constructing a road water detection model. By screening 5000 road traffic waterlogging images on the public dataset RSCD for training, the experimental results show that the average accuracy of the method is 84.4%, which is 3.7% higher than the original YOLOv5 algorithm, and it can more accurately extract and identify the waterlogged area of the image automatically, which can pave the way for further development of related research, and provide technical support for urban waterlogging monitoring and emergency management. The method can pave the way for further related research and provide technical support for urban flood monitoring and emergency management.

**Keywords:** urban flooding · deep learning · target detection · YOLOv5 · attention mechanism

## 1 Introduction

With the increase in global climate change, extreme rainfall events have become more frequent and intense in multiple urban areas. As a result, traffic is affected, and residential and public buildings are flooded, impacting urban safety and the quality of life of inhabitants [1]. At the same time, as urbanization accelerates, the drainage system in urban areas proves insufficient, resulting in an increasing impermeable area of the city and a heightened risk of urban flooding [2]. To respond effectively to urban flood emergencies and minimize flood damage, it is urgent to monitor waterlogging in urban roads to mitigate flooding caused by urban traffic. This will improve road users' safety and reduce hidden risks [3].

H. Jin et al. (Eds.): IAIC 2023, CCIS 2059, pp. 45–58, 2024.
https://doi.org/10.1007/978-981-97-1280-9_4

At present, the methods utilized in the detection of urban flooding mainly comprise artificial and sensor detection [4]. Artificial detection involves sending individuals to regions with a high risk of flooding to conduct manual measurements. It is advantageous due to its intuitive and flexible nature, but also requires substantial human and material resources, has a low level of efficiency, and poses certain safety risks to the detection personnel. Technical advancements in sensor detection have provided an alternative solution with improved efficiency and reduced security risk. Water level sensors are utilized to detect water and collect real-time data on waterlogging. The advantages of this method are its high measurement accuracy, speed and convenience. Nevertheless, this approach has some drawbacks, including a higher cost of instruments and the requirement for routine maintenance. It is possible for external factors to interfere, thereby hindering the full detection of waterlogging on urban traffic routes. Consequently, it may fall short of the necessary requirements for maintaining city road traffic. It is, therefore, imperative to explore a reliable, cost-effective, and trouble-free monitoring method to ensure effective and secure detection of waterlogging in urban zones.

In recent years, the widespread adoption of deep learning technology has facilitated the rapid development of artificial intelligence. As a result, surveillance cameras have become ubiquitous in all areas of the city. Hence, intelligent video surveillance for monitoring waterlogging in urban road traffic has emerged as a new research focus. In this field, deep learning provides effective technical support. Among them, the YOLO algorithm has been widely applied to the target detection task due to its excellent recognition performance and remarkable results. In 2016, Redmon et al. introduced the YOLOv1 algorithm, which transformed target detection into a regression problem, leading to an improvement in detection speed [5]. The following year, Redmon et al. achieved further results. The YOLOv2 algorithm was refined from YOLOv1, where the fully connected layer for bounding box prediction led to inferior accuracy in both localization and detection. In contrast, YOLOv2 [6] uses Darknet19 [7] instead of the feature extraction network of YOLOv1, which allows for more robust feature extraction. However, challenges remain, such as suboptimal detection of group targets and detection accuracy. In 2018, Redmon et al. refined YOLOv2 and introduced the YOLOv3 algorithm [8], which incorporates FPN [9] structure to enable multi-scale prediction. However, it still faces problems in detecting targets within groups and achieving accurate detection. Although the overall performance improved significantly, the algorithm still had limitations in terms of detection accuracy and real-time performance. Subsequently, Bochkovskiy et al. proposed the YOLOv4 algorithm [10] in 2020, which used the CSPN as a backbone network for feature extraction. Using CSPDarknet53 [11] as the feature extraction backbone network and applying the SPP module to improve the feature expression capability, this model demonstrates high real-time performance and detection accuracy. However, its complexity also increases. However, the YOLOv5 algorithm has been further improved based on the YOLOv4 algorithm, which shows excellent detection speed and accuracy within the YOLO series. As such, it is more suitable for real-world engineering applications. Therefore, in this study, the YOLOv5 algorithm is used as the basic model for detecting waterlogging in road traffic. As the road traffic waterlogging image dataset is from RSCD, data enhancement techniques have been used to increase the robustness of the model and improve its accuracy. The YOLOv5 backbone network has been augmented with the CBAM attention mechanism to improve the extraction of key information during detection. In addition, the CIoU_Loss loss function was employed to improve the convergence speed. The experiments demonstrate that the enhanced YOLOv5 algorithm has improved the efficacy of waterlogging detection on roads. This confirmation verifies the algorithm's effectiveness and better satisfies the requirements of waterlogging detection in road traffic.

## 2 YOLOv5 Model Structure Analysis

YOLOv5 has four main versions, namely YOLOv5s, YOLOv5m, YOLOv5l, and YOLOv5x. For this paper, we select the YOLOv5s model in the YOLOv5 algorithm as the standard training model. YOLOv5s is the speediest version in the YOLOv5 range with the smallest network structure. Its use of the PyTorch framework is facile to deploy and it is perfectly suited for real-time detection tasks. The YOLOv5 framework comprises four primary modules: the input layer (Input), backbone network layer (Backbone), feature fusion layer (Neck), and output layer (Prediction) [12].

The initial layer of the model is the input layer, which processes the input in the pre-processing stage. The size of the input image is $608 \times 608$. Initially, the data augmentation method of mosaic is applied, which combines four images into a larger one by randomly scaling, cropping and arranging them. This larger image is then used as the training set for data augmentation to increase dataset diversity. Secondly, adaptive anchor frame calculation is utilized in the training phase to enhance the speed and precision of the model training. This is achieved by means of automatic calculation of anchor frame parameters and reducing sample impact.

The second module comprises the backbone network layer, which consists primarily of the Focus structure and the CSP structure [11]. The Focus structure serves as the initial convolutional layer of the network, playing a crucial role in slicing the input feature maps obtained from the input layer, extracting the salient valid information, as well as performing downsampling and compression operations on the feature maps. The CSP structure obtains feature extraction and improvement by splitting the input feature map into two branches and carrying out distinct operations on both branches. The convolutional operations are handled by one branch, whereas the other branch passes through numerous residual structures meant to extract more global features. The incorporation of these two branches into the CSP structure enhances the capacity of the model for effective feature extraction in the stagnant water region of road traffic on the feature map. This in turn leads to a reduction in network parameters and computations, improved extraction of features, and faster model training and inference.

The third module comprises the NECK feature fusion layer, consisting primarily of FPN + PAN architecture [13]. Its principal role involves alignment of the size of the lower-level feature map with that of the higher-level feature map, achieved through the downsampling and upsampling operations, and consequent fusion of the road traffic waterlogging functions via the feature fusion module. By doing this, the module decreases network depth and amplifies target detection.

The PREDICTION output layer is the fourth module which primarily conducts the output and prediction of the target. The anchor box predicts the size and location of the target occurrence while also outputting the predicted category and corresponding confidence level. The structure of YOLOv5 network model is illustrated in Fig. 1.

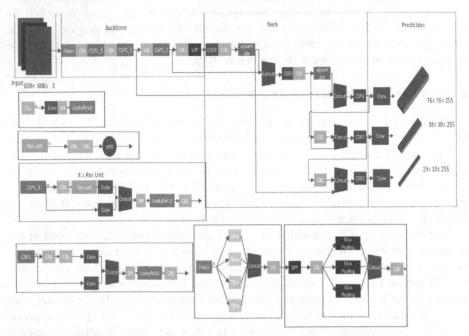

**Fig. 1.** YOLOv5s network model structure

## 3 YOLOv5s Model Structure Improvement

For the conventional YOLOv5 algorithm, this paper suggests two ways to enhance the performance of road water detection. The attention mechanism is akin to human visual attention. When processing information, the human brain will subconsciously concentrate on certain important information and eliminate other unimportant or secondary information. The attention mechanism can enhance the model's ability to suppress noisy background, focus on relevant feature information, optimize the network model, and thereby improve the detection performance of road water pooling. The loss function improvement can speed up convergence of the model and enhance detection accuracy.2.1 Introduction of CBAM Attention Mechanism.

WOO et al. introduced the CBAM [14] (Convolutional Block Attention Module) attention mechanism, illustrated in Fig. 2. The model includes both a channel attention module (SAM) and a spatial attention module (CAM). The channel attention module weighs each channel, while the spatial attention module weighs each spatial location. The employment of both the spatial and channel attention modules strengthens the feature representation of traffic clustering and reinforces the model's robustness, ultimately resulting in enhanced generalizability and elevated accuracy of the detection model.

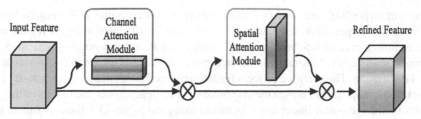

**Fig. 2.** CBAM network structure

The Channel Attention Module (SAM) is presented in Fig. 3. Its underlying process can be briefly divided into three steps: Initially, the input feature map acquires the highest value on each channel via maximum pooling, resulting in a feature vector with channel dimension 1, and the average value on each channel through average pooling, leading to a feature vector with channel dimension 1. Then, the fully connected layer and activation function process these two feature vectors to derive two separate weighting coefficients. These coefficients represent the attention weights linked to the channel maximum pooling and channel average pooling, respectively. The channel attention module then combines the input feature maps with the channel attention weights in a weighted sum by channel. This results in the feature maps that have undergone processing by the channel attention module. Through the channel attention module, the model can adjust each channel's importance and enhance significant features' representation, improving the model's performance. Equation 1 depicts the computational process.

$$M_c(F) = \sigma(MLP(AvgPool(F)) + MLP(MaxPool(F)))$$

$$= \sigma\left(W_1\left(W_0(F_{avg}^c)\right) + W_1\left(W_0(F_{max}^c)\right)\right) \tag{1}$$

In the above formula: $\sigma$ represents the Sigmoid function.

$$W_0 \in R^{C*\frac{C}{r}}, W_1 \in R^{C*\frac{C}{r}}$$

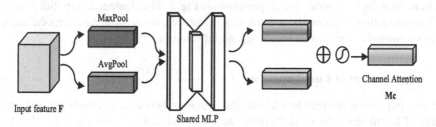

**Fig. 3.** Structure of channel attention module

The structure of the Spatial Attention Module (CAM) is portrayed in Fig. 4. The operation can loosely be divided into four main steps. Initially, the CAM module receives information from the channel attention module and optimises it by means of averaging and maximum pooling operations to extract average and maximum values in each channel direction. These operations aid in decreasing the number of channels in the feature map while retaining crucial information. The average and maximum pooling outcomes are combined with a convolutional layer to incorporate the details within the channels in the spatial dimension. The merging process further isolates the significant information in the characteristic map and decreases the computational intricacy. Next, the results of the join operation undergo a non-linear transformation using the Sigmoid activation function to approximate the output of the network layer to a non-linear function for training and predicting non-linear models. Finally, the activation function is applied to yield a fresh feature map, which is then multiplied by the input feature map of the module, resulting in the final generated feature map. This feature map serves as the CAM module's output, available for future classification, regression, or other assignments. Equation 2 displays the calculation process.

$$M_S(F) = \sigma(f^{7\times7}([AvgPool(F); MaxPool(F)]))$$

$$= \sigma \; f^{7\times7}([F_{avg}^s; F_{max}^s]))$$

(2)

In the above formula: $\sigma$ represents the Sigmoid function.
$7 \times 7$ denotes the convolution kernel size.

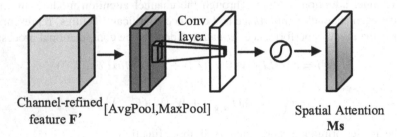

**Fig. 4.** Structure of the spatial attention module

The network structure diagram, illustrating the integration of the CBAM attention mechanism in the backbone layer, is presented in Fig. 5. The channel and spatial attention modules are utilized efficiently in extracting features from the input data, thus enhancing attention towards crucial road waterlogging information.

### 3.1 Improvement of Loss Function

The loss function is utilized to evaluate the deviation between the predicted and actual values of the model. The model's robustness is improved by choosing a loss function with a lower value and faster convergence rate during the training process. The *IoU* loss function [15, 16] possesses scalable invariant, homogeneous, and symmetrical features, but also has some drawbacks. One issue arises when the prediction frame and target frame do not intersect, as demonstrated in Fig. 6(a). The resulting *IoU* of 0 fails to reflect the actual distance between the frames. As a consequence, the loss function is not

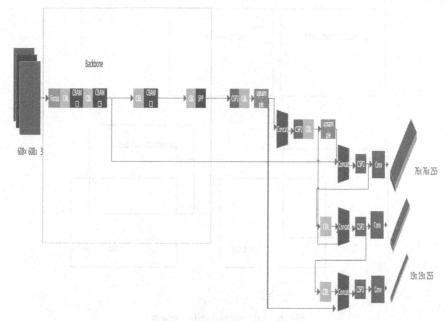

**Fig. 5.** Improved YOLOv5 network structure diagram

derivable, and $IoU$ fails to optimise cases where the frames do not intersect. Moreover, Fig. 6(b) demonstrates that when two prediction frames are of the same size, their $IoU$ values are also identical, rendering it impossible for $IoU$ to distinguish between the two intersection cases. The $GIoU$ loss function is employed in the Prediction segment of the standard YOLOv5 model, and its corresponding formula is presented in Eq. 3.

$$L_{GIoU} = 1 - IoU + \frac{|C - (A \cup B)|}{|C|} \tag{3}$$

In the above formula: $A$ is the prediction frame.

$B$ is the real frame.
$C$ is the smallest outer rectangle between the prediction frame and the true frame.
$IoU = \frac{A \cap B}{A \cup B}$, the value of the intersection of the predicted frame and the true frame compared to the concatenation.

$GIoU$ introduces the minimum outer rectangle based on $IoU$, which ensures that the prediction frame can be made as close as possible to the real frame even if there is no overlap. This solves the issue of no overlap between the detection frame and the real frame in a more objective manner. However, if the prediction frame is contained within the target frame and possesses the same dimensions, as depicted in Fig. 7, the difference sets of the prediction and target frames will be identical. Therefore, the $GIoU$ values will also be identical and in this scenario, the $GIoU$ will collapse to the $IoU$ which fails to discern the relational positioning.

Based on this, the paper utilizes the $CIoU$ [17] loss function. It is compared with the $GIoU$ loss function to measure the extent of the overlap area and aspect ratio of

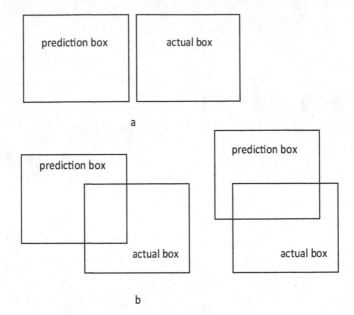

Fig. 6. IoU unavailability graph

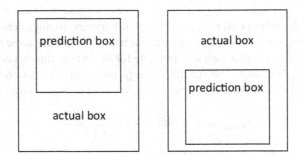

Fig. 7. GIoU unavailability map

the bounding box. This aids in enhancing the localisation accuracy of the target box, promoting the convergence speed of the model, and ultimately increasing the training efficiency.

The calculation formula is shown in Eq. 4.

$$L_{CIoU} == 1 - IoU + \frac{\rho^2(b, b^{gt})}{C^2} + \alpha v \qquad (4)$$

$$\alpha = \frac{v}{(1 - IoU) + v}$$

$$v = \frac{4}{\pi^2}\left(arctan\frac{w^{gt}}{h^{gt}} - arctan\frac{w}{h}\right)^2$$

In the above formula: $\rho$ represents the Euclidean distance between the two centre points.

c is the diagonal length of the smallest enclosing box.

$\alpha$ represents the weighting parameter.

$w$, $h$, $w^{gt}$, $w^{gt}$ represent the width and height of the prediction frame and the actual frame, respectively.

$b$ represents the centre of the projected bounding box.

$b^{gt}$ represents the central point of the actual bounding box.

## 4   Experimental Validation and Result Analysis

### 4.1   Experimental Dataset

The creation of a dataset for waterlogging detection is a crucial aspect of training and evaluating the detection system. This paper utilizes the publicly available RSCD dataset and selects 4000 images, each depicting various scenarios and conditions. From these, 3600 are used for training and the remaining 400 for validation, in a ratio of 9:1. The images are labelled with the labeling tool, as illustrated in Fig. 8. The txt format is used to store the image files, as demonstrated in Fig. 9.

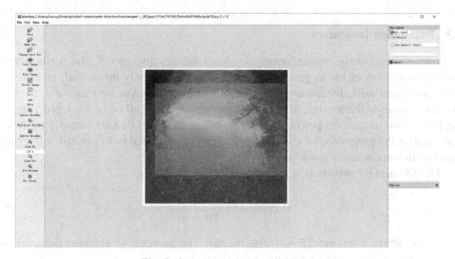

**Fig. 8.** Labelimg labelling diagram

0 0.5266532258064517 0.47830645161290325 0.8460887096774196 0.5737096774193547

**Fig. 9.** Txt format map

## 4.2 Experimental Platform and Environment

The experimental platform is a Windows 10 system with NVIDIA GeForce RTX 3060 graphics card and AMD 5800H CPU, the development framework is PyTorch, the programming language is Python 3.9, and the CUDA version is 11.3. Some of the network parameters set during the training process are shown in Table 1.

**Table 1.** Parameter settings for the training section

| training parameter | parameter value |
| --- | --- |
| momentum | 0.937 |
| weigth_decay | 0.0005 |
| lr0 | 0.01 |
| lrf | 0.2 |
| bacth-size | 10 |
| epochs | 200 |

## 4.3 Evaluation Indicators

In order to objectively and accurately evaluate the training model, the performance evaluation indices of the target detection algorithm are mainly three, such as average precision mean *mPA*, precision precision and recall recall. Among them, the average precision mean is the average of the prediction precision of all categories, precision is the proportion of samples predicted to be positive categories that are actually positive, and recall is the proportion of all samples that can be predicted to be positive among all samples that are actually positive.

The formula for precision is given by Eq. (5)

$$Precision = \frac{TP}{TP + FP} \tag{5}$$

In the above formula: *TP*——*TruePostive* The number of samples detected as positive and confirmed to be positive.

*FP*——*FalsePostive* The number of samples that detected positive but were actually negative.

The recall rate is calculated according to Eq. (6)

$$Recall = \frac{TP}{TP + FN} \tag{6}$$

In the above formula: *FN*——False*Negtive* The quantity of samples which received negative results, yet were in fact positive.

The average precision mean is calculated according to Eq. (7)

$$mPA = \frac{1}{n} \sum_{i=1}^{n} P_{A,i} \tag{7}$$

In the above formula: $P_A = \int_0^1 P(R)dR$, indicates the average precision.
$n$——— denotes the total number of categories

## 4.4 Experimental Results and Analysis

To validate the efficacy of the aforementioned enhancement techniques, this study performs ablation comparative tests on the RSCD public dataset. Table 2 depicts the outcomes. Adding the *CBAM* attention mechanism and modifying the loss function of the model can enhance the network's accuracy. The former can boost the precision rate by 2.7% in *mAP*, whereas the latter can increase it by 0.4%. Combining both of these enhancements through the improved algorithm can yield better outcomes in precision rate, recall rate and *mAP*. Figure 10 displays the precision-recall curve resulting from the training. This curve visually depicts the algorithm's performance, where a larger area enclosed by the curve implies better algorithm performance. As evident from Fig. 10, the trained model exhibits efficient performance. Combined with the ablation experiment results presented in Table 2, it is evident that this paper enhances the YOLOv5 algorithm compared to its original version, without deviating from its essence. Compared to the original YOLOv5 algorithm, the upgraded YOLOv5 algorithm outlined in this paper displays an enhanced accuracy rate of 5.5%, a heightened recall rate of 9.1%, and a boosted *mAP* of 3.7%.

**Table 2.** Results of ablation experiments (%)

| modelling | Precision | Recall | *mAP* |
|---|---|---|---|
| YOLOv5 | 82.6 | 81.3 | 80.7 |
| YOLOv5 + CBAM | 85.1 | 87.5 | 83.4 |
| YOLOv5 + CIoU | 85.8 | 82.3 | 81.1 |
| The algorithms in this paper | 88.1 | 90.4 | 84.4 |

Finally, the model underwent testing and the results are presented in Fig. 11. The leftmost image displays the original experimental figure, the middle represents the original YOLOv5 algorithm detection results, and the rightmost depicts the improved YOLOv5 detection algorithm outcomes. Following the comparison of the outcomes before and after the adjustment, the improved algorithm exhibited a significantly higher confidence level and better performance in identifying road traffic pools compared to the original YOLOv5 algorithm.

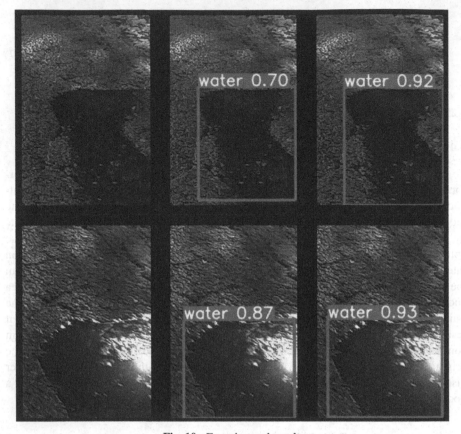

**Fig. 10.** Experimental results

## 5   Conclusion and Outlook

This paper introduces the improvement of YOLOv5 road waterlogging detection model, based on the model, for the YOLOv5 network to add the CBAM attention mechanism, so that the network will focus its attention on the region containing waterlogging; at the same time, in order to solve the original algorithm in the GIoU loss function may cause the model convergence speed is slower, the use of CIoU loss function instead of GIoU loss function to optimize the model, to improve the accuracy of detection of waterlogging on the road. The CIoU loss function is used to replace the GIoU loss function in the original algorithm to optimize the model and improve the detection accuracy of road waterlogging. Finally, the ablation experiment confirms the effectiveness of the improvement scheme proposed in this paper, but the computational volume of the network is still very large, so we will continue to optimize the model in this paper and replace it with a lightweight network to ensure the recognition accuracy while improving the recognition speed and detecting the road water more efficiently.

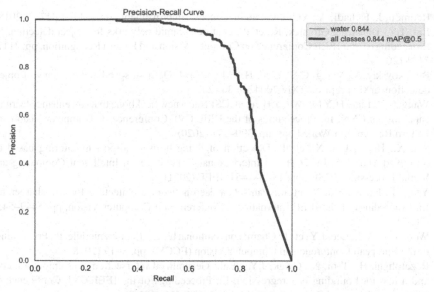

**Fig. 11.** Precision-Recall Curve

**Acknowledgments.** This work was supported by the Jiangsu Provincial College Students Innovation and Entrepreneurship Training Plan Project (grant number 202311276103Y), National Natural Science Foundation of China (grant number 41972111) and the Second Tibetan Plateau Scientific Expedition and Research Program (STEP) (grant number 2019QZKK020604).

# References

1. Lin, T., Liu, X., Song, J., et al.: Urban waterlogging risk assessment based on internet open data: a case study in China. Habitat Int. **71**, 88–96 (2018)
2. Zhang, X., Hu, M., Chen, G., et al.: Urban rainwater utilization and its role in mitigating urban waterlogging problems—a case study in Nanjing, China. Water Resour. Manage **26**, 3757–3766 (2012)
3. Jiang, J., Liu, J., Cheng, C., et al.: Automatic estimation of urban waterlogging depths from video images based on ubiquitous reference objects. Remote Sens. **11**(5), 587 (2019)
4. Liu, Y., Du, M., Jing, C., Cai, G.: Design and implementation of monitoring and early warning system for urban roads waterlogging. In: Li, D., Chen, Y. (eds.) Computer and Computing Technologies in Agriculture VIII. CCTA 2014. IFIP Advances in Information and Communication Technology, vol. 452, pp. 610–615. Springer, Cham (2015). https://doi.org/10.1007/978-3-319-19620-6_68
5. Redmon, J., Divvala, S., Girshick, R., et al.: You only look once: Unified, real-time object detection. In: Proceedings of the IEEE Conference on Computer Vision and Pattern Recognition, pp. 779–788 (2016)
6. Redmon, J., Farhadi, A.: YOLO9000: better, faster, stronger. In: Proceedings of the IEEE Conference on Computer Vision and Pattern Recognition, pp. 7263–7271 (2017)
7. Al-Haija, Q.A., Smadi, M., Al-Bataineh, O.M.: Identifying phasic dopamine releases using DarkNet-19 convolutional neural network. In: 2021 IEEE International IoT, Electronics and Mechatronics Conference (IEMTRONICS), pp. 1–5. IEEE (2021)

8. Redmon, J., Farhadi, A.: YOLOv3: an incremental improvement. arXiv:1804.02767 (2018)
9. Lin, T.Y., Dollár, P., Girshick, R., et al.: Feature pyramid networks for object detection. In: Proceedings of the IEEE Conference on Computer Vision and Pattern Recognition, pp. 2117–2125 (2017)
10. Bochkovskiy, A., Wang, C.Y., Liao, H.YM.: Yolov4: Optimal speed and accuracy of object detection. arXiv preprint arXiv:2004.10934 (2020)
11. Wang, C.Y., Liao, H.Y.M., Wu, Y.H., et al. CSPNet: a new backbone that can enhance learning capability of CNN. In: Proceedings of the IEEE/CVF Conference on Computer Vision and Pattern Recognition Workshops, pp. 390–391 (2020)
12. Shi, X., Hu, J., Lei, X., et al.: Detection of flying birds in airport monitoring based on improved YOLOv5. In: 2021 6th International Conference on Intelligent Computing and Signal Processing (ICSP), pp. 1446–1451. IEEE (2021)
13. Yang, J., Fu, X., Hu, Y., et al.: PanNet: a deep network architecture for pan-sharpening. In: Proceedings of the IEEE International Conference on Computer Vision, pp. 5449–5457 (2017)
14. Woo, S., Park, J., Lee, J.Y., et al.: Cbam: convolutional block attention module. In: Proceedings of the European Conference on Computer Vision (ECCV), pp. 3–19 (2018)
15. Rezatofighi, H., Tsoi, N., Gwak, J.Y., et al.: Generalized intersection over union: a metric and a loss for bounding box regression. In: Proceedings of the IEEE/CVF Conference on Computer Vision and Pattern Recognition, pp. 658–666 (2019)
16. Yu, J., Jiang, Y., Wang, Z., et al.: Unitbox: an advanced object detection network. In: Proceedings of the 24th ACM International Conference on Multimedia, pp. 516–520 (2016)
17. Zheng, Z., Wang, P., Liu, W., et al.: Distance-IoU loss: faster and better learning for bounding box regression. In: Proceedings of the AAAI Conference on Artificial Intelligence, vol. 34, no. 07, pp. 12993–13000 (2020)

# O&M Portrait Tag Generation and Management of Grid Business Application System Under Microservice Architecture

Dequan Gao[1]([✉]), Bing Zhang[1], Meng Yang[1], Bao Feng[2], Lei Xie[1], and Yue Shao[1]

[1] State Grid Information and Telecommunication Branch, Beijing 100761, China
gaodeq@163.com
[2] NARI Group Corporation, Nanjing 211106, Jiangsu, China

**Abstract.** With the development of microservices architecture, O&M in grid business systems is shifting from the traditional device-oriented approach to demand-oriented user experience and operational data analysis. How to achieve intelligent and demand-refined O&M has become the biggest challenge now. To solve this issue, the paper introduces an innovative approach to the automated generation of tags for time series classification through representation learning, significantly reducing tag costs associated with training. Then, focusing on the construction, management and application of portrait tags, this paper analyzes the O&M portrait indicators of grid business application system under microservice architecture, and designs and proposes a framework of portrait tag system for intelligent O&M of grid business application system to provide reference for intelligent O&M of business application system. The purpose of this system is to realize the data association and application of portrait label construction, management and application, and to provide intelligent support for the operation and maintenance of business application system. At the same time, this paper discusses the application of portrait tag in operation and maintenance decision support, anomaly detection, fault analysis and so on. The research results of this paper have important practical significance for improving the stability and security of the system and realizing the intelligent operation and maintenance of the business application system.

**Keywords:** Microservice Application System · Tag Generation · Representation Learning · Operation and Maintenance Portrait

## 1 Introduction

At present, China's power grid retains a large amount of abnormal data, which comes from the operation data of information equipment, platform software, and application systems [1]. As the amount of abnormal data increases, the scale and complexity of the service application system increase, and the operation and maintenance work becomes more and more tedious and difficult. How to realize intelligent operation and maintenance and improve the stability and security of the system has become an important issue concerned by the industry. As a comprehensive and multi-dimensional operation

H. Jin et al. (Eds.): IAIC 2023, CCIS 2059, pp. 59–69, 2024.
https://doi.org/10.1007/978-981-97-1280-9_5

and maintenance management tool, operation and maintenance portrait has become one of the important technical means of intelligent operation and maintenance of business application system. Portrait Tags can help operation and maintenance personnel to understand the system status and problems more quickly, improve the cognition and control of the system, and formulate targeted, refined, personalized service solutions through quantitative sensitivity, so as to improve service quality and service efficiency [2].

This paper aims to put forward a feasible framework around the construction, management and application of portrait tags, and provide reference for intelligent operation and maintenance of business application systems. We propose a novel mechanism to auto-generate tags. The construction and management of portrait tags can support automatic operation and maintenance, fault analysis, and improve the stability and security of the system. The research results of this paper have important practical significance for improving the stability and security of the system and realizing the intelligent operation and maintenance of the business application system.

Our prime contributions are:

(1) We propose a novel mechanism to auto-generate tags for time-series classification through representation learning. By utilizing a select amount of representative time-series data, we ensure that validation by subject experts is both efficient and cost-effective. Our primary objective is to minimize the tag costs associated with training.
(2) We propose a framework of portrait tag system for intelligent operation and maintenance of grid business application systems. The framework is based on the comprehensive analysis of the important feature indicators of the operation and maintenance portrait of the business application system in the dimensions of operation, security and business, and the corresponding tag construction, management and application methods are designed. Through comprehensive analysis of system functions, user access, run logs, and other diversity of data, we realize the data association and application of portrait tags. This gives the operations team a complete picture of how the business application is performing and potential problems.

## 2  Background

### 2.1  Portraits Technology

User portraits serve as a pivotal tool for capturing users based on a myriad of characteristic information, encompassing personal attributes, online activities, and consumer behaviors. Introduced initially by Alan Cooper [3], the pioneer of interaction design, these portraits delve deep into users' social attributes and actions. By constructing such detailed profiles [4–6], they lay the foundation for a more precise and swift analysis of user behaviors and tendencies. This not only facilitates businesses in swiftly pinpointing categorized user groups and discerning their immediate needs but also empowers users with a profound self-awareness.

The evolution of user portrait research has predominantly branched into three distinct trajectories. Firstly, the focus has been on user attributes, aiming to comprehend users by amassing feature data via social annotation systems. This approach primarily aids in understanding the user's core characteristics. Secondly, there's an emphasis on user preferences, striving to enhance the caliber of tailored recommendations by gauging users' interests. Lastly, the spotlight is on user behavior, aiming to foresee user behavioral patterns to avert customer attrition and formulate fitting strategies. For instance, in the realm of power companies, forecasting potential defaults becomes invaluable when equipped with insights into customer traits, thereby guiding decision-making processes.

Several studies have further enriched the domain of user portraits. For instance, Rosenthal et al. [7] utilized text and social attributes to categorize bloggers by age, while Mueller et al. [8] harnessed features from Twitter usernames to discern gender. Guo et al. [9] championed multi-tag classification to refine gender and age predictions. On the other hand, Chicaiza et al. [10] devised a data mining model to craft mobile user profiles using a plethora of data sources, including internet logs and user base information. Such endeavors underscore the versatility and depth of user portraits in various applications, from mobile device usage patterns to personalized service offerings.

## 2.2 Microservice Architecture

Microservice architecture, a paradigm shift in software design, has emerged as a cornerstone in the digital transformation of enterprises. Initially inspired by the need to break down monolithic systems into more manageable, scalable, and flexible components, this architecture emphasizes the decomposition of applications into loosely coupled services that function autonomously [11–13]. With the proliferation of cloud technologies, microservice architectures have facilitated the transition from traditional system-oriented operations to a more demand-driven approach, emphasizing user experience, operational data analysis, and service refinement.

The challenges posed by microservice architectures, especially in the context of power information systems, are multifaceted. These systems, characterized by their vast scale, intricate architecture, and complex business processes, often lack intuitive mechanisms to monitor resource utilization, identify business application hotspots, and assess security risks. This has led to a pressing need for innovative solutions that can enhance system optimization efficiency.

To address these challenges, researchers and practitioners have proposed various strategies. For instance, Cerny et al. [14] emphasized the importance of dynamic resource allocation in microservice environments, while Gortney et al. [15] explored the potential of using machine learning techniques to predict system bottlenecks. Blinowski et al. [16] introduced a framework for aggregating and analyzing operational data from power information systems, aiming to provide actionable insights for system optimization. These studies highlight the ongoing efforts to harness the full potential of microservice architectures, ensuring stability, scalability, and enhanced user experience.

## 3 Operation and Maintenance Portrait System Framework

### 3.1 Framework Design

This section describes the main components of the O&M portrait system framework and their corresponding capabilities. The framework is divided into three main parts, data acquisition and classification, tag generation and management and tag storage and management, the overall structure is shown in Fig. 1, and covers the entire life-cycle of profile tags from data collection to tag generation, management, and application.

The framework leverages a vast array of power system data sources, including system functionality, user access, runtime logs, and a variety of data on anomalies, hazards, alarms, failures, defects, and security vulnerabilities to provide comprehensive coverage of the operational and security complexity of business application systems.

Furthermore, a salient feature of the framework is its capability to support automated operation and maintenance. Through the integration of tagged data with state-of-the-art automated monitoring and diagnostic systems, it not only identifies anomalies and potential faults but also forecasts potential issues. This proactive approach is instrumental in curtailing downtime and optimizing the maintenance process.

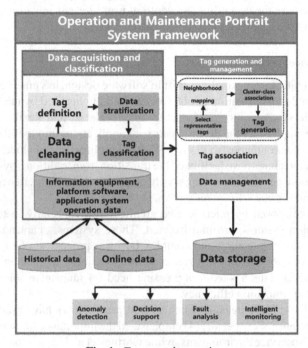

**Fig. 1.** Framework overview.

## 3.2  Data Acquisition and Classification

Data acquisition and classification are crucial steps in the framework for generating and managing profile tags. This section focuses on the processes of data cleaning, tag definition, data stratification, and tag classification.

### 3.2.1  Data Cleaning

Before generating meaningful profile tags, it is essential to perform data cleaning to ensure the quality and integrity of the collected data. Data cleaning involves preprocessing steps such as removing noise, handling missing values, resolving inconsistencies, and eliminating outliers. By applying data cleaning techniques, we can improve the accuracy and reliability of the subsequent analysis and tag generation processes.

For example, smart meters in power systems record data every 15 min. However, due to a variety of reasons, such as transmission errors, hardware failures, or temporary power outages, some meters may report unstable values or not report at all, and any consumption value that suddenly spikes to twice the average of previous readings may be flagged as a potential outlier. Similarly, if a meter has not reported data for a long period of time, interpolation techniques can be employed to estimate the missing value based on adjacent points in time or similar consumption patterns of other meters. By dealing with these anomalies, the cleaned data provides a more accurate representation of the actual electricity consumption, ensuring that the subsequent generation of configured meter labels has a more realistic classification value.

### 3.2.2 Tag Definition

Tag definition involves determining the criteria and guidelines for assigning tags to the collected data. It requires a clear understanding of the operational, security, and business aspects of the business application system. Domain experts play a vital role in defining relevant tags based on their expertise and knowledge of the system. These tags should accurately represent the characteristics, states, and potential issues of the system, enabling effective profiling and analysis.

The tag indicator system includes two categories: operation indicator and operation indicator. Running indicators include indicators at the basic platform layer, system instance layer, service layer, and user experience layer. These indicators reflect the health, stability, and reliability of the system. Operating indicators include platform operating indicators and business operating indicators, which mainly reflect the business quality and business efficiency of the system operation level and provide support for technical operations.

### 3.2.3 Data Stratification

Data stratification involves categorizing the collected data into meaningful subsets based on specific criteria. This process facilitates targeted analysis and enables the generation of more specific and contextually relevant profile tags. Stratification can be performed based on various factors, such as time periods, system components, user roles, or specific events. By stratifying the data, we can focus on different dimensions of the system's operation, security, and business aspects, leading to more accurate and insightful tag generation.

### 3.2.4 Tag Classification

Tag classification aims to assign appropriate tags to the stratified data based on predefined criteria and rules. This process involves analyzing the data and mapping it to predefined tag definitions. Classification algorithms and techniques, such as supervised learning, clustering, or rule-based approaches, can be applied to automatically assign tags to the data. The classification process ensures consistency and standardization in the tag assignment, making the profile tags interpretable and actionable for subsequent analysis and operation and maintenance decision-making.

## 3.3  Tag Generation and Management

### 3.3.1  Selection of Representative Tags

After initial data processing, a subset of time series was randomly chosen to ensure broad representation from the unlabeled data. These selected series were then annotated by domain experts to guarantee accuracy and relevance. For our experiment, the entirety of these annotated series served as the training dataset, though the size can be adjusted based on specific user requirements, ensuring adaptability to various scenarios (Fig. 2).

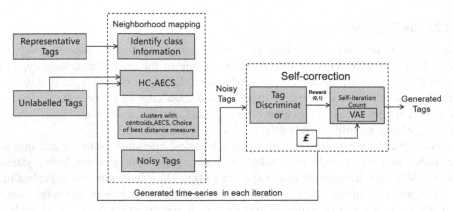

**Fig. 2.** Schematic diagram for proposed approach.

### 3.3.2  Neighborhood Mapping

Our work builds upon the foundational research conducted by Bandyopadhyay et al. [17]. By using a Seq2Seq LSTM [18] multi-layer under-complete auto-encoder, we derive a compact representation, termed AECS, of the time series $X_u$. An under-complete autoencoder is characterized by its hidden layers having fewer nodes than the input/output layers. This design ensures that the latent representation, AECS, is significantly shorter than the original time series, capturing only its most salient features. Once AECS is established, we employ Hierarchical clustering on this condensed representation to determine the optimal distance measure. The selection criterion is based on the Modified Hubert Statistic $(\mathcal{T})$, which assesses the sum of distances between each time-series pair, adjusted by the distance between their respective cluster centers. The clustering outcomes from the three distance measures are ranked according to $\mathcal{T}$, with the highest $\mathcal{T}$ value indicating the best clustering. The corresponding distance measure is then deemed the "best distance measure". Notably, the Mahalanobis distance (ML) [19] is utilized to compute $\mathcal{T}$, gauging both the distance between each time-series pair and the separation between clusters:

$$\mathcal{T} = \frac{2}{n(n-1)} \sum_{X_i \in X} \sum_{X_j \in X} d(X_i, X_j) d(c_i, c_j), \tag{1}$$

$$d(X_i, X_j) = d_{ML}(X_i, X_j); d(c_i, c_j) = d_{ML}(c_i, c_j), \tag{2}$$

where $C_i$ represents the $i^{th}$ cluster and $c_i$ is the center of cluster $C_i$, $d_{ML}(X_i, X_j)$ is the Mahalanobis distance between time-series $X_i$ and $X_j$, and $d_{ML}(c_i, c_j)$ is the Mahalanobis distance between the centres of the clusters to which two time-series $X_i$ and $X_j$ belongs.

The shortened representation diminishes the extensive computational time typically associated with hierarchical clustering, often viewed as its primary drawback. This approach is suitable for both single and multi-dimensional time series, as well as those of varying lengths.

### 3.3.3 Clustering Class Association

Clustering class association [20] groups data points based on similarities and then links these clusters to specific classes or labels. This method enhances traditional clustering by adding class associations, offering a bridge between unsupervised and supervised learning. It provides clearer insights and aids in decision-making by connecting clusters to predefined classes.

A distance matrix dist $\in R^{k \times m}$ is computed using the cluster centroids and representative time series £ exploiting the best distance measure $d_{\text{best}}$:

$$\text{dist}[i,j] = d_{\text{best}}(\text{cen}_i, \pounds_j), \tag{3}$$

where $\text{cen}_i$ is the centroid of $i^{th}$ cluster and $\pounds_j$ is the AECS of $j^{th}$ representative time-series £.

Using dist, the closest sub-group of $X_u$ for each representative instance. Define a list $rep\_clus_i$ which saves each of these associations for £. It is defined as:

$$rep\_clus_i = \underset{k}{\text{argmin dist}_i}, \forall i \in m, \tag{4}$$

where $k$ denotes the cluster for which $\text{dist}[k, i]$ is minimum for $i^{th}$ representative instance and $m$ is the total number of representative instances.

For each cluster $j$, extract the representative instances from £ nearest to $cen_j$ and their corresponding class tags in $y_{\text{ins}}$. Each unlabelled instance in cluster $j$ are labelled as $class_j$. Detailed algorithm for clustering class association is described in Algorithm 1.

---

**Algorithm 1** Clustering class Association

Input: $X_u$: unlabelled time-series, $X_u \in \mathbb{R}^{n \times t \times d}; \{£, y_r\}$ :

representative time-series, $£ \in \mathbb{R}^{m \times t \times d}$; where $m < n$

Output: $y_u$: Associated class tags for unlabelled time-

series $X_u$

Begin

1: $x_{AECS} \rightarrow AECS(£)$

2: $k \rightarrow$ No. of unique elements in $y_r$ (Num of classes)

3: $X \rightarrow X_u \cup £$

4: $X_{AECS}$, Clus, $d_{best}$, Cen $\rightarrow HCAECS(X)$

5: Compute distance of AECS of each representative timeseries to each cluster centroid using best distance measure $d_{best}$ in matrix dist $\in R^{k \times m}$

  6: dist $\rightarrow d_{best}(x_{AECS},$ Cen )

7: Find the cluster whose centroid has minimum distance for each representative time −series using dist matrix. rep_clus $_i \rightarrow \text{argmin}_k$ dist $_i$, $\forall i \in m$

8: for $j = 1,.., k$  do

9:        Find the representative instances nearest to cluster $j$

10:    ins $\rightarrow \{i \mid$ rep_clus $s_i = j\}$

11:    Find the class labels of instances ins in $y_{ins}$

12:    $y_{ins} \rightarrow y_l[$ ins $]$

13:        Class having maximum representative instances nearest to centroid of cluster $j$  is associated to cluster $j$.

14:    class$_j \rightarrow$ Mode$(y_{ins})$

15: **end for**

---

## 3.4  Tag Storage and Management

By defining the tag life-cycle, we can ensure that tags are effectively managed and applied throughout the operation and maintenance process. This definition of life cycle enables tags to be adjusted and updated as the system changes and requirements evolve, thereby improving the efficiency of operation and maintenance and system stability. Our tag life cycle is defined as follows.

At this stage, the tag should be able to accurately describe the characteristics and status of the business application system in the dimensions of operation, security and business. In addition, the tags must have the ability to capture system exceptions, hidden dangers, alarms, and faults.

Operation and maintenance personnel evaluate the usage rate and effect of the tag to determine whether to improve or downgrade the tag.

Our Tag system needs to carry relatively large C-terminal traffic, and the real-time requirements are also relatively high. Therefore, Tags storage should support high-performance queries to cope with large-scale C-end traffic, and should support SQL to facilitate data analysis scenarios.

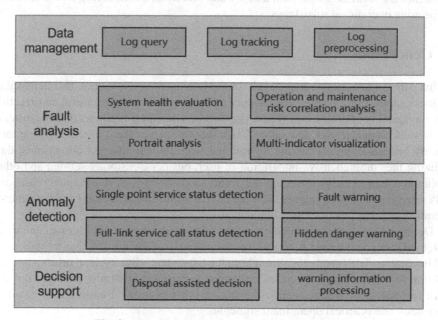

**Fig. 3.** The management framework of portrait tag.

Historical data is mainly stored in HIVE, and basic Tags are imported to Doris. Real-time data is also stored in Doris. Joint query of HIVE and Doris is performed based on Spark, and the calculated results are stored in Redis. After the improvement of this version, the real-time offline engine storage is unified, and the performance loss is within the tolerable range.

Our management framework is shown in Fig. 3. Within the ambit of fault analysis, the framework undertakes a rigorous assessment of system health, leveraging advanced algorithms to discern operation and maintenance risk correlations. Through an in-depth portrait analysis, it elucidates latent vulnerabilities and delineates areas necessitating intervention. The integration of multi-indicator visualizations facilitates a nuanced interpretation of complex data patterns, enabling stakeholders to derive cogent insights.

In the context of anomaly detection, the framework's robustness is underscored by its adeptness in identifying nuanced disruptions in single-point services, achieved through the deployment of state-of-the-art algorithms. Its capacity to continuously monitor full-link service call statuses ensures comprehensive oversight of the service continuum, mitigating potential system inefficiencies. The framework's proactive stance is further accentuated by its dual-tiered warning mechanism, which not only provides immediate alerts for extant faults but also proactively identifies latent threats, facilitating preemptive interventions.

Conclusively, the decision support facet of the framework augments its utility. It not only capacitates administrators with data-driven disposal recommendations but also optimizes the assimilation and response mechanism for warning information. By proffering actionable insights underpinned by empirical evidence, the framework ensures that anomalies are addressed with both alacrity and precision, underscoring its commitment to fostering operational robustness.

## 4 Conclusion

In this paper, we propose a framework for the intelligent operation and maintenance of business application systems. The essence of the framework is the careful construction and management of profile tags. These tags serve as concise descriptors that encapsulate key system characteristics and states. Their design enables them to skillfully capture, classify and associate data across operational, security and business dimensions, thus realizing the comprehensive integration of user, business, customer service and other application data, greatly enhancing the ability of data analysis, sharing and reuse, and fully releasing the management efficiency of power information system and the value of operation and maintenance operation data assets.

Delving deeper into the implications, the framework stands as a beacon of enhanced system stability and robust security. By meticulously analyzing key feature indicators and integrating data correlation throughout the tag system's life-cycle, it offers a holistic solution to the challenges faced in contemporary business system operations. Beyond mere technological advancements, the framework represents a harmonious blend of innovation and practical operational strategies.

While our framework offers a promising approach to the intelligent operation and maintenance of business application systems, it is not without its limitations. One potential constraint lies in the scalability of the profile tag system when faced with exponentially growing data sets or rapidly evolving business environments. Additionally, the reliance on generative modeling, though effective, might not capture all nuances in highly heterogeneous data landscapes.

**Acknowledgment.** This work was supported by the Foundation of State Grid Information & Telecommunication Brach Science and Technology Program under Grant No. 52993920002H.

## References

1. Liang, H., Ma, J.: Data-driven resource planning for virtual power plant integrating demand response customer selection and storage. IEEE Trans. Ind. Inf. **18**, 1833–44 (2021)
2. Rahdari, F., Movahhedinia, N., Khayyambashi, M., Valaee, S.: QoE-aware power control and user grouping in cognitive radio OFDM-NOMA systems. Comput. Networks **189**, 107906 (2021)
3. Cooper.: The Inmates are running the asylum. In: Publishing House of Electronics Industry (2006)
4. Gu, H., Wang, J., Wang, Z., et al.: Modeling of user portrait through social media. In: IEEE International Conference on Multimedia, pp. 1–6 (2018)

5. Huang, K.H., Deng, Y.S., Chuang, M.C.: Static and dynamic user portraits. Adv. Hum. Comput. Interact. **123725**, 1–6 (2012)
6. Xiong, R., Donath, J.: PeopleGarden: creating data portraits for users. In: ACM Symposium on User Interface Software and Technology (1999)
7. Rosenthal, S., McKeown, K.: Age prediction in blogs: a study of style, content, and online behavior in pre- and post-social media generations. In: Annual Meeting of the Association for Computational Linguistics (2011)
8. Mueller, J., Stumme, G.: Gender inference using statistical name characteristics in Twitter. In: Proceedings of the 3rd Multidisciplinary International Social Networks Conference on SocialInformatics, Data Science (2016)
9. Guo, N., Wei, R.K., Shen, Y.P.: Abnormal feature extraction method in large data environment based on user portrait. Comput. Simul. **37**(8), 332–336 (2020)
10. Chicaiza, J., Díaz, P.V.: A comprehensive survey of knowledge graph-based recommender systems: technologies, development, and contributions. Information **12**, 232 (2021)
11. Zhang, J., Huang, W., Ji, D., et al.: Globally normalized neural model for joint entity and event extraction. Inf. Process. Manag. **58**, 102636 (2021)
12. Cerný, T., Donahoo, M., Trnka, M.: Contextual understanding of microservice architecture: current and future directions. ACM Sigapp Appl. Comput. Rev. **17**, 29–45 (2018)
13. Cerný, T., Abdelfattah, A.S., Bushong, V., et al.: Microservice architecture reconstruction and visualization techniques: a review. In: IEEE International Conference on Service-Oriented System Engineering, pp. 39–48 (2022)
14. Tetiana, Y., Bagge, A.H.: Overcoming security challenges in microservice architectures. In: 2018 IEEE Symposium on Service-Oriented System Engineering (SOSE), IEEE (2018)
15. Gortney, M.E., Harris, P.E., Cerný, T., et al.: Visualizing microservice architecture in the dynamic perspective: a systematic mapping study. IEEE Access **10**, 119999–20012 (2022)
16. Blinowski, G., Ojdowska, A., Przybyłek, A.: Monolithic vs. microservice architecture: a performance and scalability evaluation. IEEE Access **10**, 20357–20374 (2022)
17. Bandyopadhyay, S., Datta, A., Pal, A.: Automated label generation for time series classification with representation learning: reduction of label cost for training. arXiv preprint arXiv: 2107.05458 (2021)
18. Tang, R., Zeng, F., Chen, Z., et al.: The comparison of predicting storm-time ionospheric TEC by three methods: aRIMA, LSTM, and Seq2Seq. Atmosphere (2020)
19. McLachlan, G.J.: Mahalanobis distance. Resonance **4**(6), 20–26 (1999)
20. Mattiev, J., Kavšek, B.: CMAC: clustering class association rules to form a compact and meaningful associative classifier. In: International Conference on Machine Learning, Optimization, and Data Science (2020)

# An Erase Code Based Collaborative Storage Method in RFID Systems

Feng Lin[1,2], Yu Liu[2], Yunke Yang[2], Bin Peng[2], Xuemei Cui[2], Chengjiang Qiu[3(✉)], Heng Zhang[1], Ling Li[1], and Shaochen Su[1]

[1] Yunnan Transport Engineering Quality Inspection Co., LTD., Kunming, China
[2] Yunnan Traffic Science Research Institute Co., LTD., Kunming, China
[3] West Yunnan University of Applied Sciences, Dali, China
qiucjfp@wyuas.edu.cn

**Abstract.** Ensuring the safe and reliable storage of important data under limited storage resources is an important challenge faced by RFID systems. Cooperative storage technology based on network coding can divide the transmitted data into blocks and store them redundantly on multiple RFID tags. This paper designs an erase-code-based cooperative storage scheme for important information in a large-scale RFID system. Moreover, the optimization of scheme parameters is studied according to different scenarios to ensure that key data can be successfully recovered under different RFID tag loss probabilities. Finally, the simulation platform is designed to verify the performance of the proposed scheme. Through testing, this study found that when the loss rate of RFID tags is 5%, 10%, 15%, and 20%, respectively, different data block sizes and data redundancies will have different probabilities of successfully restoring the original data.

**Keywords:** Collaborative Storage · Erase Code · RFID System

## 1 Introduction

Radio Frequency Identification (RFID) is a wireless communication technology that enables automatic identification and information acquisition of target objects through radio frequency signals [1,2]. Each RFID tag contains a unique electronic code that is used to identify the target object, and also has the function of storing information that needs to be identified and transmitted. As an important component of the Internet of Things (IoT), RFID systems have been widely used in industries such as intelligent logistics, intelligent transportation, and intelligent healthcare [3–6]. However, due to the limited storage and computing power of RFID tags, and the vulnerability of RFID tags to message tampering, node loss, and other issues [7–11], the security of the Internet of Things (including RFID systems) faces many challenges, although research on IoT security is also constantly evolving [12–15]. Therefore, how to ensure the safe and reliable storage of important data under limited storage resources is an important challenge faced by RFID systems.

Network coding is a method of encoding information into multiple data streams for transmission and storage over a network [16]. It enables encoding and decoding

H. Jin et al. (Eds.): IAIC 2023, CCIS 2059, pp. 70–79, 2024.
https://doi.org/10.1007/978-981-97-1280-9_6

operations on data streams at various nodes in the network, enabling reliable transmission and collaborative processing of information. The principle of network coding is to divide information into small pieces and add some redundant information to each piece, so that errors during transmission can be repaired through these redundant information. Meanwhile, network coding can also improve the throughput and reliability of the network, avoid network congestion and data loss and other problems. Its principle is to divide information into small pieces and add some redundant information to each piece, so that errors during transmission can be repaired through these redundant information. The earliest network coding was based on linear algebraic theory, mainly used to solve the problems of link packet loss and bottleneck issues, but the complexity of encoding and decoding is high, not suitable for large-scale networks [17–19]. Subsequently, in order to solve the complexity problem of basic network coding, researchers proposed distributed network coding, applying linear algebraic theory to distributed networks, achieving reliable and efficient data transmission [20]. With the development of wireless communication technology [21], cloud edge collaborative computing [22] and big data, distributed storage has become an important research direction. Network coding-based distributed storage technology applies network coding to distributed storage, achieving efficient, reliable and secure storage and transmission of data [23].

Cooperative storage technology based on network coding can divide the transmitted data into blocks and store them redundantly on multiple RFID tags. By performing operations such as XOR, matrix multiplication, and Reed-Solomon encoding on the data [24], the fault tolerance of RFID systems is improved, and reliable storage of RFID system data is achieved. The basic idea of RFID system collaborative storage based on network coding is to coordinate multiple RFID tags and achieve distributed storage and processing of information using network coding methods. When RFID tags carrying encoded blocks enter the reading range of a reader, the tags send their stored encoded blocks to the reader. If enough encoded blocks are collected from the tags, the reader can use network coding technology to decode these encoded blocks into the original information. In a large number of existing works, people focus on applying network coding technology to computer networks, but it is rarely applied to the Internet of Things represented by RFID system. This study abstracts RFID tags as distributed storage nodes, and applies the relevant theory of network coding in distributed storage to RFID system collaborative storage, utilizing relevant matrix theory to achieve RFID system collaborative storage and distribute information to multiple RFID tags. In a RFID system composed of a large number of RFID tags, if some tags fail or lose data, we can use erasure code encoding and decoding technology to recover the data from storage of other tags, achieving efficient and reliable storage and recovery of data. The main contribution of this article is proposing a collaborative storage scheme for RFID systems based on erasure codes and designing and implementing a simulation platform to verify the scheme.

## 2  Erasure Code

Erasure Code is a type of coding that can detect and correct data loss or damage, mainly used in distributed storage, cloud storage, and high-speed networks. Erasure Codes can encode original data into multiple coding blocks. In the process of data storage or transmission, even if data loss or damage occurs, it can still be recovered through the encoded blocks, thereby ensuring the integrity and reliability of the data. The principle of Erasure Codes is to divide a file into equal-sized k pieces, then encode these k pieces of data into n encoded fragments using maximum distance separable codes, and select n nodes to store one piece of data respectively. Then, by collecting any k fragments for decoding, the source file can be restored. The general Erasure Codes include the following:

*Redundant Array of Independent Disks Code (RAID Code).* RAID code is the earliest form of Erasure Codes [25]. It divides data into multiple data blocks, encodes these data blocks to generate multiple parity blocks, and then stores the data blocks and parity blocks on different disks separately. When a disk is damaged or lost, it can be recovered through the parity blocks to ensure the integrity and reliability of the data.

*Reed-Solomon Code.* Reed-Solomon Code is a commonly used Erasure Code that uses polynomial calculation for encoding and decoding, which can detect and correct the loss or damage of multiple data blocks [26]. The main advantage of Reed-Solomon Code is that the encoding and decoding complexity is low, which is suitable for large-scale storage and transmission scenarios.

*Fountain Code.* Fountain Code is an Erasure Code based on random coding [27]. It divides the original data into multiple data blocks, randomly encodes these data blocks to generate multiple encoded blocks, and then transmits and stores the encoded blocks. When the receiver receives a sufficient number of encoded blocks, it can obtain the original data through decoding. The main advantage of Fountain Code is its efficient encoding and decoding speed, which is suitable for high-speed networks and real-time transmission scenarios.

*Luby Transform Code (LT Code):* LT Code is an Erasure Code based on random coding [28]. It divides the original data into multiple data blocks, randomly encodes these data blocks to generate multiple encoded blocks, and then transmits and stores the encoded blocks. When the receiver receives a sufficient number of encoded blocks, it can obtain the original data through decoding. The main advantage of LT Code is its efficient encoding and decoding speed, which is suitable for high-speed networks and real-time transmission scenarios.

## 3    Erasing Code Scheme for RFID Systems

The basic principle of erasure codes is to divide a given original file into $k$ equal-sized data blocks, add m redundant check blocks, and form a whole consisting of $k + m$ data blocks. Using the erasure code to encode the whole, when the number of missing data blocks in the whole does not exceed $m$, the lost data blocks can be recovered through computation and the original file can be restored. Take the k+m erasure code storage strategy as an example (where $k = 5$ and $m = 3$): five blocks of raw data are $k_1$, $k_2$, $k_3$, $k_4$, and $k_5$, and three redundant check blocks are $m_1$, $m_2$, and $m_3$.

*The Encoding Process.* Establish the equation between the original data block and the redundancy check block as follows:

$$
\begin{aligned}
m_1 &= k_1 + k_2 + k_3 + k_4 + k_5 \\
m_2 &= k_1 + 2k_2 + 3k_3 + 4k_4 + 5k_5 \\
m_3 &= k_1 + 4k_2 + 9k_3 + 16k_4 + 25k_5
\end{aligned}
\tag{1}
$$

*Stored Procedure.* Stores the above eight blocks of data to different nodes as shown in Fig. 1.

*Recovery Process.* Suppose that any three of the original data $k_i$ are lost, which can be determined by solving the above equation; If the redundant data is lost, you can choose to calculate the above formula again to recover the redundant data.

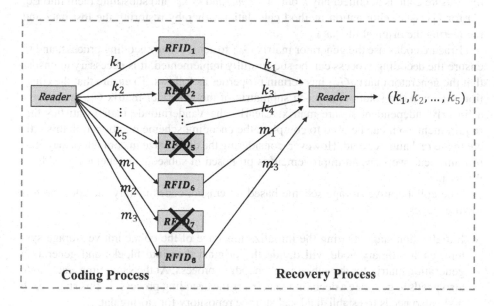

**Fig. 1.** The encoding and recovery process of erasing code.

In order to facilitate the design and discussion of the relationship equation between the original data block and the redundant data block during the encoding process, the coefficients of the first equation in Eq. (1) remain unchanged, while the coefficients of the second equation are changed to $1, 2, 2^2, 2^3, 2^4$, and $2^5$, and the coefficients of the third equation are changed to $1, 3, 3^2, 3^3, 3^4$, and $3^5$. And the example is extended to the case of k+m, so that we can obtain the $k + m$ erasure code encoding matrix in matrix form, as shown below:

$$
\begin{bmatrix} m_1 \\ m_2 \\ m_3 \\ \vdots \\ m_m \end{bmatrix} = \begin{bmatrix} 1 & 1 & 1 & & 1 \\ 1 & 2 & 2^2 & \cdots & 2^{k-1} \\ 1 & 3 & 3^2 & & 3^{k-1} \\ & \vdots & & \ddots & \vdots \\ 1 & m & m^2 & \cdots & m^{k-1} \end{bmatrix} * \begin{bmatrix} k_1 \\ k_2 \\ k_3 \\ \vdots \\ k_k \end{bmatrix} \tag{2}
$$

We can use $M$ to represent the left redundant data block column vector, $K$ to represent the rightmost original data column vector, and $G_{mk}$ to represent the erasure code encoding matrix in the middle, also known as the generator matrix, where each row vector composed of elements is called an encoding vector. In this article, the Vandermonde matrix is selected as the encoding matrix for the erasure code. The decoding process of the erasure code is to collect any $k$ out of $m_1$ $m_m$ and $k_1$ $k_k$, and substitute them into Eq. (2) for Gaussian elimination method calculations, thereby restoring the lost data and recovering the original file data.

Erasure codes use the generator matrix $G_{mk}$ to achieve the encoding process, and to ensure the decoding process can be successfully implemented, it is necessary to ensure that the generator matrix $G_{mk}$ has certain properties as follows. To ensure that the equation system has a solution, any $i$-th submatrix of the generator matrix $G_{mk}$ must form a linearly independent square matrix. Clearly, the Vandermonde matrix satisfies this requirement, so it can be used to complete the encoding scheme mentioned in this article in the real number field. However, considering the digital size in computers may lead to numerical overflow, an improvement is proposed in subsequent schemes to address this issue.

The collaborative storage scheme based on erasure codes mainly includes the following stages.

- Initialization stage: During the initialization stage of the collaborative storage system, each ordinary node will divide the original data into blocks and generate a generation matrix for the subsequent encoding process. At the same time, each node also needs to generate a decoding matrix for the decoding process. In this stage, each node also needs to establish a local storage repository for storing data;
- Encoding stage: In this stage, reader uses the previously generated encoding generation matrix to encode its own data blocks. The encoded data will be sent to the RFID tags of other clusters, and then the cluster header nodes will forward the data to other nodes for storage;
- Data storage stage: In this stage, each ordinary node will send the encoded data blocks to other clusters through the cluster header nodes. This way, even if some nodes fail or data is lost, it can be retrieved from other clusters. Additionally, each node also needs to send the checksum of the data blocks to the cluster header node so that the cluster header node can detect and repair any errors;
- Decoding stage: In this stage, if a data block is lost or corrupted, the previously generated decoding matrix can be used to regenerate this data block. Specifically, you only need to multiply the row vector corresponding to the lost or corrupted data block with the decoding matrix to obtain the original data block.

Now we introduce the details of each process as follows.

In the initialization stage, The Reader first divides the file $A$ into $k$ equal parts $(A_1, A_1, \cdots, A_k)$. If the last piece of data is not long enough, it will be supplemented with zeros to make up the length. The most important task in the initialization stage is for nodes to generate logarithm tables and exponent tables. In the encoding stage, multiplication and division operations in matrix operations are typically converted to logarithm and power operations.

In the encoding stage, by using a Cauchy matrix as the generator matrix, the encoding process is as follows:

$$
\begin{bmatrix} A_1 \\ A_2 \\ A_3 \\ \vdots \\ A_k \\ B_1 \\ B_2 \\ B_3 \\ \vdots \\ B_m \end{bmatrix} = \begin{bmatrix} 1 & 0 & & 0 \\ 0 & 1 & \cdots & 0 \\ 0 & 0 & & 0 \\ \vdots & & \ddots & \vdots \\ 0 & 0 & & 1 \\ \frac{1}{x_1+y_1} & \frac{1}{x_1+y_2} & & \frac{1}{x_1+y_k} \\ \frac{1}{x_2+y_1} & \frac{1}{x_2+y_2} & \cdots & \frac{1}{x_2+y_k} \\ & \vdots & & \vdots \\ \frac{1}{x_m+y_1} & \frac{1}{x_m+y_2} & & \frac{1}{x_m+y_k} \end{bmatrix} * \begin{bmatrix} A_1 \\ A_2 \\ A_3 \\ \vdots \\ A_k \end{bmatrix}
\tag{3}
$$

where the right $k + m$ blocks $(A_1, A_1, \cdots, A_k, B_1, B_2, \cdots, B_m)$ are the encoded data, and the generator matrix is the Cauchy matrix. Then, each of the blocks $(A_1, A_1, \cdots, A_k, B_1, B_2, \cdots, B_m)$ can be transmitted to different RFID tags.

In the decoding stage, when the reader obtains enough RFID tag's encoded blocks, it can recover the original file by generating a decoding matrix. The strategy is as follows: remove the corresponding number of rows of missing data from the encoding matrix, and then remove the corresponding number of rows of missing data from the encoding matrix to supplement the decoding matrix, ensuring that the decoding matrix is a $k$-order square matrix, with each row corresponding to an original data block, as follows:

$$
\begin{bmatrix} A_1 \\ A_2 \\ A_4 \\ \vdots \\ A_{k-1} \\ B_1 \\ B_2 \end{bmatrix} = \begin{bmatrix} 1 & 0 & & 0 \\ 0 & 1 & \cdots & 0 \\ 0 & 0 & & 0 \\ \vdots & & \ddots & \vdots \\ 0 & 0 & & 0 \\ \frac{1}{x_1+y_1} & \frac{1}{x_1+y_2} & \cdots & \frac{1}{x_1+y_k} \\ \frac{1}{x_2+y_1} & \frac{1}{x_2+y_2} & & \frac{1}{x_2+y_k} \end{bmatrix} * \begin{bmatrix} A_1 \\ A_2 \\ A_3 \\ \vdots \\ A_k \end{bmatrix}
\tag{4}
$$

$$
\begin{bmatrix} A_1 \\ A_2 \\ A_3 \\ \vdots \\ A_k \end{bmatrix} = \begin{bmatrix} 1 & 0 & & & 0 \\ 0 & 1 & \cdots & & 0 \\ 0 & 0 & & & 0 \\ & \vdots & & \ddots & \vdots \\ 0 & 0 & & & 0 \\ \frac{1}{x_1+y_1} & \frac{1}{x_1+y_2} & \cdots & & \frac{1}{x_1+y_k} \\ \frac{1}{x_2+y_1} & \frac{1}{x_2+y_2} & & & \frac{1}{x_2+y_k} \end{bmatrix}^{-1} * \begin{bmatrix} A_1 \\ A_2 \\ A_4 \\ \vdots \\ A_{k-1} \\ B_1 \\ B_2 \end{bmatrix} \tag{5}
$$

In Eq. (4), it is assumed that the original data blocks $A_3$ and $A_k$ are missing, so the third and $k$-th rows are removed from the original encoding matrix when calculating, and the $k+1$-th and $k+2$nd rows of the original encoding matrix are added. Then, Eq. (5) performs the inverse matrix calculation on the intermediate generator matrix and performs matrix multiplication to recover the original data.

## 4   Experimental Verification

In this section, we conduct extensive experiments to verify the performance of the proposed method on a PC of CPU with Intel(R) Core(TM)i5-8265H CPU@ 1.60GHz2.30GHz, Memory with 8G, Hard disk with 20GB, Operating system Ubuntu 18.04, Build environment python 3.7, and Virtual machine with VMware® Workstation 14 Pro.

In the experiment, 300 RFID tags and a RFID reader constitute a system. Based on the different working environments, we set a loss rate of 5% to 20% for RFID tags to represent the severity of the RFID system's working environment. Four parallel experiments were conducted with four different loss rates of 5%, 10%, 15%, and 20%. We control the number of encoded data segments to 15, which means the sum of the number of data segments and the number of data redundancies is equal to 15. The number of data segments represents the total number of segments of the original data, and the number of data redundancies represents the redundant data segments added during the encoding process. The encoding parameters for the four experimental scenarios are shown in Table 1. The experimental results are shown in Table 2. We can see that, as the node loss rate continues to increase, it can be seen that the probability of system errors and fragment loss thresholds significantly increases.

**Table 1.** Experimental Scenarios.

| Scenarios | Node Loss Rate | Cod. Parameter 1 | Cod. Parameter 2 | Cod. Parameter 3 |
|-----------|----------------|------------------|------------------|------------------|
| Scenario 1 | 5% | (8, 7) | (10, 5) | (12, 3) |
| Scenario 2 | 10% | (8, 7) | (10, 5) | (12, 3) |
| Scenario 3 | 15% | (8, 7) | (10, 5) | (12, 3) |
| Scenario 4 | 20% | (8, 7) | (10, 5) | (12, 3) |

**Table 2.** Experimental Scenarios.

| Scenarios | Node Loss Rate | Redundancy | Error Rate | Recovery Rate | Threshold |
|---|---|---|---|---|---|
| Scenario 1 | 5% | 87.5% | $2.123*10^{-26}$ | $1-0.1^{20}$ | $5.817*10^{-24}$ |
| | 5% | 50% | $5.279*10^{-14}$ | $1-0.1^{13}$ | $1.446*10^{-16}$ |
| | 5% | 36.3% | $1.752*10^{-10}$ | $1-0.1^{9}$ | $4.798*10^{-13}$ |
| | 5% | 25% | $4.263*10^{-7}$ | $1-0.1^{6}$ | $1.167*10^{-9}$ |
| Scenario 2 | 10% | 87.5% | $2.714*10^{-19}$ | $1-0.1^{18}$ | $7.432*10^{-22}$ |
| | 10% | 50% | $1.687*10^{-12}$ | $1-0.1^{11}$ | $4.606*10^{-15}$ |
| | 10% | 36.3% | $5.139*10^{-9}$ | $1-0.1^{8}$ | $6.514*10^{-12}$ |
| | 10% | 25% | $3.406*10^{-6}$ | $1-0.1^{5}$ | $9.322*10^{-9}$ |
| Scenario 3 | 15% | 87.5% | $4.613*10^{-18}$ | $1-0.1^{17}$ | $1.268*10^{-20}$ |
| | 15% | 50% | $8.716*10^{-15}$ | $1-0.1^{14}$ | $2.399*10^{-17}$ |
| | 15% | 36.3% | $1.217*10^{-11}$ | $1-0.1^{10}$ | $3.502*10^{-14}$ |
| | 15% | 25% | $1.148*10^{-5}$ | $1-0.1^{4}$ | $3.140*10^{-9}$ |
| Scenario 4 | 20% | 87.5% | $3.463*10^{-17}$ | $1-0.1^{16}$ | $9.484*10^{-20}$ |
| | 20% | 50% | $4.916*10^{-14}$ | $1-0.1^{13}$ | $1.346*10^{-16}$ |
| | 20% | 36.3% | $5.382*10^{-11}$ | $1-0.1^{10}$ | $1.473*10^{-13}$ |
| | 20% | 25% | $2.717*10^{-5}$ | $1-0.1^{4}$ | $7.430*10^{-8}$ |

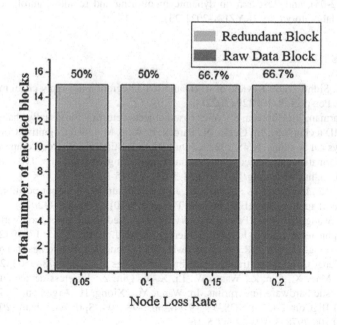

**Fig. 2.** Optimization results of encoding parameters under different loss rates.

To further optimize the encoding parameters, we have added a new set of data redundancy rate parameters for experimentation. Figure 2 depicts the optimization results of encoding parameters under different loss rates. As shown in Fig. 2, when the node loss rate is below 10%, to ensure that the system has a reliability of $1 - 10^{-9}$ and above, a coding scheme with a redundancy of 50% can be used; When the node loss rate is between 15% and 20%, to ensure that the system has a reliability of $1 - 10^{-11}$ or higher, a coding scheme with a redundancy of 66.7% can be used.

## 5    Conclusion

This paper proposes a collaborative storage scheme for RFID systems based on erasure codes. Through testing, when using the erasure code encoding technique, this study found that when the loss rate of RFID tags is 5%, 10%, 15%, and 20%, respectively, different data block sizes and data redundancies will have different probabilities of successfully preserving and restoring the original data. Not only that, on the basis of ensuring a high success rate in restoring the original data, we try to reduce the storage and communication overhead of the system, which is mainly reflected in the system reliability and redundancy parameters in the experiment. The larger the first parameter and the smaller the second parameter, the smaller the additional storage overhead required to achieve reliable storage, and the more reliable the system.

**Acknowledgements.** This work is supported by Research on the construction of intelligent full-process laboratory system based on RFID (Radio frequency identification) technology (No. YCIC-YF-2022-05), and Research on dynamic monitoring and remote control technology of environmental laboratory (No. JKYZLX-2021-23).

## References

1. Tan, W.C., Sidhu, M.S.: Review of RFID and IoT integration in supply chain management. Oper. Res. Perspect. **9**, 100229 (2022)
2. Gayatri Sarman, K., Gubbala, S.: Voice based objects detection for visually challenged using active RFID technology. In: Gupta, N., Pareek, P., Reis, M. (eds.) Cognitive Computing and Cyber Physical Systems. IC4S 2022. Lecture Notes of the Institute for Computer Sciences, Social Informatics and Telecommunications Engineering, vol. 472, pp. 170–179. Springer, Cham (2022). https://doi.org/10.1007/978-3-031-28975-0_14
3. Farahsari, P.S., Farahzadi, A., Rezazadeh, J., et al.: A survey on indoor positioning systems for IoT-based applications. IEEE Internet Things J. **9**(10), 7680–7699 (2022)
4. Chen, D., Wang, H., et al.: Privacy-preserving encrypted traffic inspection with symmetric cryptographic techniques in IoT. IEEE Internet Things J. **9**(18), 17265–17279 (2022)
5. Yang, Y., Wang, H., Jiang, R., et al.: A review of IoT-enabled mobile healthcare: technologies, challenges, and future trends. IEEE Internet Things J. **9**(12), 9478–9502 (2022)
6. Chen, D., Mao, X., Qin, Z., Wang, W., Li, X.-Y., Qin, Z.: Wireless device authentication using acoustic hardware fingerprints. In: Wang, Yu., Xiong, H., Argamon, S., Li, X.Y., Li, J.Z. (eds.) BigCom 2015. LNCS, vol. 9196, pp. 193–204. Springer, Cham (2015). https://doi.org/10.1007/978-3-319-22047-5_16
7. Zuo, Y.: Survivability experiment and attack characterization for RFID. IEEE Trans. Dependable Secure Comput. **9**(2), 289–302 (2011)

8. Shuyu, C., Limin, Y.: A low-overhead PUF for anti-clone attack of RFID tags. Microelectron. J. **126**, 105497 (2022)
9. Chen, D., Zhao, Z., et al.: MAGLeak: a learning-based side-channel attack for password recognition with multiple sensors in IIoT environment. IEEE Trans. Industr. Inf. **18**(1), 467–476 (2022)
10. Wang, Y., Wang, Q., Chen, X., et al.: ContainerGuard: a real-time attack detection system in container-based big data platform. IEEE Trans. Industr. Inf. **18**(5), 3327–3336 (2020)
11. Gao, M., Lu, Y.B.: URAP: a new ultra-lightweight RFID authentication protocol in passive RFID system. J. Supercomput. **78**(8), 10893–10905 (2022)
12. Chen, D., et al.: Audio-based security techniques for secure device-to-device (D2D) communications. IEEE Network **36**(6), 54–59 (2022)
13. Abdulghani, H.A., Nijdam, N.A., Konstantas, D.: Analysis on security and privacy guidelines: RFID-based IoT applications. IEEE Access **10**, 131528–131554 (2022)
14. Sun, J., et al.: A privacy-aware and traceable fine-grained data delivery system in cloud-assisted healthcare IIoT. IEEE Internet Things J. **8**(12), 10034–10046 (2021)
15. Chen, D., Jiang, S., et al.: On message authentication channel capacity over a wiretap channel. IEEE Trans. Inf. Forensics Secur. **17**, 3107–3122 (2022)
16. Wijethilaka, S., Liyanage, M.: Survey on network slicing for internet of things realization in 5G networks. IEEE Commun. Surv. Tutorials **23**(2), 957–994 (2021)
17. Cai, N., Yeung, R.W.: Network coding and error correction. In: Proceedings of the IEEE Information Theory Workshop, pp. 119–122 (2002)
18. Wu, Y.: Network coding for wireless networks. Adaptation and Cross Layer Design in Wireless Networks, CRC Press, pp. 213–242 (2018)
19. Matsuda, T., Noguchi, T., Takine, T.: Survey of network coding and its applications. IEICE Trans. Commun. **94**(3), 698–717 (2011)
20. Usman, M., Yang, N., Jan, M.A., et al.: A joint framework for QoS and QoE for video transmission over wireless multimedia sensor networks. IEEE Trans. Mob. Comput. **17**(4), 746–759 (2017)
21. Zhang, N., Yang, P., et al.: Synergy of big data and 5G wireless networks: opportunities, approaches, and challenges. IEEE Wirel. Commun. **25**(1), 12–18 (2018)
22. Ale, L., Zhang, N., et al.: Online proactive caching in mobile edge computing using bidirectional deep recurrent neural network. IEEE Internet Things J. **6**(3), 5520–5530 (2019)
23. Xie, D., Peng, H., Li, L., et al.: An efficient privacy-preserving scheme for secure network coding based on compressed sensing. AEU Int. J. Electron. Commun. **79**, 33–42 (2017)
24. Tang, Y.J., Zhang, X.: An efficient parallel architecture for resource-shareable reed-solomon encoder. In: IEEE Workshop on Signal Processing Systems, pp. 152–157 (2021)
25. Ramkumar, M.P., Balaji, N., Emil Selvan, G.S.R., Jeya Rohini, R.: RAID-6 code variants for recovery of a failed disk. In: Nayak, J., Abraham, A., Krishna, B.M., Chandra Sekhar, G.T., Das, A.K. (eds.) Soft Computing in Data Analytics. AISC, vol. 758, pp. 237–245. Springer, Singapore (2019). https://doi.org/10.1007/978-981-13-0514-6_24
26. Dimakis, A.G., Godfrey, P.B., Wainwright, M.J., et al.: The benefits of network coding for peer-to-peer storage systems. In: Third Workshop on Network Coding, Theory, and Applications, pp. 1–9 (2007)
27. Mallick, A., Chaudhari, M., Joshi, G.: Fast and efficient distributed matrix-vector multiplication using rateless fountain codes. IEEE ICASSP **2019**, 8192–8196 (2019)
28. Fang, B., Han, K., Wang, Z., et al.: Latency optimization for Luby transform coded computation in wireless networks. IEEE Wireless Commun. Lett. **12**(2), 197–201 (2022)

# Authentication that Combines rPPG Information with Face Detection on the Blockchain

Maoying Wu[1], Wu Zeng[2(✉)], Ruochen Tan[3], Yin Ni[1], and Lan Yang[1]

[1] Department of Eletrical and Electronic Engineering, Wuhan Polytechnic University, Hubei 430023, China
[2] School of Mathematics and Computer Science, Wuhan Polytechnic University, Hubei 430023, China
zengwu@whpu.edu.cn
[3] Computer Science and Engineering, University of California, San Diego, La Jolla, CA 92093, USA

**Abstract.** Blockchain is a decentralized distributed ledger, through a unique chain block structure to verify and store data, so as to ensure the privacy and security of all kinds of information, but the blockchain personal identity authentication link is not complete, relying only on the key can operate personal accounts there are serious security problems. Therefore, a signature scheme is proposed, which takes face biometric as input, uses convolutional neural network (CNN) facenet to encode the face feature information, and then uses homomorphic encryption scheme to encrypt the face encoding and compare it with the face template in the database. At the same time, in order to ensure the authenticity and security of opeation, the user's rPPG signal are detected in real time during the authentication in considred successfully. Finally, using the information mixing algorithm, biometric and RSA key are fused to form a combined key for signature. The experiment shows that, under the condition of obtaining biometric information, the user's identity is verified correctly, and the contract is signed correctly within 2 s. In the whole scheme, the template creation time is 5.62 s, the encryption time of the input biometric information is 0.52 s, the heart rate detection time (including the camera time) is 5.59 s, and the user can be fully identified within four times, with an accuracy of 98%. The scheme improves the security of the blockchain transaction and signing process.

**Keywords:** Blockchain · Face recognize · Fully homomorphic encryption · Heart rate detector · Information fusion · RSA

## 1 Introduction

Blockchain is essentially a distributed ledger, by using encryption technology, transaction records are stored in hashed blocks to maintain and protect the transaction ledger, and data is stored on the network in a decentralized way. For each member of the network, a ledger is maintained containing a copy of every transaction and its corresponding

H. Jin et al. (Eds.): IAIC 2023, CCIS 2059, pp. 80–94, 2024.
https://doi.org/10.1007/978-981-97-1280-9_7

hash value for data integrity [1]. In decentralized storage, it is difficult for any intruder to change the data stored in most locations, so decentralized storage offers higher security compared to centralized storage [2]. Blockchain technology emerged as a solution to the longstanding issue of user trust [3] by providing a decentralized system architecture that enables nodes to conduct credit-based peer-to-peer transactions, coordination, and collaboration in a distributed system without requiring central authority. When a user registers on the blockchain for the first time, a fixed-length private key string is randomly generated by the system to uniquely identify their identity in transactions. The mathematical proof of security ensures that each private key is unique and secure within the blockchain system [4], and if you want to attack the blockchain, at least 51% of the system nodes need to be controlled, but the cost of attack is very expensive, and the attack is meaningless. Therefore, the security of users in the blockchain system only relies on the private key to ensure that once the private key is lost or stolen by others, it will inevitably cause property losses of users, so users should take good care of their private keys.

While distributed electronic trading methods have been successful, authentication of users and the "real identity" of the owner of each transaction is equally important. Currently, in blockchain-based electronic currency transactions, there is little concern about whether the trader is the account owner himself, which is very bad for the security of users and the system. If the user's own biometric characteristics are applied to transaction verification [5], it can effectively solve the problem of easy loss of private keys in the blockchain system and theft by others. Face recognition is now the most popular way of human-computer interaction [6], it can determine identity information through facial features, can be used as an identity in the field of blockchain. The purpose of this biometric facial recognition is to check the identity of the account owner and allow the transaction to proceed only when the identity of the account owner is verified [7], which helps to curb the illegal behavior of network hackers using private keys to steal account funds. However, the recognition process of face recognition system is in clear text state, facing the hidden danger of personal facial information disclosure. Homomorphic encryption technology [8] can perform arithmetic operations in ciphertext state, and the result is the same as that obtained by plaintext operations. Combining homomorphic encryption with face recognition can not only provide data security, but also provide privacy protection for users.

Face recognition systems often face fraud attacks of various fake faces [9], recently, DeepFake [10] has also become the focus of social attention, which is a rapidly developing AI face exchange technology with very realistic visual effects. Therefore, live detection is also a key issue that needs to be considered. This is about the authenticity of the identification. The commonly used live detection method [11] is to make corresponding actions according to the system prompts, such as blinking, turning the head, etc., but this method is cumbersome and time-consuming, and is not suitable for the transaction process, and in the face of challenging attack types such as DeepFake, it often shows low robustness. Related studies [12] have shown that fake faces do not show normal human physiological functions, such as the periodic changes in face skin color caused by heartbeat. Remote photoplethysmography (rPPG) technology is capable of detecting subtle changes in skin color through a camera and analyzing the periodic components of the signal to obtain various physiological parameters related to the human body. When applying the technique to video sequences featuring synthetic faces, the estimated heart rate signal displays substantial deviation compared to that extracted from authentic human faces [13], so the heart rate signal in the video is a powerful tool for detecting fake faces and preventing malicious users from using accounts illegally.

In order to confirm the "real identity" of traders in blockchain transactions, this paper proposes a new solution that applies face recognition technology to blockchain, taking into account the problem of live detection and face information leakage, combines remote photoplethysmography and homomorphic encryption algorithm, and carries out identity verification while ensuring security. The vector sequence with physiological characteristics information is fused with the user key to sign the transaction. This scheme not only determines whether the current trader is the user himself, but also ensures the privacy of the user and improves the security of the transaction.

## 2 Related Work

Security authentication on blockchain is a serious problem. In 2017, LinZhang et al. [14] proposed to design a completely distributed user authentication scheme using blockchain technology, and Abdul Ghafoor Abbasi et al. [15] proposed VeidBlock. Verification can be done without interacting with the identity owner and revealing their personal information. Lee et al. [16] designed a Blockchain-based Distributed biometric Authentication System (BDAS), which provides a distributed mechanism for processing biometric identity authentication. Adopt the blockchain mechanism to provide reliable identity authentication. Rafael Paez [17] proposed an architecture for a blockchain-based biometric electronic identity document (e-ID) system. Ching Sheng Hsu [18] et al. proposed a digital diploma system based on Hyperledger Fabric, which verifies diplomas through facial recognition, prevents others from impersonating them, and links ownership to private keys held by individuals. The use of blockchain technology for identity authentication services, but the identity information is stored outside the blockchain, increasing the risk of information disclosure, and does not solve the identity of the private key is easy to forget and lost, stolen and other problems. Storing the collected user biometric information on the block can solve the problem of information leakage existing in the centralized database, greatly improve the accuracy of identity authentication, and achieve the purpose of protecting the security of user information.

Based on the decentralization and tamper-proof characteristics of the alliance blockchain architecture, Xu et al. [19] proposed a cross-domain dual-unload biometric authentication scheme that uses blockchain to store public biometric information, realized cross-domain biometric authentication of users in local and remote environments, and solved the threat problem of fuzzy extraction technology that is easy to be actively attacked. Bao et al. [20] used blockchain technology to solve the problem of data centralization of traditional identity authentication, and combined with fuzzy extractor, proposed an identity authentication scheme to realize two-factor identity authentication. Simply using blockchain as a tool to store users' identity information will not solve the authentication problem in blockchain-based systems.

YananQi et al. [21] proposed a blockchain transaction system based on fingerprint recognition, randomly selecting a user's fingerprint to generate a private key, and creating a local database to identify the user during sensitive transactions, which could not only prevent others from logging in and using their account, but also handle a large number of concurrent transactions. But fingerprint-based recognition is very easy to forge, which is very unfriendly to users; CarmenBisogni et al. [22] proposed a new signature scheme, which encodes face information and fuses it with RSA key to generate a new combined key, and uses biometric characteristics to sign transactions, which does not play a necessary role in identity verification in ordinary transactions. The original owner of the biological information can be queried only when necessary.

In order to verify the identity by using the characteristics of the user himself on the blockchain, face recognition is introduced, which is already a very mature computer recognition technology, mainly considering the security of its face template database.Yatao Yang et al. apply homomorphic encryption technology to face recognition [23], facenet neural network learns face ciphertext, realizes face classification, and finally realizes face recognition. This scheme ensures the privacy and security of face information in the process of face recognition.

Deepfakes can produce the same appearance as real faces, but cannot mimic their internal physiological features. As early as 2014, V Conotter et al. [24] proposed to use small changes in the face caused by blood flow to determine whether a human face is computer-generated. Hua Qi [25] believes that the facial physiological information of real faces will be destroyed in the process of computer synthesis, and remote photoelectric volume pulse map (ppg) can be used to monitor subtle changes in facial color. Therefore, DeepRhythm is proposed to detect deepfake videos by monitoring heart rate changes. Umur Aybars [26] Ciftci introduced a deepfake detector, FakeCatcher, to extract real physiological information hidden in videos of faces and use it as the reason for the determination. Based on the Deepphys [27] framework, Javier Hernandez-Ortega [12] made changes that are more suitable for detecting Deepfake videos. He proposed DeepFakesON-Phys to study the influence of rPPG on videos, extract and analyze temporal and spatial information from video frames, and detect fake videos. The area under the curve (AUC) and equal error rate (EER) can be compared for the various assays, as shown in Table 1. These results are derived from [12], it can be seen that the performance of DeepFakesON-Phys framework is better.

**Table 1.** Comparison of different deepfake detectors

| Study | Classifiers | Performance | Dataset |
| --- | --- | --- | --- |
| Conotter et al | – | Acc. = 100.0% | Own |
| DeepRhythm | CNN + Attention Mechanism | Acc. = 100.0% | FF + + (FaceSwap) |
| | | Acc. = 100.0% | FF + + (DeepFake) |
| | | Acc. = 64.1% | DFDC Preview |
| FakeCatcher | SVM/CNN | Acc. = 94.9% | FF + + (DeepFakes) |
| | | Acc. = 91.5% | Celeb-DF |
| DeepFakesON-Phys | CAN | AUC = 99.9% | Celeb-DF v2 |
| | | AUC = 98.2% | DFDC Preview |

In consideration of identity verification and the counterfeiting problem, face recognition technology is combined with homomorphic encryption and rPPG to be applied to the blockchain system. Face recognition completes identity verification in the process of blockchain transactions, and heart rate detection and homomorphic encryption are used as face detection in vivo to prevent false face deception and correct face matching under ciphertext. The authenticity of the transaction and the privacy of the user are guaranteed, and the design focuses on user friendliness and privacy. In order to make the transaction information more real and reliable, the information fusion algorithm is introduced, which fuses the biometric characteristics with the user key to generate a combined key to sign the transaction information.

## 3   Methodology

We propose a system framework, as shown in Fig. 1, that combines facial feature information, rPPG, and homomorphic encryption to provide identity authentication for transactions in the blockchain, demonstrating how to ensure the identity of transaction performers while ensuring the privacy information of individuals' faces. Combined with face feature information and RSA key, the combined key is generated for transaction signature.

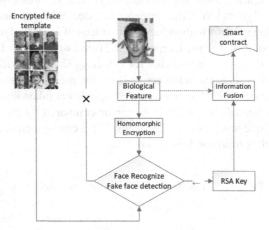

**Fig. 1.** Transaction signature framework

### 3.1   Face Recognition

The Facenet [28] algorithm was proposed by Schroff et al. in Google in 2015. The algorithm uses the image mapping technology of deep learning neural network and triplet loss function to train the neural network, and maps the face image to multidimensional space, whose distance corresponds to the measure of face similarity. The biggest innovation of Facenet is the use of triplet loss function, that is, the square Euclide distance of two image vectors (x, y) with the same identity label is as close as possible. On the contrary, the distance between the two image vectors with different identity tags is as far as possible. The square Euclidean distance is defined as follows:

$$\text{Distance}(x, y) = \sum (\mathbf{xi} - \mathbf{yi})^2 \tag{1}$$

## 3.2  Fully Homomorphic Encryption

In order to protect the face information from threats, a homomorphic encryption algorithm is used to encrypt the face feature vector, which allows the result of specific algebraic operation on ciphertext to be the same as the result of the same operation on plaintext after decryption, which can keep the data in ciphertext form. Total homomorphic encryption is mainly composed of four algorithms: key generation algorithm KenGen(), encryption algorithm Encrypt(), Decrypt(), ciphertext computing algorithm Evaluate().

KenGen ():key generation algorithm, is used to generate encryption private key (pk, sk) and calculating the key Evk (computing key not all solutions must be used).

Encrypt ():encryption algorithm, the use of pk user's public key to Encrypt plaintext message to m, using Encrypt () new cipher encryption cipher get called, namely the cipher text does not have calculated, is an initial ciphertext.

Decrypt (): the use of user private key to Decrypt the ciphertext information c sk corp, to Decrypt the new cipher decryption algorithm can not only, also be able to Decrypt the cipher text after homomorphism calculation.

Evaluate (): cipher computing, said ciphertext information to operation, the cipher text input to the function f is calculated.

Fully homomorphic encryption scheme to data privacy and data using the convenience of clever unifies in together, on the third party service providers don't know, without the decryption key of users' personal data for the corresponding processing and get the correct result, ensure user data privacy and security in the process in a third party.

## 3.3  Fake Face Detection Based on rPPG

Javier Hernandez-Ortega proposed DeepFakesON-Phy [13], a deep fake face detection framework based on heart rate characteristics, which uses remote photoelectric volumetry (rPPG) to detect subtle color changes under skin tissue, in order to analyze heart-rate related information in video sequences and detect fake faces.

The convolutional neural network is used to estimate the heart rate feature from the video. Since our goal is to detect fake faces, we do not calculate the heart rate directly, but use its time derivative as the weight of the classification layer to evaluate the score of each frame in the video, and the score is directly related to the probability of whether the face is real. Figure 2 shows our proposed fake face detection method. After the video is preprocessed in the time domain, the convolutional attention network (CAN) composed of two parallel convolutional neural networks (CNNS) is input respectively. The network can extract spatiotemporal information from the video frame for detecting the DeepFake video synthesized by deep learning.

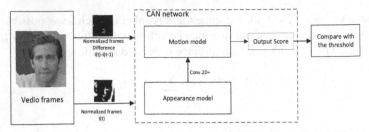

**Fig. 2.** DeepFakesON-Phys process

The convolutional attention Network (CAN) consists of two distinct CNN branches:

Motion Model: Detect changes between consecutive video frames, analyze the video for a brief duration to identify spurious facial images, and the input at time t comprises the normalized difference between the current frame and its preceding frame.

Appearance model: The focus is on analyzing the static information in each video frame, with the goal of providing the motion model with feature points that may contain DeepFake information, a batch of attention masks shared across different layers of the CNN. The input to this branch at time t is the original frame of the video I(t), normalized to zero mean and one standard deviation.

The attention mask from the appearance model is shared with the motion model, taking the output layer of the motion model as the final output of the entire CAN.

The weights learned from the heart rate estimation are used in the detection task of DeepFakes, set a detection threshold to assess whether the input face is a real face, and the faces that appear in the transaction must be detected not only by identity matching, but also by DeepFakesON-Phys scores.

### 3.4 Information Fusion Algorithm

Information fusion algorithm [29] starts with face biometric information and account key to calculate the combined key. An important step of the algorithm is to convert the two vectors containing biometrics and keys into a square matrix, and the number of the two components determines the order of the matrix. The calculation steps are shown in Fig. 3.

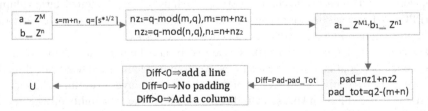

**Fig. 3.** Matrix calculation process

Contain biometric information vector a, b contains key information, $nz_1$, $nz_2$ says the number of each vector to be filled. Define vector $a_1 \in Z^{m_1}$ and $b_1 \in Z^{n_1}$, and the first element of the vector a, b and the first element of the same, the last element respectively $nz_1$, $nz_2$ added.

Padding is performed by components identified by the private key of the same vector, with the first component of the key serving as the index of the first value to be filled; A second component is added to the previous index and gives the index of the second filling value during the loop. In addition, each vector is partitioned and inserted into the matrix U in an appropriate way as a row of U:

$$nblocc\_a_1 = m_1/q \tag{2}$$

$$nblocc\_b_1 = n_1/q \tag{3}$$

The two vectors are computed and filled.

When constructing the matrix U, the insertion order of the blocks is given by the private key. Vector has an index of "0" and vector has an index of "1". In addition, the value of the first component of the private key vector also determines the number of blocks of the vector to be inserted into the matrix, and the value of the next component determines the number of blocks of the other vector, so that if all blocks of one of the two vectors are inserted, all blocks of the other vector will be inserted in turn.

According to the cryptography of the algorithm, the private key selection matrix is composed of 6 $3 \times 3$ permutation matrices, which are obtained from the same matrix:

$$P_0 = \begin{bmatrix} 1 & 0 & 0 \\ 0 & 1 & 0 \\ 0 & 0 & 1 \end{bmatrix} \quad P_1 = \begin{bmatrix} 1 & 0 & 0 \\ 0 & 0 & 1 \\ 0 & 1 & 0 \end{bmatrix} \quad P_2 = \begin{bmatrix} 1 & 0 & 0 \\ 1 & 0 & 0 \\ 0 & 0 & 1 \end{bmatrix}$$

$$P_3 = \begin{bmatrix} 1 & 0 & 0 \\ 1 & 0 & 0 \\ 0 & 1 & 0 \end{bmatrix} \quad P_4 = \begin{bmatrix} 0 & 1 & 0 \\ 0 & 0 & 1 \\ 1 & 0 & 0 \end{bmatrix} \quad P_0 = \begin{bmatrix} 0 & 0 & 1 \\ 0 & 1 & 0 \\ 1 & 0 & 0 \end{bmatrix} \tag{4}$$

The product of two matrices is defined as:

$$F = UP \tag{5}$$

The resulting fusing F matrix is then decomposed and the output along the row builds the V vector, which is the hybrid face coding.

BNIF encoding is closely linked to key and face encoding when implemented, the information it contains is random to external and potential intruders, and the original owner of the biometric information can be identified only through the private key. The FIF algorithm is completely reversible (provided that there is a physiological signature), the original information is obtained by tracing the steps of the FIF algorithm, the private key allows easy transmission of the permutation matrix P, considering the transposition of a single diagonal block, and remembering:

$$P_3 = P_4^T \tag{6}$$

and:

$$P_4 = P_3^T \tag{7}$$

When i = 0,1,2,5 is a symmetric matrix:

$$P_i = P_i^T \tag{8}$$

The union matrix U consists of:

$$U = FP^T \tag{9}$$

By eliminating possible padding, two vectors are obtained: the biometric component vector and the key vector.

### 3.5  User Identification and Contract Signing

To authenticate contract operators in transactions using face recognition and sign transactions using mixed codes, the contract needs to be identified before the contract is instantiated, authentication and fake face detection of the user are performed, and the associated mixed codes are calculated.

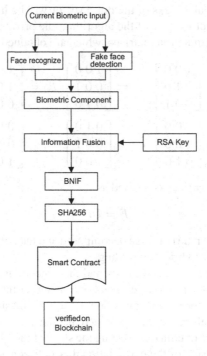

**Fig. 4.** Matrix calculation process

The contract signing process is shown in Fig. 4, which is divided into the following stages:

a, Biometric acquisition: Obtain the current face vector coding and rPPG signal of the user who signs the contract. If the user uses face recognition to sign the contract for the first time, the face biometric template is created for the user. In order to ensure the security of the information, the feature vector in the template is encrypted and stored using homomorphic encryption.

b, identification and fake face detection: will obtain biological characteristics comparing with the template and is used to detect the fake face, authenticate users and their authenticity. If the recognition passes, a facial code with heart rate information is used for the hybrid calculation.

c, the RSA key generation: using the RSA key generation algorithm, the length of 2048 bits key generated.

d, information fusion, will face encoding and RSA key is calculated according to the information fusion algorithm, generating a composite key.

e, contract signing, the combination of the generated by SHA256 encryption key, in the end for the contract signature, instantiate the smart contracts.

The transfer process in the blockchain system is shown in Fig. 5. And is performed using smart contracts:

**Fig. 5.** Transfer trading system

Biometrics are obtained to generate BNIF vector for users to sign transactions. Figure 6 is the user's RSA key and Fig. 7 is the user's BNIF vector. BNIF vector signature ensures the security of transactions and the safe transfer of funds.

16992898292048228782605906527866182154960311866152530931201333966648758733802991780590978579
38258116893471795735633246263799481984639056381897089590893422083470693909363528516129363600
96124972340541488079240926806938038158274127433064295895790046582863561984781173269174618382
67424621726314260352845136149718883312998226703645394079871322828705700909802361051342727240
61053013506901614242814453663820819681549428337540303969527000186721014240415491724122764417
08588508213759190938107713942077080884954098450024310366690637989823988591902442009050466677
0971693103846936110188003354169734246185238279254405351

**Fig. 6.** User's RSA

97998086710086289650874379990707661099682928904228278486205965078266128141595630018166125509
31321033936668449705873832901970850998747579832581611984377197356532364266379491984869306351
88790098590938224803740693930396653285611932366009126947327450414808927049628690830834185274
27133460492595897040768528656398174811726319764913882642762471236126430582815413649118788313
29892267860634534099871732281820750770099803620513147222476015083103509066112448214546368362
80196158944823735034309960572000861217104420154941276421276414087858058132579918039810717943
02707888094504794850043203166609679398832998859102942400950066476790076193138069463101180830
34541967344628513287295420455391609992699992999244700664749044766707946876808760994490

**Fig. 7.** User's BNIF coding

## 4   Experiment and Result

In this section, the experimental results of face detection and contract signing in system design are described.

### 4.1   Face Detection Result

The face detection part mainly includes face recognition and in vivo detection, using the facenet algorithm described in Sect. 3 and the DeepFakesON-Phys framework respectively. Before adding in vivo detection, it can be recognized only by relying on pictures.

The data set used in the face recognition module for testing is the publicly available celeba data set of the Chinese University of Hong Kong, including 202,599 face pictures and 10,177 personal identification marks, all of which are well marked with relevant features, which is very suitable for face-related training. In this dataset, we selected 1000 identities, each identity selected 4 images, and used 4000 images for testing.

We first test whether homomorphic encryption has an impact on the accuracy rate of identity recognition. The results are shown in Table 2. The addition of homomorphic encryption algorithm has no impact on the accuracy rate of face recognition, only there is a difference in time.

**Table 2.**  Face recognition comparison.

| face recognition algorithm | accuracy | Recognization time(s) |
|---|---|---|
| Transactional | 76% | 0.5 |
| Interactive | 76% | 1.03 |

Due to hat occlusion and only side face, face recognition is carried out several times, the results are shown in Table 3. The accuracy rate of the first recognition is 70%. After adjusting the face direction of the unsuccessful user, it can be successfully recognized for up to four times, and the accuracy rate is 98%.

**Table 3.**  Multiple test.

|  | First time | Second time | Third time | Forth time |
|---|---|---|---|---|
| Accuracy | 0.70 | 0.88 | 0.92 | 0.98 |
| Error rate | 0.30 | 0.12 | 0.08 | 0.02 |

After the addition of live detection, it is impossible to rely on "false faces" such as photos and videos for identification, and only the account owner can successfully identify them. Using rPPG information for in-vivo detection is essentially video analysis, and the system records the operator's video and analyzes it during authentication.

The fake face detection experiment used the Celeb-DF v2 public database, which contains 590 raw videos and 5,639 synthetic Deepfake videos, covering different ages, races and genders, as well as a very rich video footage in different lighting and scenes. Enough images are provided to calculate weights for iPPG feature estimation, suitable for DeepFake detection.

As shown in Fig. 8, the real face evaluation score is 1 most of the time and is always higher than the detection threshold, while the fake face score is constantly changing, if only with the naked eye observation, it is difficult to distinguish the true and false video.

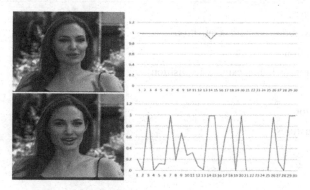

**Fig. 8.** Real and fake face scores

Different values are used as the thresholds for judging the authenticity of video faces, and the experimental results as shown in Fig. 9 are obtained. When the thresholds are distributed between 0.52 and 0.60, the accuracy rate of judging the authenticity of faces is more than 80%. When the threshold is 0.59, the correct rate is 86%, so 0.59 is a reasonable threshold for judgment.

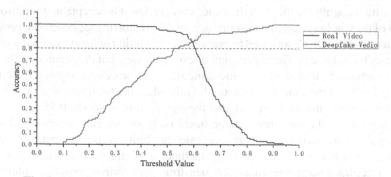

**Fig. 9.** The Accuracy rate of judgment under different threshold values

## 4.2   Results of Cintract Signature

As for the experiment on the execution time of the signature process, as shown in Table 4, the stage that takes the longest is the creation stage of the face template (5.62 s), but this stage only occurs once when the transaction is signed for the first time. When the user signed the contract, it took a total of 8 s, including the biometric encryption time of 0.52 s, the fake face detection time of 5.59s, the identity identification time of 1.03 s, and the RSA key generation time of 0.27 s.

**Table 4.** Time test of contract signing process

| Process | Execution time (s) |
|---|---|
| Template creation | 5.62 |
| Identity check(excluding camera acquisition) | 1.03 |
| Biometric encryption | 0.52 |
| Fake face detection(including camera acquisition) | 5.59 |
| RSA Key generation | 0.27 |
| Information Fusion | 0.139 |
| Transaction Signatures | 0.05 |

### 4.3   Discussion

The complete framework of the signing process and the technologies and algorithms used in each module are presented in detail in this paper. Celeba data set published by the Chinese University of Hong Kong was used to test whether homomorphic encryption would prolong the time of face recognition. The results show that homomorphic encryption has no effect on the accuracy of face recognition. The facenet algorithm combined with homomorphic encryption is 0.5s slower than facenet, and can reach 98% accuracy in the fourth recognition, although the time is longer, but it is acceptable to improve the security of the whole system and user privacy. In the prosthetic face detection, prosthetic face datasets Celeb-DF v2 was used to evaluate the feasibility of heart rate characteristics as the basis for judgment. The experimental results showed that the accuracy values was 86%, and achieved good results. At the same time, the process of signing the contract was described, and the time was measured. Only when the heart rate was detected, the camera spent more time (at least 5 s), and the signing time was only 0.75 s.

In future work, the signature scheme based on heart rate can be replaced by other physiological characteristics, such as heart rate variability, respiratory rate, blood pressure, etc., and a better signal processing method or neural network can be proposed to extract accurate heart rate characteristics from face videos, avoiding interference from light and motion. This allows the detection of deeper fake face videos (generating adversarial network synthesis video and facial deformation, etc.), while a more humane detection method can also improve the user experience, such as reminding the user to change the environment when the light is unstable, rather than directly declaring the detection failure.

## 5  Conclusion

In this paper, we propose a digital transaction authentication scheme based on biometric information. On the basis of protecting user privacy, identity authentication is carried out to sign transactions, and the security of accounts and transactions is improved. This paper presents a complete framework of the signing process, as well as the techniques and algorithms used in it. The combination of face recognition, homomorphic encryption and rPPG can maximize the verification of the account operator, and use the fusion information algorithm to merge the physiological characteristics and RSA key into a hybrid key, making it difficult for malicious users to utilize the signature. Based on the proposed framework, we conducted relevant experiments, and the results show that homomorphic encryption has no impact on the accuracy of face recognition, and the accuracy can reach 98% when repeated for four validations, and the heart rate detection can accurately determine whether the current face is a fake face. The process of signing the contract is described, and the time is measured. Only when the heart rate is detected, the camera takes more time (at least 5 s), and the signing time is only 0.75 s.

## References

1. Chao, L., Debiao, H., Xinyi, H., Khan, K.: DCAP: a secure and efficient decentralized conditional anonymous payment system based on blockchain. IEEE Trans. Inf. Forensics Secur. **15**, 2440–2452 (2020)
2. Pal, O., Alam, B., Thakur, V., Singh, S.: Key management for blockchain technology. ICT Express **7**(1), 76–80 (2021)
3. Johar, S., Ahmad, N., Asher, W., Cruickshank, H., Durrani, A.: Research and applied perspective to blockchain technology: a comprehensive survey. Appl. Sci. **11**(14), 6252 (2021)
4. Swan, M.: Blockchain: Blueprint for a New Economy. O'Reilly Media Inc., Sebastopol (2015)
5. Medikonda, A.K., Padmatti, Y., Kosuru, V.B., Thudumu, R.: Biometric authentication: a holistic review. In: 2018 2nd International Conference on I-SMAC (IoT in Social, Mobile, Analytics and Cloud) (I-SMAC)I-SMAC (IoT in Social, Mobile, Analytics and Cloud) (I-SMAC), 2018 2nd International Conference on, Palladam, India, pp. 428–433 (2018)
6. Sharma, V.K.: Designing of face recognition system. In: 2019 International Conference on Intelligent Computing and Control Systems (ICCS), Madurai, India, pp. 459–461 (2019)
7. Shamini, B.P., Nithish, S.H., Surendar, N.: Bank transaction using face recognition. In: 2022 International Interdisciplinary Humanitarian Conference for Sustainability (IIHC), Bengaluru, India, pp. 772–774 (2022)
8. Cheon, J.H., Kim, A., Kim, M., Song, Y.: Homomorphic encryption for arithmetic of approximate numbers. In: Takagi, T., Peyrin, T. (eds.) Advances in Cryptology – ASIACRYPT 2017. Lecture Notes in Computer Science(), vol. 10624, pp. 409–437. Springer, Cham (2017). https://doi.org/10.1007/978-3-319-70694-8_15
9. Agarwal, A., Yadav, D., Kohli, N., Singh, R., Vatsa, M., Noore, A.: Face presentation attack with latex masks in multispectral videos. In: 2017 IEEE Conference on Computer Vision and Pattern Recognition Workshops (CVPRW), Honolulu, HI, USA, pp. 275–283 (2017)
10. Citron, D.K.: How deepfakes undermine truth and threaten democracy. TED (2019)
11. Galbally, J., Marcel, S., Fierrez, J.: Biometric antispoofing methods: a survey in face recognition. IEEE Access **2**, 1530–1552 (2014)
12. Hernandez-Ortega, J., Tolosana, R., Fierrez, J., Morales, A.: Deepfakeson-phys: Deepfakes detection based on heart rate estimation. arXiv preprint: arXiv:2010.00400 (2020)

13. Erdogmus, N., Marcel, S.: Spoofing face recognition with 3D masks. IEEE Trans. Inf. Forensics Secur. **9**(7), 1084–1097 (2014)
14. Zhang, L., Li, H., Sun, L., Shi, Z., He, Y.: Poster: towards fully distributed user authentication with blockchain. In: 2017 IEEE Symposium on Privacy-Aware Computing (PAC), Washington, DC, USA, pp. 202–203 (2017)
15. Abbasi, A.G., Khan, Z.: Veidblock: verifiable identity using blockchain and ledger in a software defined network. In: Companion Proceedings of the10th International Conference on Utility and Cloud Computing, pp. 173–179 (2017)
16. Lee, Y.K., Jeong, J.: Securing biometric authentication system using blockchain. ICT Express **7**(3), 322–326 (2021)
17. Páez, R., Pérez, M., Ramírez, G., Montes, J., Bouvarel, L.: An architecture for biometric electronic identification document system based on blockchain. Future Internet **12**(1), 10 (2020)
18. Hsu, C.S., Tu, S.F., Chiu, P.C.: Design of an e-diploma system based on consortium blockchain and facial recognition. Educ. Inf. Technol., 1–25 (2022)
19. Xu, Y., Meng, Y., Zhu, H.: An efficient double-offloading biometric authentication scheme based on blockchain for cross domain environment. Wireless Pers. Commun. **125**(1), 599–618 (2022). https://doi.org/10.1007/s11277-022-09567-4
20. Bao, D., You, L.:Two-factor identity authentication scheme based on blockchain and fuzzy extractor. Soft Comput., 1–13 (2021)
21. Qi, Y., Fu, Y., Wang, T., Lv, H.: Pftom: a blockchain based on fingerprint. In: 20Chinese Automation Congress (CAC), Shanghai, China, pp. 5338–5344 (2020)
22. Bisogni, C., Iovane, G., Landi, R.E., Nappi, M.: ECB2: A novel encryption scheme using face biometrics for signing blockchain transactions. J. Inf. Secur. Appl. **59**, 102814 (2021)
23. Yang, Y., Zhang, Q., Gao, W., Fan, C., Shu, Q., Yun, H.: Design on face recognition system with privacy preservation based on homomorphic encryption. Wireless Pers. Commun. **123**(4), 3737–3754 (2021). https://doi.org/10.1007/s11277-021-09311-4
24. Conotter, V., Bodnari, E., Boato, G., Farid, H.: Physiologically-based detection of computer generated faces in video. In: IEEE International Conference on Image Processing (ICIP), pp. 248–252. IEEE (2014)
25. Qi, H., et al.: Deeprhythm: exposing deepfakes with attentional visual heartbeat rhythms. In: Proceedings of the 28th ACM International Conference on Multimedia, pp. 4318–4327 (2020)
26. Ciftci, U. A., Demir, I., Yin, L.: Fakecatcher: detection of synthetic portrait videos using biological signals. In: IEEE Transactions on Pattern Analysis and Machine Intelligence (2020)
27. Chen, W., McDuff, D.:DeepPhys: video-based physiological measurement using convolutional attention networks. In: Ferrari, V., Hebert, M., Sminchisescu, C., Weiss, Y. (eds.) Computer Vision – ECCV 2018. Lecture Notes in Computer Science(), vol. 11206, pp. 349–365. Springer, Cham (2018). https://doi.org/10.1007/978-3-030-01216-8_22
28. Schroff, F., Kalenichenko, D., Philbin, J.: FaceNet: a unified embedding for face recognition and clustering. In: 2015 IEEE Conference on Computer Vision and Pattern Recognition (CVPR), Boston, MA, USA, pp. 815–823 (2015)
29. Iovane, G., Bisogni, C., De Maio, L., Nappi, M.: An encryption approach using information fusion techniques involving prime numbers and face biometrics. IEEE Trans. Sustain. Comput. **5**(2), 260–267 (2018)

# A Memoryless Information Sharing RFID Tag Anti-Collision Protocol

Yan Liu[1], Qi Tang[1], Gang Li[3], Zhong Huang[2], Zihan Huang[1], Xiaochuan Fang[4], and Guangjun Wen[1(✉)]

[1] School of Information and Communication Engineering, University of Electronic Science and Technology of China, Chengdu, Sichuan, People's Republic of China
wgj@uestc.edu.cn
[2] School of Physics and Electronic Engineering, Sichuan Normal University, Chengdu, Sichuan, China
[3] School of Information and Software Engineering, University of Electronic Science and Technology of China, Chengdu, Sichuan, China
[4] School of Electronic Engineering and Computer Science, Queen Mary University of London, London E1 4NS, UK

**Abstract.** Traditional anti-collision algorithms in radio frequency identification (RFID) systems primarily target single-channel, short-distance, point-to-point systems. They often rely on time division multiplexing to avoid tag collisions, resulting in low system efficiency. This article introduces a memoryless information-sharing tag anti-collision protocol. The protocol exploits the multi-channel attributes of a passive cellular IoT system, utilizing relay nodes to access known tag information. Additionally, it employs a bit tracking mechanism, proposing a Shared Intelligent Traversal Tree (BSSTT) protocol based on bit tracking. BSSTT protocol effectively reduces the number of invalid time slots. Simulation results, based on the application of various memoryless protocols in the presented scenario, show that the BSSTT protocol achieves a system recognition efficiency of 86.8% when recognizing 4000 tags, which is 59.1% higher than the existing QT protocol. Moreover, compared to the latest memoryless protocols, BSSTT significantly reduces both the total recognition time and overall energy consumption.

**Keywords:** RFID · Anti-collision · no memory protocol · multi-channel · system efficiency

## 1 Introduction

Radio Frequency Identification (RFID) is a non-contact technology for radio frequency identification implemented through backscatter communication. It is widely used due to its advantages of low cost, easy implementation, and rapid recognition in various fields such as inventory management, intelligent transportation, and indoor positioning [1,2]. RFID systems typically consist of a reader and multiple tags, which can be categorized into active tags (powered

© The Author(s), under exclusive license to Springer Nature Singapore Pte Ltd. 2024
H. Jin et al. (Eds.): IAIC 2023, CCIS 2059, pp. 95–109, 2024.
https://doi.org/10.1007/978-981-97-1280-9_8

by batteries) and passive tags (powered by signals from the reader). Passive tags are widely applied in wireless sensor networks, passive cellular IoT, and other fields due to their benefits of low power consumption, cost-effectiveness, and long lifespan [3]. However, in backscatter communication systems, where RFID readers and receivers operate at the same frequency, the self-interference signal generated by the reader can potentially interfere with the reflected signal from the tags. This reduces the reception sensitivity of RFID readers, thereby limiting the uplink communication distance of the system.

Kimionis et al. proposed a dual RFID Reader (dual base station) backscatter system in 2014 [4], which adopts a separation of transmitter and receiver to avoid the self interference problem of the RFID Reader and effectively improve the system coverage. Based on dual base station backscatter technology, a large number of carrier activators (also known as relay nodes) are deployed to form a passive cellular Internet of Things system architecture as shown in Fig. 1, and its communication link is shown in Fig. 2. In this system, the RFID Reader adopts a frequency division multiple access method, broadcasting query commands to the carrier exciter within the coverage range through the downlink. The carrier exciter selects different frequency channels and forwards the required commands to the labels within the coverage range. Passive cellular IoT systems can provide energy for passive tags through carrier activators. The cost and data complexity of carrier activators are much lower than those of readers and writers. Therefore, this system not only improves the communication distance between traditional readers and tags, but also has the advantages of reducing system power consumption, cost, easy maintenance, and wide access. This article mainly studies efficient tag access protocols under this system.

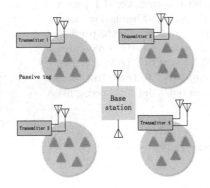

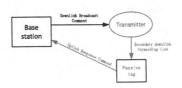

**Fig. 1.** Passive Cellular IoT System Architecture [5,6].    **Fig. 2.** System communication link.

RFID systems primarily utilize two tag access protocols: Aloha-based and tree-based methods [5–11]. The Aloha protocol's efficiency, although economical, peaks at only 0.368 due to its susceptibility to the "label hunger" issue [12,13].

In contrast, tree-based protocols, such as Query Tree (QT) and Binary Partitioning (BS), tackle this problem more effectively [8,14,15]. Memoryless tags, which rely on environmental energy, are emerging as vital components in passive cellular IoT systems. However, they face challenges like data collisions and prolonged idle times, especially in larger systems. The QT protocol, representative of memoryless methods, is known to be time-consuming with increasing tag numbers. Recent research has focused on enhancing QT. Notably, the intelligent traversal tree protocol adjusts query commands based on label density, although it has its trade-offs [11,16]. Bit tracking methods, such as the collision tree protocol, employ collision bits for better query prefixes [10,17,18]. The optimal query tracking tree protocol, using a bit estimation algorithm, offers improved label recognition but with added overheads [19]. Some protocols manage label bitstream lengths, balancing recognition and energy efficiency, albeit with some efficiency trade-offs [20].

Addressing the limitations of traditional protocols and leveraging the characteristics of passive cellular IoT systems, this paper introduces the Shared Intelligent Traversal Tree (BSSTT) protocol. It employs a multi-channel approach and bit tracking mechanism to enhance data throughput. By categorizing tags under multiple carrier activators and recording recognized tag information, the protocol significantly minimizes idle slots. Simulation results indicate that the BSSTT protocol holds advantages in system efficiency, recognition time, and energy conservation.

The rest of this article is organized as follows: Sect. 2 delves into relevant work and common memoryless protocols. Section 3 explicates the proposed system model and BSSTT algorithm. Performance evaluations are presented in Sect. 4, culminating in conclusions in Sect. 5.

## 2 Related Work

Firstly, provide some definitions for correctly understanding the protocol proposed in this article:

- **Time slot**: A duration for tag identification, encompassing RFID Reader commands and tag responses. Time slots are classified into readable (only one tag response on a unique frequency channel, successfully recognized by the RFID Reader), collision (multiple tag responses on the same channel causing recognition failures), and idle (no tag responses to the reader's inquiry on a particular channel) categories.

- **Query**: The bit string sent by the reader to the tag, and the tag ID will only respond if it matches the query. Based on the type of protocol, respond with either the remaining ID or all IDs.

- **Identification cycle**: The time period during which all tag IDs are identified, consisting of different time slots.

Based on the above definition, we introduce some assumptions: assuming that the transmission channel is ideal, there is no interference between frequency channels, and there is no erroneous transmission; The response of the carrier exciter and all tags is synchronized. These assumptions have been widely used to analyze existing anti-collision protocols [8, 10, 19].

## 2.1   Query Tree Protocol (QT)

In the QT protocol, the RFID Reader queries labels, which must match the query with their unique IDs. Successful matches result in a response with the remaining ID. Labels form a binary tree in space. The reader maintains a query stack, starting with an empty string $\epsilon$. It updates the stack based on label responses, navigating the tree. No response leads to an idle slot, a single response to a readable slot, and multiple responses to a collision slot. Collisions prompt the reader to create new queries by adding '0' and '1' to the current query. An empty stack indicates successful identification of all labels. This protocol experiences many idle and collision slots, reducing performance, as seen in Fig. 3.

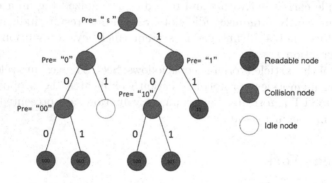

**Fig. 3.** QT algorithm.

## 2.2   Intelligent Traverse Tree Protocol (STT)

The Intelligent Traverse Tree Protocol (STT) uses Query Traverse Path (QTP) to update query prefixes, dynamically adjusting QTP to reach readable nodes through online learning of label distribution, thereby reducing the number of empty and collision nodes. The RFID Reader controls the traversal speed through two adjustable parameters r and w, where r represents the minimum number of layers moved and w represents the maximum number of consecutive collisions or voids recorded. STT reduces a large number of collision nodes at the cost of the number of empty nodes, thereby improving the efficiency of the original QT. However, the appearance of a large number of empty nodes wastes the system's recognition time and energy. To address this issue, this article proposes an information sharing mechanism (skipping known queries) and elaborates on its principle in the third section, effectively reducing the number of empty nodes in the STT algorithm and improving system efficiency.

## 2.3   Bit Tracking Technology

Based on Manchester coding principle, the bit tracking technology defines the value of tag bit as the voltage change within the bit window, as shown in Fig. 4. In the RFID system, tags transmit signals based on the Manchester encoding method, with bit '0' defined as positive conversion encoding and bit '1' defined as negative conversion encoding. When two tags transmit different values at the same time, the positive and negative values of the receiving bits cancel out each other, and the reader cannot recognize it as 0 or 1, thus determining it as a collision bit. As shown in Fig. 4, the second, fourth, and fifth bits collide. When the tag transmission values are the same, it can be correctly received, as shown in the other bits in Fig. 4.

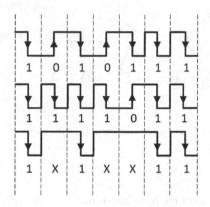

Fig. 4. Example of Manchester encoding.

# 3   Shared Intelligent Traversal Tree Protocol Based on Bit Tracing(BSSTT)

## 3.1   Algorithm Ideas

### 3.1.1   Dynamically Adjusting Query Prefix

A Query Traversal Path (QTP) refers to a sequence of query prefixes, defined as $Q = q_1 q_2 q_3 ... q_n$. An ideal QTP consists of a query prefix set that only contains singleton nodes. Due to the dynamic distribution of tags, this is highly unlikely to achieve. Our goal is to find an optimal QTP with the minimum cost. This requires the algorithm to predict which segmentation layer is most likely to contain the next singleton node. During the label recognition process, when a query node encounters a collision, it indicates that the current level is too high and should move down one level. When an empty node is encountered, it signifies that the current level is too low and should move up one level. When a query node is a singleton node, it suggests that the current level is appropriate, and no movement is needed. The QTP is updated using online learning of the tag distribution, dynamically adjusting it to reach readable nodes, thus reducing the number of empty and collision nodes. Figure 5 provides an example of the dynamic adjustment process. While interacting with different nodes along the traversal path, the behavior varies. Upon reaching node A and encountering a collision, the level descends. At node C, a singleton node, the level stays unaltered. However, at node J, which represents an empty node, the level ascends.

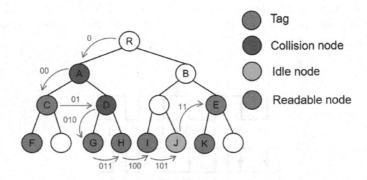

**Fig. 5.** Dynamic query adjustment process.

### 3.1.2   Skip Known Query Mechanism

In the coverage area of a base station, each tag possesses a unique ID, and these tag IDs, under each transceiver, collectively form a spatial binary tree. Each node within this binary tree represents a distinct query command. Figure 5 illustrates a binary tree under a single transceiver, where the green leaf nodes represent tags. The process of querying these tags can be likened to traversing this binary tree. At the outset of the traversal, all nodes within the binary tree are considered unknown (i.e., their node types remain uncertain). However, as the queries progress, certain unknown nodes transition into known nodes, their node types confirmed. The traversal of the binary tree depicted in Fig. 5 and the update process of the known query set are detailed in Table 1. Due to the unique nature of tag IDs, if a new query generated by the transceiver is already present in the query set, accessing that node will inevitably lead to an empty node. Therefore, it becomes evident that the node is, in fact, an empty node, allowing for the generation of a new query. This, in turn, minimizes slot wastage.

**Table 1.** Tag response status and updates to the known query set.

| Serial number | Query prefix | Response status | Known query set |
|---|---|---|---|
| 1 | 0 | Collision time slot | *null* |
| 2 | 00 | Readable time slot | 000 |
| 3 | 01 | Collision time slot | 000 |
| 4 | 010 | Readable time slot | 000, 010 |
| 5 | 011 | Readable time slot | 01, 000, 010, 011 |
| 6 | 100 | Readable time slot | 01, 000, 010, 011, 100 |
| 7 | 101 | Idle time slot | 01, 000, 010, 011, 100 |
| 8 | 11 | Readable time slot | 01, 000, 010, 011, 100, 110 |

## 3.2 Algorithm Flow and Examples

### 3.2.1 Algorithm Flow

The algorithm flowchart is shown in Fig. 6. The algorithm begins by generating a default query prefix set to 0 and updating the query prefix collection as null. It broadcasts the query command to the tags, resulting in three types of tag responses:

- **Collision time slot**: When there are two or more tag responses under this transceiver, the next query command sent by the transceiver under this condition adds '0' after the collision prefix $q_c$, *i.e.*, $q_n = q_c 0$.

- **Readable time slot**: When there is one tag response under this transceiver, or two tag responses with only one collision bit, the next query command sent by the base station to this transceiver is the value of the readable prefix $q_c + 1$, i.e., $q_n = q_c + 1$.

- **Idle time slot**: When there is no tag response under this transceiver, the query length needs to be reduced. If the idle prefix $q_c$ ends with '0', it becomes '1'. If the last bit of the idle prefix value $q_c$ is '1', assuming the bit length of the idle prefix $q_c$ is n, and the number of consecutive '1' bits at the rightmost end of $q_c$ is m, we first flip the bits from n-m to n-1 of $q_c$ ('1' becomes '0', '0' becomes '1'), and then remove the last bit to obtain the new query $q_n$.

In each response time slot, the base station needs to check whether the new query $q_n$ is in the known query set. If it is included, a new query prefix is generated based on the known query mechanism.

### 3.2.2 Algorithm Example

Table 2 provides an example of communication using BSSTT. Assume there are two transceivers, A and B, numbered 1 and 2, respectively, within the coverage range of the base station. The tag IDs under A are "00000", "00010", "00011", "00101", and "01000", while the tag IDs under B are "00001", "00100", "00110",

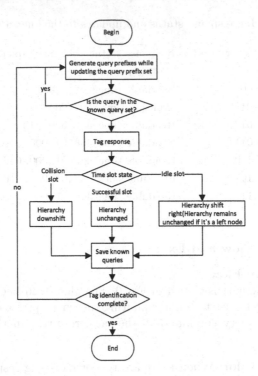

**Fig. 6.** BSSTT Algorithm Flowchart.

"00111", and "01001". The base station initially broadcasts a query command with a prefix of '0', and each transceiver forwards its respective command. Tags respond synchronously on different frequency channels. Throughout the identification process, a total of 14 time slots are consumed to identify 10 tags.

## 3.3    System Energy and Commands

### 3.3.1    System Energy Model

Query commands are sent by the base station and forwarded to the tags within the coverage area by transceivers. Upon receiving a query command, the tags send response signals back to the base station. In this system, tags within the coverage areas of different transceivers respond on different frequency channels without interfering with each other. Therefore, when tags on different frequency channels simultaneously respond to the query, the base station can correctly receive and identify tags from different transceivers.

Figure 7 provides a detailed description of the BSSTT's operational timing, including three types of time slots: collision, readable, and idle. The transmission model follows the EPC C1 Gen2 standard. The query command sent by the base station to the transceiver is referred to as 'Command' and takes time $t_R$. After receiving the 'Command' from the base station, the transceiver forwards

**Table 2.** BSSTT communication example.

| Number | Base station | A | Identify tags | B | Identify tags | Known queries |
|--------|-------------|---|--------------|---|--------------|---------------|
| 1 | (1, 0, 2, 0) | (0) | *null* | (0) | *null* | *null* |
| 2 | (1, 00, 2, 00) | (00) | *null* | (00) | *null* | *null* |
| 3 | (1, 000, 2, 000) | (000) | *null* | (000) | 00001 | 00001 |
| 4 | (1, 0000, 2, 001) | (0000) | 00000 | (001) | *null* | 0000, 00000, 00001 |
| 5 | (1, 0001, 2, .0010) | (0001) | 00010, 00011 | (0010) | 00100 | 000, 0000, 0001, 00000, 00001, 00010, 00011, 00100 |
| 6 | (1, 0010, 2, 0011) | (0010) | 00101 | (0011) | 00110, 00111 | 00, 000, 001, 0000, 0001, 0010, 0011, 00000, 00001, 00010, 00011, 00100, 00101, 00110, 00111 |
| 7 | (1, 01, 2, 0100) | (01) | 01000 | (0100) | 01001 | 00, 000, 001, 0000, 0001, 0010, 0011, 0100, 00000, 00001, 00010, 00011, 00100, 00101, 00110, 00111, 01000, 01001 |

the signal and continuously transmits a CW (continuous wave) for the tags to harvest energy during the time $T_{exc}$. The forwarded query command is referred to as 'Query'. After receiving the 'Query' from the transceiver, the tags generate responses to the base station during time $T_1$, and during time $T_2$, the base station receives and processes all tag responses. If the base station does not receive any tag responses during time $T_3$, it considers the time slot as an idle time slot and no longer waits for tag responses. Additionally, the response time of the tags is $T_t$.

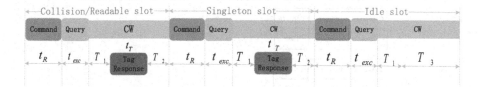

**Fig. 7.** Link timing for collision time slots, readable time slots, and idle time slots.

Let E represent the total energy consumed by the identification cycle of the system. The power for transmitting commands from the base station is denoted as $P_{tx}$, the power for the transceiver to receive the base station's signal is denoted as $P_{cmd}$, the power for the transceiver to excite the tags is denoted as $P_{exc}$, and the power for the base station to receive data returned by the tags is denoted as $P_{rx}$. Therefore, E is represented as follows:

$$E = E_c + E_s + E_i$$
$$= \sum_{j=0}^{c+s} \left[ (P_{\text{tx}} + P_{\text{cmd}}) \cdot t_R + (P_{\text{tx}} + P_{\text{exc}}) \cdot \left( T_{\text{coll}}^j + T_{\text{succ}}^j \right) + P_{\text{rx}} \cdot T_t^j \right] \tag{1}$$
$$+ \sum_{j=0}^{i} \left[ (P_{\text{tx}} + P_{\text{cmd}}) \cdot t_R + (P_{\text{tx}} + P_{\text{exc}}) \cdot T_{\text{idle}}^j \right]$$

$$T_{\text{succ}} = T_{\text{coll}} = t_{\text{exc}} + T_1 + T_t + T_2 \tag{2}$$

$$T_{\text{idle}} = t_{\text{exc}} + T_1 + T_3 \tag{3}$$

where $T_{succ}$, $T_{coll}$, and $T_{idle}$ respectively represent the time consumed in readable, collision, and idle time slots. $E_s$, $E_c$, and $E_i$ respectively represent the energy consumed in readable, collision, and idle time slots. Additionally, s, c, and i represent the number of readable, collision, and idle time slots.

### 3.3.2 Command Format

Figure 8 provides detailed command formats and tag responses. The frame header information includes header, mask, command, and address information. The base station broadcasts commands in Broad, which contains the number of each transceiver and its corresponding query command. Each transceiver receives and forwards only the query command corresponding to its number from the Broad command. Due to the varying lengths of query commands, tag response data is variable.

(a)

| Frame header | Number | Query | Number | Query | ... | Number | Query | CRC-16 |
|---|---|---|---|---|---|---|---|---|
| 37bit | 1 | Variable | 2 | Variable | ... | n | Variable | 16bit |

(b)

| Command | Query | CRC-16 |
|---|---|---|
| 4bit | Variable | 16bit |

(c)

| Preamble | Response | CRC-16 |
|---|---|---|
| 9bit | Variable | 16bit |

**Fig. 8.** Base station broadcast command, transceiver query command, and tag response(a) Base station broadcast command Broad, (b) Transceiver query command Query, and (c) Tag response command Query.

## 3.4   Algorithm Performance Analysis

In a complete binary tree, there are $2^L$ nodes on the Lth level. Assuming there are M tags within the transceiver's coverage range, and tag responses are independent, the probability of multiple tags being assigned to the same node follows a binomial distribution. Let P be the probability of a specific tag selecting the same node among the $2^L$ nodes. We can deduce that on the Lth level, the probability of K tags simultaneously selecting the same node is:

$$P(M,K)_L = \binom{M}{K} P^k (1-P)^{M-K} \tag{4}$$

In accordance with the Eq. 4, the probabilities of empty nodes, readable nodes, and collision nodes are as follows:

$$P(M, K \geq 2)_L = 1 - \left( \binom{M}{0} P^0 (1-P)^{M-0} + \binom{M}{1} P^1 (1-P)^{M-1} \right) \tag{5}$$

With tags uniformly distributed and the prior condition for a node to be queried being that its parent node is a collision node, where each node has an equal collision probability, let $h(i)_L$ represent the probability of the ith node in the Lth layer being queried, and $a(i)_L$ represent the probability of the ith node in the Lth layer being a collision node. Thus, we can derive:

$$a(i)_L = a_L = P(M, L \geq 2)_L \tag{6}$$

Hence, the probability of a node being queried is:

$$h(i)_L = h_L = \begin{cases} 1, & \text{if } L = 0 \\ a(i)_{L-1}, & \text{if } L \geq 1 \end{cases} \tag{7}$$

The probability of all nodes being queried is:

$$\begin{aligned} T(M) &= 1 + \sum_{L=1}^{\infty} 2^L a_{(L-1)} \\ &= 1 + 2 \sum_{L=0}^{\infty} 2^L \left[ 1 - \left(1 - 2^{-L}\right)^M - M \cdot 2^{-L} \left(1 - 2^{-L}\right)^{M-1} \right] \end{aligned} \tag{8}$$

The total number of collision time slots during the tag recognition process is:

$$C(M) = \sum_{L=0}^{\infty} \sum_{i=1}^{2^L} a(i)_L = \sum_{L=0}^{\infty} 2^L \cdot a_L = \frac{1}{2} \cdot (T(M) - 1) \tag{9}$$

The number of idle time slots obtained is:

$$I(M) = T(M) - C(M) - M = \frac{1}{2} \cdot T(M) + \frac{1}{2} - M \qquad (10)$$

Using bit tracking and skipping known nodes, we calculate the total number of BSSTT time slots based on the collision bit count (N) and the skipped time slots (S). N varies from M/2 to 0, and S ranges from M to 0.

$$T(M)_{BSSTT} = T(M) - 2N - S, \quad 0 \le N \le \frac{M}{2}, \quad 0 \le S \le M \qquad (11)$$

The number of collision time slots in BSSTT and the number of idle time slots are:

$$C(M)_{BSSTT} = C(M) - 2N, \quad 0 \le N \le \frac{M}{2} \qquad (12)$$

$$I(M)_{BSSTT} = I(M) - S, \quad 0 \le S \le M \qquad (13)$$

## 4    Simulation Results Analysis

We applied the BSSTT protocol and compared its performance with existing memoryless protocols in the scenario presented in this paper. We conducted evaluations using MATLAB in a simulated environment consisting of one base station and two transceivers. In this setup, tags were uniformly distributed around each transceiver, ensuring that all tags could receive commands from the transceivers without errors, and tag IDs were uniformly distributed. We assumed ideal communication with no transmission errors, and there was no interference between frequency channels. The relevant simulation parameters are defined in Table 3, and they comply with EPC standards. All simulation results represent an average of 1000 iterations to ensure the fairness of the experiments.

**Table 3.** Simulation Parameter Definitions.

| Parameters | Values | Parameters | Values | Parameters | Values |
|---|---|---|---|---|---|
| B | 12 bit | Rate | 160 kbps | $T_1$ | 25 $\mu$s |
| $T_2$ | 25 $\mu$s | $T_3$ | 30 $\mu$s | $P_{exc}$ | 825 mW |
| $P_{rx}$ | 125 mW | $P_{tx}$ | 825 mW | $P_{cmd}$ | 125 mW |

This paper compares several representative memoryless protocols, including QT, STT, CWT, and QWT. All protocols were simulated using the simulation parameters defined in Table 3 and the same simulated scenario. The number of tags varied from 100 to 4000, with IDs uniformly distributed. Tags were uniformly distributed around two transceivers. The protocol performance was evaluated in terms of slot count, system efficiency, total recognition time, and energy consumption, among others. The simulation results, as shown in Fig. 9, indicate that the performance of BSSTT is dependent on the density of tags. With a higher number of tags and increased density, the algorithm's performance improvement becomes more pronounced. The protocol performance was evaluated, including slot count, system efficiency, total recognition time, and energy consumption, among other aspects.

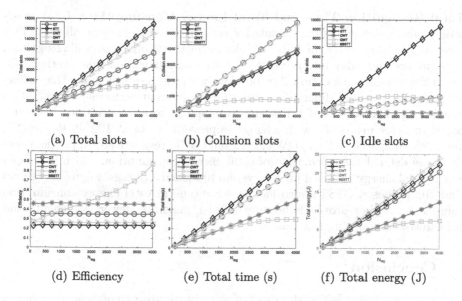

(a) Total slots     (b) Collision slots     (c) Idle slots

(d) Efficiency     (e) Total time (s)     (f) Total energy (J)

**Fig. 9.** Simulation results with uniformly distributed tags

**Total Slot Count:** As observed in Fig. 9, the five schemes consume total slot counts in descending order: STT, QWT, QT, CWT and BSSTT. BSSTT outperforms other memoryless protocols in terms of total slot consumption, requiring the fewest slots to identify the same number of tags. Specifically, compared to STT, QWT, QT, and CWT, BSSTT saves 72.3%, 69.9%, 59.1%, and 46.8% of slots when the tag count is 4000, respectively. From Fig. 9(c), it can be seen that BSSTT reduces idle slots to some extent compared to STT. At the same time, as shown in Fig. 9(b), BSSTT has an advantage in collision slots compared to other protocols.

**System Efficiency:** According to existing work, system efficiency is widely considered an important metric for evaluating RFID system performance, especially in assessing RFID anti-collision protocol performance. System efficiency, as defined in relevant anti-collision literature, is the ratio of the number of successfully identified tags to the total slot count consumed for their identification. Figure 9(d) shows that BSSTT outperforms other memoryless protocols in terms of system efficiency, with average improvements of 52.8%, 48.9%, 31.8%, and 11.4% compared to STT, QWT, QT, and CWT, respectively. However, considering the differences in protocol duration, a comprehensive comparison of actual performance should also consider factors such as total recognition time and energy consumption.

**Total Recognition Time and Total Energy Consumption:** Total recognition time is defined as the cumulative time required to identify all tags. This measure can help reduce the impact of slot count variations on overall recognition performance and takes into account the total communication time overhead for the entire system, as mentioned in previous work. Figure 9(e) shows that different algorithms exhibit different rankings under different performance evaluation metrics. Specifically, BSSTT has an advantage in total recognition time compared to other protocols, with average improvements of 52.1%, 46.1%, 46.2%, and 9.8% compared to STT, QWT, QT, and CWT, respectively. In previous sections, we defined the calculation of total energy consumption, which is defined as the total energy required for the system to identify all tags. Figure 9(f) shows that the proposed BSSTT also has an advantage in total energy consumption compared to other protocols because BSSTT generates fewer slot counts and transmitted bits during tag reading.

## 5   Conclusion

The passive cellular IoT system can effectively improve communication range. In this paper, a collision avoidance protocol for passive RFID tags with no memory sharing is proposed based on this scenario. BSSTT employs a skip-known query mechanism and a bit tracking mechanism, reducing a significant number of idle and collision slots, enhancing system efficiency, and reducing identification time and energy consumption. Finally, through a performance comparison with existing memoryless protocols, it is demonstrated that the proposed BSSTT can effectively enhance the efficiency of passive cellular IoT systems. In summary, BSSTT is a candidate solution for improving system efficiency and reducing system identification time and energy consumption, with a positive impact.

## References

1. Yuan, L.: Research and practice of RFID-based warehouse logistics management system. In: Proceedings of the 2019 International Conference on Smart Grid and Electrical Automation (ICSGEA), IEEE (2019)
2. Lu, J., Yang, L., Qiu, Y., et al.: An improved RFID anti-collision algorithm and its application in food tracking. In: Proceedings of the 2019 14th International Conference on Computer Science & Education (ICCSE), IEEE (2019)
3. Landaluce, H., Arjona, L., Perallos, A., et al.: A review of IoT sensing applications and challenges using RFID and wireless sensor networks. Sensors **20**(9), 2495 (2020)

4. Kimionis, J., Bletsas, A., Sahalos, J.N.: Increased range bistatic scatter radio. IEEE Trans. Commun. **62**(3), 1091–1104 (2014)

5. Chen, W.-T.: An accurate tag estimate method for improving the performance of an RFID anticollision algorithm based on dynamic frame length ALOHA. IEEE Trans. Autom. Sci. Eng. **6**(1), 9–15 (2008)

6. Knerr, B., Holzer, M., Angerer, C., et al.: Slot-wise maximum likelihood estimation of the tag population size in FSA protocols. IEEE Trans. Commun. **58**(2), 578–585 (2010)

7. Su, J., Sheng, Z., Leung, V.C., et al.: Energy-efficient tag identification algorithms for RFID: survey, motivation and new design. IEEE Wirel. Commun. **26**(3), 118–124 (2019)

8. Law, C., Lee, K., Siu, K-Y.: Efficient memoryless protocol for tag identification. In: Proceedings of the 4th International Workshop on Discrete Algorithms and Methods for Mobile Computing and Communications, IEEE (2000)

9. Myung, J., Lee, W., Srivastava, J., et al.: Tag-splitting: adaptive collision arbitration protocols for RFID tag identification. IEEE Trans. Parallel Distrib. Syst. **18**(6), 763–775 (2007)

10. Jia, X., Feng, Q., Yu, L.: Stability analysis of an efficient anti-collision protocol for RFID tag identification. IEEE Trans. Commun. **60**(8), 2285–2294 (2012)

11. Pan, L., Wu, H.: Smart trend-traversal protocol for RFID tag arbitration. IEEE Trans. Wireless Commun. **10**(11), 3565–3569 (2011)

12. Nithya, S., Jerlin, M., Priya, S.: Tag starvation and tag collisions in RFID system-A solution. Int. J. Eng. Technol. **5**(3), 2519–2522 (2013)

13. Barletta, L., Borgonovo, F., Cesana, M.: A formal proof of the optimal frame setting for dynamic-frame Aloha with known population size. IEEE Trans. Inf. Theory **60**(11), 7221–7230 (2014)

14. Myung, J., Lee, W., Shih, T.K.: An adaptive memoryless protocol for RFID tag collision arbitration. IEEE Trans. Multimedia **8**(5), 1096–1101 (2006)

15. La Porta, T.F., Maselli, G., Petrioli, C.: Anticollision protocols for single-reader RFID systems: temporal analysis and optimization. IEEE Trans. Mob. Comput. **10**(2), 267–279 (2010)

16. Pan, L., Wu, H.: Smart trend-traversal: a low delay and energy tag arbitration protocol for large RFID systems. In: Proceedings of the IEEE INFOCOM 2009, IEEE (2009)

17. Chen, W-C., Horng, S-J., Fan, P.: An enhanced anti-collision algorithm in RFID based on counter and stack. In: Proceedings of the 2007 Second International Conference on Systems and Networks Communications (ICSNC 2007), IEEE (2007)

18. Su, J., Chen, Y., Sheng, Z., et al.: From M-Ary query to bit query: a new strategy for efficient large-scale RFID identification. IEEE Trans. Commun. **68**(4), 2381–2393 (2020)

19. Lai, Y.-C., Hsiao, L.-Y., Chen, H.-J., et al.: A novel query tree protocol with bit tracking in RFID tag identification. IEEE Trans. Mob. Comput. **12**(10), 2063–2075 (2012)

20. Landaluce, H., Perallos, A., Onieva, E., et al.: An energy and identification time decreasing procedure for memoryless RFID tag anticollision protocols. IEEE Trans. Wireless Commun. **15**(6), 4234–4247 (2016)

# Prudent Promotion, Steady Development: Capability and Safety Considerations for Applying Large Language Models in Medicine

Sheng Xu[1], Shuwen Chen[1,2], and Mike Chen[1,2(✉)]

[1] School of Physics and Information Engineering, Jiangsu Second Normal University, Nanjing 211200, China
mc277509@ohio.edu
[2] Jiangsu Province Engineering Research Center of Basic Education Big Data Application, Nanjing 211200, China

**Abstract.** The powerful capabilities of large language models (LLMs) in the medical field have been affirmed by various benchmark tests. However, safety assessments are indispensable for high-risk medical applications. We systematically analyzed LLMs, including their technical principles, applicability in healthcare, medical capability assessment, potential security risks, and countermeasures. The study found that LLMs demonstrate strong capabilities in medical text processing, decision support, and text generation, but also have risks like "hallucination". To ensure safe and effective applications of LLMs, we analyzed the causes of hallucination and proposed using output detection and prompting techniques to mitigate hallucinations generated by LLMs. The results affirm that LLMs can advance AI innovation in healthcare, but need to be introduced prudently without comprehensive safety assessments.

**Keywords:** Large language models · medical application · safety

## 1 Introduction

In recent years, pretrained language models have achieved remarkable progress in natural language processing. The emergence of large language models (LLMs) like OpenAI's GPT series and Baidu's Ernie Bot have demonstrated powerful capabilities in text generation and language understanding tasks [1, 2]. The technical breakthroughs brought by these models also provide new opportunities for innovation in healthcare [3–5].

Specifically, LLMs like ChatGPT show promising application potential in medical question answering [6], clinical decision support [7], patient note generation [8], drug discovery [9], and more. Researchers are exploring the transfer learning and fine-tuning capabilities of these pretrained models to adapt them to specific medical text processing

---

S. Xu and S. Chen—These authors contributed equally to this work. S. Xu and S. Chen—Co-first authors

© The Author(s), under exclusive license to Springer Nature Singapore Pte Ltd. 2024
H. Jin et al. (Eds.): IAIC 2023, CCIS 2059, pp. 110–123, 2024.
https://doi.org/10.1007/978-981-97-1280-9_9

tasks like entity recognition [10], relation extraction [11], disease classification [12]. Meanwhile, the medical text generated by ChatGPT also contains significant quality issues, including misleading information, biases, plagiarism, lack of originality, inaccuracy, and risks of hallucination [13]. Moreover, ethical and safety challenges exist in the medical application of these models [14].

Therefore, this study aims to systematically assess the capacities and limitations of LLMs in the medical domain, by analyzing their performance on tasks like medical information processing, decision support, and text generation, as well as the challenges they face. We conduct quantitative and qualitative analyses to examine different models' effects based on public medical corpora and real patient cases. Our goal is to provide insights for further optimization and safe and effective application of these powerful models in clinical practice. This study holds significance for guiding responsible and controllable use of large language models in healthcare.

## 2 Review of Related Work

### 2.1 Large Language Models

With the advancement of computational capabilities, the emergence of LLMs has presented new opportunities for natural language processing in the healthcare domain. The striking feature of these models is their gigantic scale, pretraining in an unsupervised manner on massive text data to learn powerful language representational capabilities. For instance, BERT was pretrained on around 30 million words from books, while GPT-3 utilized approximately 200 billion words from web data [15, 16]. The pretraining enables fast adaptation to downstream tasks when supervised fine-tuning [17].

In terms of parameter size, as shown in Fig. 1, early LLMs like BERT-Large have 0.34B parameters, while more recent models have reached hundreds of billions of parameters, like InstructGPT, the predecessor of ChatGPT with 175B parameters. The huge parameter space allows language models to learn intricate and sophisticated language representations, as more parameters provide greater flexibility for the model to capture subtle nuances and complex structures of language.

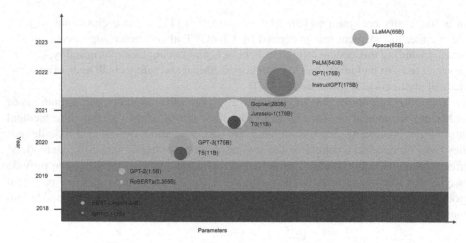

**Fig. 1.** Parameter Size and Release Timeline of LLMs

This is greatly attributed to the neural network architecture of Transformer [18] for sequential data processing. As shown in Fig. 2, It employs an attention mechanism that allows the model to weigh the importance of each word in relation to other words. This facilitates capturing long-range dependencies and contextual information, enhancing the model's ability to generate coherent text. By increasing the number of layers and parameters, Transformers can expand to handle more complex language tasks and generate more coherent and contextually relevant text.

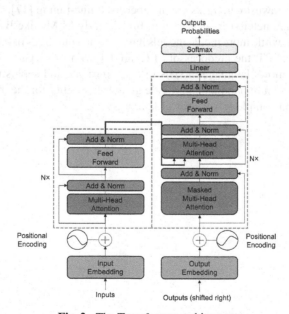

**Fig. 2.** The Transformer architecture

In the field of Natural Language Processing (NLP), these LLMs have introduced powerful capabilities such as context awareness, transfer learning, continuous text generation, and zero-shot learning. These capabilities have revolutionized traditional NLP practices [19], which were heavily reliant on manual feature engineering and rule-based approaches, requiring human intervention to capture various patterns and rules within language. The emergence of LLMs has enhanced the effectiveness, flexibility, and efficiency of NLP tasks, leading to significant technological innovation in the NLP domain. The immense potential of LLMs has sparked boundless imagination about their breakthroughs in health-related fields.

## 2.2 Applications of Large Language Models in Healthcare

As the medical domain starts to incorporate these models into various NLP tasks, their contextual awareness enables better understanding of terminology and entity relationships in medical texts, leading to outstanding performance in text classification, named entity recognition, etc. The transfer learning capacity allows the models to learn rich language knowledge from general corpora first, and then adapt to the particular context of the medical domain through fine-tuning, such as clinical notes [20].

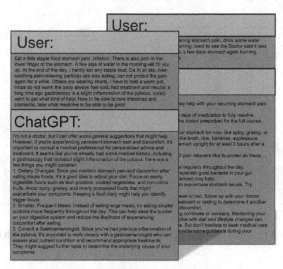

**Fig. 3.** We use ChatGPT (GPT-3.5) as a medical assistant, and consult with it for medical advice. It is able to take the patient's past medical history as background, and provide further medical recommendations in a continuous conversation.

With continuous evolution, the application of large pretrained language models in healthcare is expanding to more complex tasks like clinical diagnosis support, drug relation extraction, text generation, etc. As shown in Fig. 3, the models can provide personalized diagnostic suggestions based on patients' medical history to assist clinicians in making more accurate decisions [21, 22]. They can also extract drug-drug interactions from literature to facilitate drug development [23] and treatment design. In addition, the continuous text generation ability allows generating medical text like discharge summaries, clinical reports, etc., thereby reducing documentation burden for physicians [24].

## 3   Analysis of Safety and Application Examples

Before exploring the safety of LLMs, we first need to determine the medical capabilities inherent in LLMs, and then look for potential safety issues in the scenarios where LLMs can be applied. Therefore, in this chapter, we start from the test results of LLMs' medical capabilities, analyze the application examples of LLMs, and gradually look for potential safety issues.

### 3.1   Medical Capability Test of LLMs

As shown in Fig. 4, the excellent performance of GPT-4 in medical competency exams and benchmark datasets indicates that it can contribute to the daily workflow of clinical reasoning and medical practice [25].

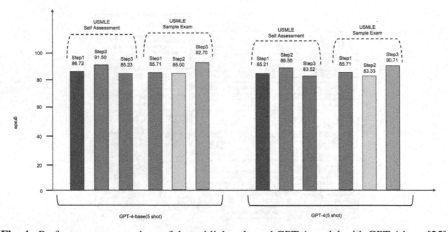

**Fig. 4.** Performance comparison of the publicly released GPT-4 model with GPT-4-base [25].

We have read some publicly published articles on the medical competency tests for LLMs, and all conclusions are that LLMs have extremely strong medical capabilities. However, as shown in Table 1, their understanding of safety is rather vague.

**Table 1.** Considerations of safety in different experiments on clinical text generation

| Experiment contents | Considerations of safety | Means of judging safety |
| --- | --- | --- |
| Experiment on evaluating ChatGPT's answers to medical questions [26] | Define accuracy and completeness as criteria for judging safety. Inaccurate or incomplete answers may mislead doctors or patients | Answers are scored for accuracy and completeness by professional physicians |
| Experiments on GPT series models' performance on medical exams [27] | Pay attention to the "hallucination" problem. Overconfident but erroneous assertions by models may jeopardize medical safety | Calculate the incidence of hallucinations and compare between different models |
| Experiment on GPT-4's performance on medical licensing exams [25] | Consider risks of incorrect medical advice and information | Discuss need for cautious use and expert verification of model outputs |
| Experiments on GPT series models' performance in Korean medicine [28] | Need to correct errors to ensure safe and reliable application | Model optimization to reduce error rates |

Considering the considerations for safety in the table above, cautious clinical integration and expert supervision are necessary at this stage. Accuracy, completeness, avoiding omission of key information, and reducing "hallucinations" are all key indicators for determining the safety of clinical text generation. It is important to adopt appropriate judgment methods, have professional medical personnel check the results, and continuously optimize the model to improve safety.

## 3.2 Medical Image Analysis and Reporting

After ChatGPT demonstrated unprecedented potential, ChatCAD [29] and GPT4MIA [30] have emerged, showcasing the tremendous potential of leveraging ChatGPT for medical image analysis and reporting. This brings new opportunities for AI in healthcare [31]. In Table 2, we compare the technical contents of the two papers and provide corresponding commentary.

While advancing the application of LLMs in medicine, we should realize from the content of these two articles that the lack of interpretability must be addressed first, enhancing the transparency and interpretability of the model through technical means. At the same time, a standardized safety and ethical evaluation system needs to be established to conduct a comprehensive risk assessment before using LLMs. From the perspective of model safety, a comprehensive safety audit of large models is required to identify and eliminate potential biases or errors. From the perspective of data safety, patient privacy must be strictly protected to prevent misuse of medical data. From the perspective of system safety, an end-to-end security protection mechanism needs to be established to prevent attacks and intrusions. From the perspective of process safety, safety assessments and verifications should be carried out at every stage of development and deployment. In

**Table 2.** ChatCAD and GPT4MIA Key points comparison

| Comparison Point | ChatCAD [29] | GPT4MIA [30] |
|---|---|---|
| Main Contribution | Proposes combining LLMs with medical image CAD, utilizing the logical reasoning abilities of LLMs to improve image classification, segmentation, etc. | Proposes using GPT-3 for transductive inference, converting image information into text through prompt design to improve medical image classification accuracy |
| Experimental Setup | Trains report generation models on MIMIC-CXR dataset, trains image classification models on CheXpert dataset, finally evaluates report quality on MIMIC-CXR test set | Trains ResNet classification models on MedMNIST dataset, constructs prompts on validation set, finally evaluates GPT-3's transductive inference effects on test set |
| Experimental Results | Significantly improves diagnostic metrics of report generation, increasing by 16.42% | Outperforms traditional inductive and transductive methods in both use cases of detecting prediction errors and improving classification accuracy |
| Personal Opinions | Provides a new idea of utilizing LLMs' knowledge for multi-model output integration. But safety and interpretability need consideration | Demonstrates the powerful logical reasoning capabilities of LLMs, but usage cost is high and practicality needs evaluation |

addition, ethical issues should be actively considered, assessing the impact of artificial intelligence on different groups and establishing a partnership between humans and machines.

## 4   Security Issues and Solutions

Early discussions focused on issues like bias and interpretability [32], and with the application of these LLMs in sensitive areas like healthcare, more rigorous regulations and governance have become a focus [33]. In Chapter 3, the introduction of LLMs has brought about new security issues, which we will explain and propose solutions for in this chapter."

### 4.1   Hallucination

The two major current issues with LLMs are hallucination and toxic content generation [29]. Hallucination refers to the problem of models generating non-existent or inaccurate information. The limitations of training data make it hard to avoid such biases in models. Toxic content generation involves models potentially producing racist, hateful or otherwise objectionable content. Both these issues can impair user experience and ethical norms in model applications. However, at present, commercially available Language Models (LLMs) pay more attention to the regulation of generating toxic information. Although there have been instances of prompting attacks that cause LLMs to generate toxic information [34], these are not common in regular medical scenarios. Therefore, we focus on discussing the phenomenon of hallucinations.

Although there are comprehensive explanations for the generation of hallucinations in language models [35], in traditional tasks, hallucinations usually refer to Faithfulness. However, for LLMs, the hallucinations we usually consider are Factualness. Moreover, LLMs are very likely to need to produce information in medical tasks that is not in the input source but conforms to the content of the information.

In order to study the potential hallucination scenarios that the most advanced LLMs may encounter today, we have chosen ChatGPT, which is currently leading in the industry, as the test model. For the dataset, we have chosen PubMedQA [36]. The PubMedQA dataset consists of research questions answered using corresponding abstracts. Each instance contains a question along with a context, a long answer, and a short answer that is either 'yes', 'no', or 'maybe'. This aligns well with common medical consultation scenarios (Table 3).

**Table 3.** Comparison of GPT's Judgements and Explanations Under Different Prompting Situations, "Accuracy" refers to the correct judgments made by ChatGPT, "Indecisive" refers to situations where ChatGPT cannot determine an answer, "Direct Wrong Judgment" refers to incorrect judgments made by ChatGPT, and "Provide Explanation" refers to instances where ChatGPT provides an explanation for its judgment.

| Situation | No Limited Options | Limited Options (Yes/No/Maybe) |
|---|---|---|
| Accuracy | 68.16% | 73.52% |
| Indecisive | 0 | 24.8% |
| Direct Wrong Judgement | 31.84% | 8.79% |
| Provide Explanation | 24.29% | 8.66% |

Overall, both scenarios demonstrate the limitations of GPT in making judgments without clear evidence. Although limiting options can reduce the errors made by GPT in direct judgments, it also reduces the provision of explanations and increases the uncertainty of judgments. Moreover, we found that ChatGPT made mistakes such as giving fictional content explanations for incorrect judgments and giving incorrect explanations for correct judgments when analyzing specific answer situations. After statistically analyzing these errors, we referred to existing papers [37–39] on the hallucination of LLMs and classified them into four categories, as shown in Fig. 5.

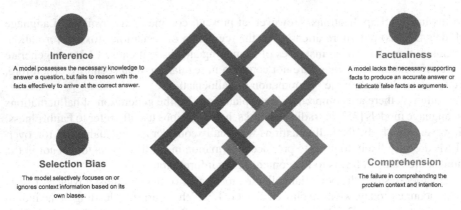

**Inference**
A model possesses the necessary knowledge to answer a question, but fails to reason with the facts effectively to arrive at the correct answer.

**Factualness**
A model lacks the necessary supporting facts to produce an accurate answer or fabricate false facts as arguments.

**Selection Bias**
The model selectively focuses on or ignores context information based on its own biases.

**Comprehension**
The failure in comprehending the problem context and intention.

**Fig. 5.** Four Types of Errors

According to research [1, 37], the reasons for the hallucinations currently experienced by LLMs are diverse, mainly divided into the following types (due to space limitations, this article does not delve into the causes): 1. False information in the training data 2. LLM has learned incorrect relationships 3. Transformers tend to generate freely rather than infer logically 4. LLM lacks understanding of the physical world's common sense 5. Misleading prompts.

## 4.2 Output Detection and Prompting

Considering the feasibility of actual operation, among the above reasons for hallucinations, points 1 and 2 cannot be changed after training the LLMs. Therefore, we only seek to alleviate the hallucinations of LLMs by improving the last three points.

**Transformers Tend to Generate Freely Rather than Infer Logically.** To solve this problem, improvements have been made in two aspects: output detection and prompt design. In terms of output detection [40], as shown in Fig. 6, a rule-based fact verification module has been designed to analyze the logical consistency of the output using a relation extraction model. Additionally, a knowledge graph is used to check for consistency and identify baseless digressions.

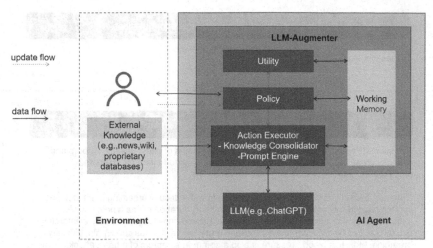

**Fig. 6.** LLM-AUGMENTER architecture showing how its plug-and-play modules interact with the LLM and the user's environment [40].

*In terms of prompt design* [25, 41]. LLMs are explicitly required to perform logical reasoning and provide complete information to reduce ambiguity. The generation of explanatory statements is also required to clarify the basis and guide the model to generate more rational inferences. It is worth noting that while Chain-of-Thought(CoT) has been proven to have positive effects, it also has the potential to create an hallucination snowball effect [42].

**LLM Lacks Understanding of the Physical World's Common Sense.** Due to the limitations of training data, LLMs generally lack a basic understanding of common sense about the physical world, which can result in generated content that violates common sense constraints.

*In terms of output detection* [40]. To improve its understanding of common sense, a common sense knowledge base can be constructed and used to check the reasonableness of outputs. Additionally, a pre-trained common sense reasoning model can be used for judgment.

*In terms of prompt design.* Incorporating common sense premises and providing background common sense knowledge can guide the generation process. This can constrain the scope of the model's discussion and reduce instances of violating common sense.

**Misleading Prompts.** Incomplete or biased prompts can potentially mislead the model generation. To avoid this, some articles [43–45] use prompt annealing to regenerate and compare outputs to identify sources of misinformation. Additionally, we can optimize prompt formulation by avoiding ambiguous vocabulary, providing complete contextual information, and adding explicit instructions to clarify the expected interaction, thereby reducing the impact of misleading prompts on the generated output (Fig. 7).

---

**Chain-of-Thought Prompt template for questions**

Question: {{question}}
Let's think step by step.

---

**Prompt template with common sense constraints.**

Question: {{Patient is male, 45 years old. He complains of chest pain for 2 hours and difficulty breathing. Physical examination reveals elevated blood pressure and electrocardiogram shows ST segment elevation. }}

Common Sense Assumptions :Acute chest pain with difficulty breathing in adult males is often caused by cardiovascular diseases. A short duration of chest pain usually indicates angina or myocardial infarction. Background knowledge: Myocardial infarction is often caused by coronary artery blockage and can manifest as chest pain, difficulty breathing, etc. High blood pressure is also a common symptom of a heart attack.

**Fig. 7.** Prompt Template.

Through the above improvement methods, we conducted re-experiments using CoT prompting. From Table 4 and Fig. 8, we can see that incorporating CoT prompts can effectively reduce the occurrences of hallucinations generated by LLMs.

From the Table 4, we can see that compared to regular open-ended prompting, CoT prompting can effectively reduce the wrong judgment rate (direct wrong judgment) and indecisiveness of language models. This demonstrates that CoT prompting helps improve the logical reasoning of language models.

**Table 4.** Comparison of GPT's Judgements and Explanations Under Different Prompting Situations.

| Situation | No Limited Options | Limited Options (Yes/No/Maybe) | Limited Options (Chain-of-Thought Prompt) |
|---|---|---|---|
| Accuracy | 68.16% | 73.52% | **80.27%↑** |
| Indecisive | 0 | 24.8% | 20.15% |
| Direct Wrong Judgement | 31.84% | 8.79% | **5.25%↑** |
| Provide Explanation | 24.29% | 8.66% | 100% |

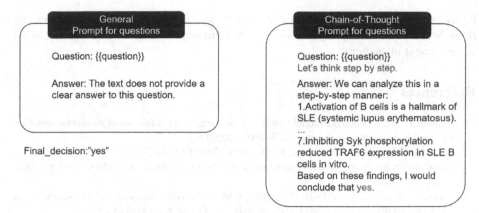

**Fig. 8.** Comparison of results between General prompting and CoT prompting

Overall, the reasonable use of CoT prompting can significantly reduce the incorrect outputs of language models, and is an effective method to promote their safe application in the medical field. However, targeted optimization is still needed based on different tasks.

## 5    Conclusion and Outlook

In summary, this study conducted a systematic analysis of the capabilities and limitations of large language models in the medical domain. Our experiments and evaluations, while again demonstrating the formidable strengths of LLMs in medical text processing, decision support, and text generation tasks, also revealed significant challenges and risks that need addressing before deploying these models. Models like GPT-3 and ChatGPT exhibit huge potential in assisting clinicians through patient data reasoning and medical text generation, but also uncover notable pitfalls like hallucination that could generate inaccurate or misleading medical information, compromising patient safety. Issues like lack of explainability, biases, and harmful content generation also need redressing.

Therefore, we emphasize the importance of rigorous testing, risk assessment, and governance of these models before real-world application in healthcare, and provide detailed suggestions on several methods to mitigate hallucination in LLMs which we demonstrate to be effective.

While LLMs present new opportunities for AI innovation in medicine, they require careful and responsible implementation. Our study provides insights and guidance for future research directions. Key priorities going forward include: 1) Improving model transparency and explainability; 2) Building benchmark datasets to comprehensively evaluate capabilities and limitations; 3) Establishing standardized protocols for risk assessment and safeguards; 4) Optimizing training and prompting techniques to reduce hallucination and biases; 5) Implementing robust human-in-the-loop frameworks to enable safe and effective LLM utilization.

By integrating principles of ethics and safety precautions, large language models can become valuable assistants to human experts, augmenting medical intelligence for better healthcare. But this requires ongoing research to address their current flaws and risks. We hope our analysis provides a meaningful step towards safe and beneficial AI advancement in medicine.

# References

1. Brown, T., et al., Language models are few-shot learners. In: Advances in Neural Information Processing Systems, vol. 33, pp. 1877–1901 (2020)
2. OpenAI, GPT-4 Technical Report. ArXiv: arXiv:2303.08774 (2023)
3. Bommasani, R., et al.: On the opportunities and risks of foundation models. arXiv preprint: arXiv:2108.07258 (2021)
4. Wang, S.-H., et al.: COVID-19 classification by CCSHNet with deep fusion using transfer learning and discriminant correlation analysis. Inf. Fusion **68**, 131–148 (2021)
5. Zhang, Y., et al.: Deep learning in food category recognition. Inf. Fusion, 101859 (2023)
6. Muftić, F., et al.: Exploring medical breakthroughs: a systematic review of ChatGPT applications in healthcare. Southeast Europe J. Soft Comput. **12**(1), 13–41 (2023)
7. Liu, S., et al.: Using AI-generated suggestions from ChatGPT to optimize clinical decision support. J. Am. Med. Inform. Assoc. **30**(7), 1237–1245 (2023)
8. Liu, J., Wang, C., Liu, S.: Utility of ChatGPT in clinical practice. J. Med. Internet Res. **25**, e48568 (2023)
9. Sharma, G., Thakur, A.: ChatGPT in drug discovery (2023)
10. Gu, Y., et al.: Domain-specific language model pretraining for biomedical natural language processing. ACM Trans. Comput. Healthc. (HEALTH) **3**(1), 1–23 (2021)
11. Si, Y., et al.: Enhancing clinical concept extraction with contextual embeddings. J. Am. Med. Inform. Assoc. **26**(11), 1297–1304 (2019)
12. Huang, K., Altosaar, J., Ranganath, R.: Clinicalbert: modeling clinical notes and predicting hospital readmission. arXiv preprint: arXiv:1904.05342 (2019)
13. Sallam, M.: ChatGPT utility in healthcare education, research, and practice: systematic review on the promising perspectives and valid concerns. In: Healthcare. MDPI (2023)
14. Wang, C., et al.: Ethical considerations of using ChatGPT in health care. J. Med. Internet Res. **25**, e48009 (2023)
15. Devlin, J., et al.: BERT: pre-training of deep bidirectional transformers for language understanding. arXiv preprint: arXiv:1810.04805 (2018)
16. Kaplan, J., et al.: Scaling laws for neural language models. arXiv preprint: arXiv:2001.08361 (2020)
17. Qiu, X., et al.: Pre-trained models for natural language processing: a survey. Sci. China Technol. Sci. **63**(10), 1872–1897 (2020)
18. Vaswani, A., et al.: Attention is all you need. In: Advances in Neural Information Processing Systems, vol. 30 (2017)
19. Sun, X., et al.: Pushing the Limits of ChatGPT on NLP Tasks. arXiv preprint: arXiv:2306.09719 (2023)
20. Lee, J., et al.: BioBERT: a pre-trained biomedical language representation model for biomedical text mining. Bioinformatics **36**(4), 1234–1240 (2019)
21. Cheng, K., et al.: The potential of GPT-4 as an AI-powered virtual assistant for surgeons specialized in joint arthroplasty. Ann. Biomed. Eng., 1–5 (2023)
22. He, Y., et al.: Will ChatGPT/GPT-4 be a lighthouse to guide spinal surgeons? Ann. Biomed. Eng., 1–4 (2023)
23. Peng, C., et al.: A study of generative large language model for medical research and healthcare. arXiv preprint: arXiv:2305.13523 (2023)

24. Adams, L.C., et al.: Leveraging GPT-4 for post hoc transformation of free-text radiology reports into structured reporting: a multilingual feasibility study. Radiology **307**(4), e230725 (2023)
25. Nori, H., et al.: Capabilities of GPT-4 on medical challenge problems. arXiv preprint: arXiv: 2303.13375 (2023)
26. Johnson, D., et al.: Assessing the accuracy and reliability of AI-generated medical responses: an evaluation of the Chat-GPT model (2023)
27. Ali, R., et al.: Performance of ChatGPT, GPT-4, and Google bard on a neurosurgery oral boards preparation question bank. Neurosurgery, 10–1227 (2022)
28. Jang, D., Kim, C.-E.: Exploring the potential of large language models in traditional Korean medicine: a foundation model approach to culturally-adapted healthcare. arXiv preprint: arXiv:2303.17807 (2023)
29. Wang, S., et al.: ChatCAD: interactive computer-aided diagnosis on medical image using large language models. arXiv preprint: arXiv:2302.07257 (2023)
30. Zhang, Y., Chen, D.Z.: GPT4MIA: Utilizing Geneative Pre-trained Transformer (GPT-3) as A Plug-and-Play Transductive Model for Medical Image Analysis. arXiv preprint: arXiv:2302. 08722 (2023)
31. Zhang, Y.-D., et al.: Advances in multimodal data fusion in neuroimaging: overview, challenges, and novel orientation. Inf. Fusion **64**, 149–187 (2020)
32. McCradden, M.D., et al.: Ethical limitations of algorithmic fairness solutions in health care machine learning. Lancet Digit. Health **2**(5), e221–e223 (2020)
33. Morley, J., et al.: The ethics of AI in health care: a mapping review. Soc. Sci. Med. **260**, 113172 (2020)
34. Zhao, S., et al.: Prompt as triggers for backdoor attack: examining the vulnerability in language models. arXiv preprint: arXiv:2305.01219 (2023)
35. Ji, Z., et al.: Survey of hallucination in natural language generation. ACM Comput. Surv. **55**, 1–38 (2022)
36. Jin, Q., et al.: PubMedQA: a dataset for biomedical research question answering (2019). arXiv:1909.06146. https://doi.org/10.48550/arXiv.1909.06146
37. Zheng, S., Huang, J., Chang, K.C.-C.: Why does ChatGPT fall short in answering questions faithfully? arXiv preprint: arXiv:2304.10513 (2023)
38. Griffin, L.D., et al.: Susceptibility to influence of large language models. arXiv preprint: arXiv:2303.06074 (2023)
39. Bang, Y., et al.: A multitask, multilingual, multimodal evaluation of ChatGPT on reasoning, hallucination, and interactivity. arXiv preprint: arXiv:2302.04023 (2023)
40. Peng, B., et al.: Check your facts and try again: Improving large language models with external knowledge and automated feedback. arXiv preprint: arXiv:2302.12813 (2023)
41. Kim, S., et al.: The CoT collection: improving zero-shot and few-shot learning of language models via chain-of-thought fine-tuning. arXiv preprint: arXiv:2305.14045 (2023)
42. Zhang, M., et al.: How language model hallucinations can snowball. arXiv preprint: arXiv: 2305.13534 (2023)
43. Manakul, P., Liusie, A., Gales, M.J.: SelfcheckGPT: zero-resource black-box hallucination detection for generative large language models. arXiv preprint: arXiv:2303.08896 (2023)
44. Mündler, N., et al.: Self-contradictory hallucinations of large language models: evaluation, detection and mitigation. arXiv preprint: arXiv:2305.15852 (2023)
45. Huang, K.-H., Chan, H.P., Ji, H.: Zero-shot faithful factual error correction. arXiv preprint: arXiv:2305.07982 (2023)

# Blockchain-Enhanced Device to Device Network Identity Verification Based on Zero Knowledge Proof

Xingyu Shang[(✉)], Meiling Dai, and Xiaoou Liu

China Telecom Research Institute, Beijing, China
{shangxy1,daiml1,liuxo}@chinatelecom.cn

**Abstract.** With the large-scale application of device to device(D2D) network, the authentication process between devices is complicated and needs to be optimized. Identity authentication also needs to protect the privacy of users. Therefore, this paper proposes a D2D network identity authentication method based on blockchain and zero knowledge proof, which realizes automatic authentication of devices in the network through blockchain-based smart contracts, and improves the efficiency of authentication between devices. The zero knowledge proof method is used to protect the user's identity privacy and realize the anonymous authentication of the identity. The effectiveness of the proposed method is proved by experiments.

**Keywords:** Device to device · Blockchain · Zero knowledge proof

## 1 Introduction

With the continuous development of the Internet of Thing (IoT), 5G Advanced further improves the data transmission rate of communication by using higher frequency bands based on 5G. In this process, many directions have been explored. To solve the problem of poor coverage of high-frequency bands, scholars have proposed a new network deployment scheme named Device to Device (D2D) communication [1]. The base station(BS) relay communication structure is further extended to the communication between user equipment by D2D, providing a deployment scheme with low maintenance cost and high coverage. It is complementary to the existing high-frequency schemes which can better extend the coverage radius and fill the coverage hole in high-frequency mode, so as to find the balance between input-output ratio and high-quality network construction for operators.

The expanded BS structure has access to a large number of heterogeneous devices. Different types of devices will directly establish contact when connecting, which often leads to identity authentication difficulties and complex authentication processes. Low-efficiency identity authentication has become a difficult problem that cannot be ignored in the application of D2D network. Blockchain can synchronize the rules in all device nodes by consensus algorithm through smart contract, so blockchain and smart contract as communication relay access to identity management tools become the key to make

H. Jin et al. (Eds.): IAIC 2023, CCIS 2059, pp. 124–134, 2024.
https://doi.org/10.1007/978-981-97-1280-9_10

D2D achieve efficient identity authentication. The smart contracts can also be executed to incentivize users to utilize smart devices to act as D2D relay nodes.

There are some problems in identity privacy protection based on D2D network. As a relay node, the connected device must frequently obtain user location information. Combined with the registered identity information, user privacy may be compromised. Since the blockchain network can synchronize ledger data across all nodes, a smartphone that joins the D2D network as a relay node has access to all the data in the blockchain. When a user accesses the D2D network using a device, the relay node can get user's private information, such as identity and transaction, which causes the risk of privacy disclosure. Zero-knowledge proof is a widely used privacy protection protocol, which has been studied in the fields of Internet of Things [2], wireless network [3], Internet of Vehicles [4] and so on. Zero-knowledge proof can verify the identity information of both parties when the privacy data is hidden. The method of zero-knowledge proof can hide the user's identity privacy information while synchronizing the identity in the blockchain, and realize the identity authentication in the D2D network at the same time.

At present, traditional identity authentication schemes are difficult to meet the needs of node privacy protection in D2D, and lack of effective privacy-preserving policies to protect the identity information of the nodes. For some problems existing in the identity authentication and privacy protection of D2D network, we propose two methods:

The proposed solution suggests an automatic identity authentication method based on smart contracts to address the challenges posed by a large number of D2D network access devices and their inefficient identity authentication process. This approach aims to achieve automated user identity management, thereby enhancing device access and facilitating efficient network information updating.

All participating nodes synchronize the identity authentication information based on smart contracts in the D2D network, causing the corresponding identity information to be transparent to all participating nodes, which may disclose user identity privacy. Therefore, a privacy protection method based on non-interactive zero-knowledge proof is proposed to realize the privacy protection of user information. The relay node in the blockchain is unable to access the user's identity or transaction information, but it can verify the user's identity through the generated zero-knowledge proof.

The structure of the paper is as follows: The Sect. 2 introduces the related work of identity authentication privacy protection based on blockchain; The Sect. 3 introduces our proposed D2D network identity authentication architecture and privacy protection mechanism. The Sect. 4 introduces the D2D network privacy protection method based on zero-knowledge proof. The Sect. 5 is experiment and analysis; Sect. 6 is the conclusion and prospect.

## 2  Related Work

### 2.1  Blockchain in Authentication and Privacy Preserving

[5] based on efficient pair and Certificate Free batch signature (CLBS), identity-based prefix encryption and China Residual Theorem (CRT), a unified and efficient mechanism for anonymous access device discovery and batch authentication of heterogeneous D2D terminals is proposed, which can be applied to all 5G heterogeneous access scenarios of

D2D communication Liu et al. [6] realized the Internet of Things information sharing scheme in a zero-trust environment through blockchain, got rid of the dependence on trusted third parties, and realized the identity authentication and privacy protection of Internet of Things devices while ensuring the reliability of data;Li and Shi [7] proposed an alliance chain enhanced IoT architecture, which increased the scalability, and on this basis provided a privacy protection scope query based on alliance chain, improving the query efficiency of iot data Goyat et al. [8] used blockchain and cloud storage in the Internet of Things composed of wireless sensor networks to ensure the privacy of user identity authentication and resist various internal and external attacks, but the system needs to treat BS as a trusted entity.

### 2.2  Zero Knowledge Proof in Privacy Preserving of Blockchain

With the rapid development of blockchain, privacy protection has become a key issue to be solved in the field. Zero Knowledge Proof (ZKP), as an efficient privacy protection technology, can prove to other parties that they have a secret or perform a certain operation without disclosing the original information.

In recent years, many researchers have explored the use of zero-knowledge proof in blockchain privacy protection. Literature [9] proposes a block privacy protection scheme based on zero-knowledge proof, which uses zero-knowledge proof to verify the authenticity of blockchain transactions while protecting the privacy of transactions. But there is a problem of low efficiency. Literature [10] proposes a layered privacy protection scheme based on zero-knowledge proof to achieve different levels of privacy protection. Literature [11] studies the application of zero-knowledge proof in the privacy protection of smart contracts, and proposes a smart contract privacy protection scheme based on zero-knowledge proof, but this scheme is only limited to the field of smart contracts and does not consider other application scenarios.

## 3  Proposed Model

### 3.1  System Architecture

In order to realize blockchain-based D2D identity authentication, a blockchain-based near-domain wireless access system architecture, as shown in Fig. 1, is constructed, which consists of the D2D resource layer and the blockchain identity management layer.

D2D resource layer: including consensus node and relay node, the user terminal nodes and BS node four types of entities. BS is the public mobile communication BS, which is the interface equipment for the user's mobile terminal to access the Internet. BS set is defined as $S = \{1, 2, \cdots, w\}$; The D2D relay is formed by the idle intelligent terminal of the user and is responsible for the transfer of user signals. The relay node set is defined as $R = \{1, 2, \cdots, n\}$; User mobile terminals are devices with communication requirements such as smartphones, and the user mobile terminal set is defined as $U = \{1, 2, \cdots, m\}$; Consensus node is a full node that provides blockchain consensus and saves all data of the blockchain network. At the same time, the operator node will also provide identity certificates for nodes accessing the D2D network. Consensus node set is defined as $C = \{1, 2, \cdots, v\}$.

Blockchain identity management layer: The construction of the blockchain network in the blockchain management layer, based on mature technologies like Ethereum, is completed by configuring various types of blockchain nodes for all participating nodes in the D2D resource layer. During the initial stage of network construction, basic blockchain nodes are established to accomplish the primary network setup. The edge server, located near BS, is configured as a full node with complete blockchain functionalities including full-chain data storage and consensus. D2D relays and user mobile terminals are configured as lightweight blockchain nodes or clients, depending on their storage capacity. This enables essential functions like accessing the blockchain network and initiating transactions.

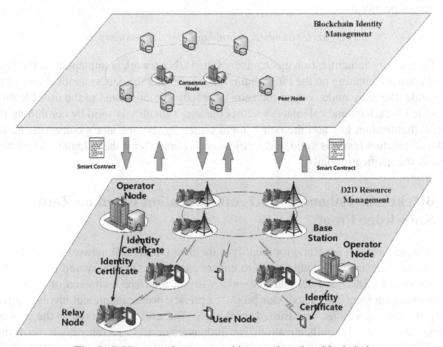

**Fig. 1.** D2D network system architecture based on blockchain

## 3.2   Registration and Authentication in D2D Network

To realize the registration and authentication of relay nodes and user nodes, the D2D network identity authentication process is defined as follows: First, a user or relay node registers with the D2D network to obtain the identity of providing or using D2D services. The relay node needs to upload the relevant information required for registration to the provincial company operator. The relay node can obtain the identity certificate through the nearest consensus node (that is, the provincial company operator in the region), and complete the steps of identity linking through the consensus node. The user node can know the location and service information of the relay node through broadcasting. The user first checks whether the user has registered in the D2D system through BS. If the registration information is not checked, the identity information is encrypted through the provincial company operator consensus node, and the identity is registered in the blockchain network through the smart contract. At the same time, the provincial company operator node issues the identity certificate Cert. The relay node needs to provide the corresponding registration information to operator node:

$$Regis\_re(ID\,number,\ phone\,number,\ equipment\,type,\ resource,\ time\,stamp,\ initial\,score)$$

Operators evaluate the quality of service they can provide by the type of equipment of relay nodes; The registration information submitted by the user to the provincial company operator is

$$Regis\_ur(ID\,number,\ phone\,number,\ time\,stamp)\circ$$

The identity authentication mechanism of the D2D network is implemented through smart contracts running on the blockchain. After receiving the access application of the user node, the relay node obtains the zero-knowledge proof related to the user identity from the blockchain and calculates whether the user's identity is valid by combining its own authentication key and the public input issued by the operator's consensus node. If the calculation result is valid, the verification is complete; if the calculation result is invalid, the verification fails.

## 4   Blockchain-Enhanced D2D Authentication Based on Zero Knowledge Proof

This chapter presents a method for identity authentication in D2D networks, which uses non-interactive zero knowledge proof to ensure user node anonymity and privacy. We assume that the D2D and cellular networks will not interfere with each other on the communication frequency band. Our proposed privacy protection model involves three roles: the prover, verifier, and trusted third party. The user node serves as the prover, providing evidence for authentication. The authenticator acts as a relay node, verifying the evidence provided by the prover in a zero knowledge condition to confirm their identity. The operator node serves as the trusted third party, endorsing the credibility of the verification process, and providing the relevant keys and public input required for verification.

### 4.1   Automatic Registration of Users and Relay Nodes

We propose an automatic registration method based on smart contract in D2D network. Once registration is completed, identity credentials and keys are provided to users via a zero knowledge system. The specific methods are as follows:

In initialization, the operator selects the appropriate arithmetic circuit and public input according to the encryption method, and generates the proof key and verification key through the algorithm; The new relay node submits the registration information and initiates the registration application to BS; BS verifies the relay registration request and associates the registration information with the blockchain account/wallet; BS packages and writes the registration information into the smart contract; The registration information is broadcast in a blockchain network that includes each BS light node and operator consensus node; After the carrier node obtains the synchronized registration information, it issues the authentication key and identity certificate required for authentication to the relay node. BS confirms the registration information to the new relay node; The relay node is registered. The new user submits the registration information to the operator node and applies for registration; After verifying the user's registration request, the operator will associate the registration information with the blockchain account/wallet; By inputting the public input and user registration information into the arithmetic circuit, the operator obtains zero-knowledge proof which hides the user identity privacy through calculation. The operator node packages the zero-knowledge proof that does not contain privacy information onto the chain; Zero-knowledge proofs are broadcast in the blockchain network (each BS light node and operator consensus node); After the zero-knowledge proof broadcast, the operator consensus node issues identity credentials and proof keys to new users. The user node is registered.

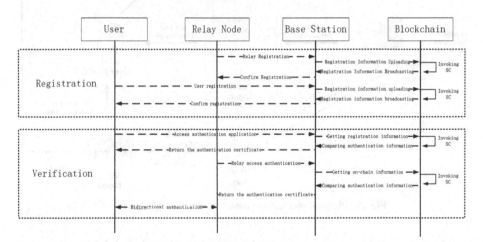

**Fig. 2.** Identity authentication process based on smart contract

## 4.2   Identity Authentication Process Based on Zero Knowledge Proof

The registered user can select the relay node for network access as required, submit the key and prove the zero-knowledge identity authentication, and realize the authentication while ensuring their privacy information. The specific process is as follows:

Users select relay nodes that meet their own requirements according to D2D network direct and indirect excitation algorithms. After the relay node is selected, the user initiates an access service application to the relay node, and submits the identity certificate and proof key required for identity authentication. The relay node obtains zero-knowledge proof related to user identity from the blockchain network by invoking smart contracts;

The relay node calculates the verification key, zero-knowledge proof and public input, and obtains the calculation result. The relay node determines whether the computing result is valid. If the result is valid, the authentication is completed. If the result is invalid, authentication fails. After the authentication is completed, the relay node provides its own identity information to the user. The user gets identity information of the relay node on blockchain by invoking the smart contract; Through the relay node registration information obtained from blockchain to complete the relay identity authentication; The user node and relay node are authenticated bidirectional (Fig. 3).

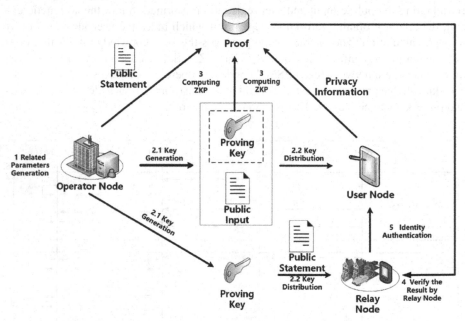

**Fig. 3.** Identity verification based on zero knowledge proof

The proposed privacy protection model is shown in Fig. 2, and the key operations required are as follows:

a) $Setup(1^{\lambda}, C) \rightarrow pubpara$:proposed security parameters, generating a set of public parameters $pubpara = (P; e; G_1; P_1; G_2; P_2; G_t; F_p)$, where p is a prime number, $G_1$ and $G_2$ are two cyclic group of order p, corresponding to generator P1 and P2, e is a bilinear map: $e : G_1 \times G_2 \rightarrow G_t$. It follows that $(G_1, G_2, G_t)$ are three cyclic groups of order P.$F_p$ is defined as a finite field. All algorithms use pubpara as the default public input parameter. $C$ is a given arithmetic circuit.

b) $Keygen(C, pubpara) \rightarrow (pk, vk)$: In our proposed zero-knowledge encryption scheme, the operator is required to execute the algorithm to generate the proof key and the authentication key. In order to make the whole process credible, the operator must not let the participant know the coordinates of the random points it chooses. Through the given arithmetic circuit $C$ and generated parameter $pubpara$, it can be calculated that the final outputs of the key generation process are the proof key $pk$ and the verification key *vk*, *pk* is the proof key used for generating the proof and is kept by the user;*vk* is the verification key used for proof and is saved by the authenticator's relay node.

c) *Proofgen(pk, ε, m)* → *π*: The proof generation algorithm takes the proof key *pk*, the public declaration ε(as the public input of the circuit *C*)and the privacy information m of the prover (the auxiliary input of the circuit C) as inputs. The final output of this part is π, which is a zero-knowledge proof that proves the relationship between ε and m built by the circuit C. ε and π are publicly released and can be used by anyone; m is the privacy information provided by the prover, that is, the identity information provided by the user registration.

d) *Verify(vk, ε, π)* → *boolean*: The verification algorithm takes (*vk, ε, π*) as input, vk is the verification key held by the verifier, ε is the public statement in Proofgen used to generate π, π is the zero-knowledge proof generated in Proofgen(); Generates a boolean output indicating whether proving π is valid (boolean = true) or not. The relay node receives the message to verify that the relay node does not have any direct interaction with the user. This non-interactive scheme ensures that no identity information of the user node is leaked to the relay node, so the relay node cannot forge proofs.

e) Calculation result: If boolean = true, the relay node completes anonymous identity authentication and allows the user node to access; If boolean = false, validation fails and no service can be provided.

## 5   Security and Privacy Analysis

### 5.1   Threat Model

When a device accesses a D2D network, due to the openness and randomness of the network, any two devices with communication requirements can establish a D2D communication link. Therefore, the devices in the D2D communication system are more likely to become the target of malicious attacks.

**Anti-man-in-the-Middle-Attack:**   Because the operators in the D2D network involved in this study are trusted third parties to grant the identity of the nodes, the participation of malicious users can be excluded;

**Anti-sybil Attack:**   In sybil attack, the attacker takes over a specific network by creating multiple false identities. However, in the D2D network, the cost of such operation is high, and the zero-knowledge proof will hide the user's identity information to ensure the confidentiality and integrity of user information.

**Anti-replay Attack:**   In a replay attack, an attacker replays an intercepted message to the recipient to create a false impression or interfere with normal communication. Our proposed zero-knowledge proof method can hide the knowledge in the authentication process, even if the authentication information is intercepted, the enemy can not change the original authentication content, and can not get the real identity information;

**Single Point of Failure:**   Due to the distributed featrue of the blockchain, the ledger is synchronized across all nodes, so our decentralized architecture can effectively resist the problem of a single point of failure.

## 5.2 Privacy Analysis

In D2D communication, the communication link established between terminals can avoid data processing by network core devices, thus reducing network load and delay, and improving spectrum utilization. However, this method of communication can also lead to privacy issues such as impersonation of the communication entity and disclosure of the real identity.

In a D2D network, the user's identity information is sensitive. Attackers can carry out various attacks by obtaining the user's identity information. Therefore, the user's identity information needs to be encrypted and protected. The zero-knowledge proof method we propose can hide the user's real identity.

# 6 Experiment Result

We simulated the communication between blockchain nodes on a device configured with Intel i5-1135G7 CPU, 16GB RAM, and Ubuntu18.04. Assuming that every node on the blockchain is a device in the D2D network, we use the communication and authentication between nodes to simulate the zero-knowledge verification process in the D2D network: the prover node generates proof, passes it to the verifier node through the communication between nodes, and the verifier node verifies the authentication between different nodes to complete the identity authentication.

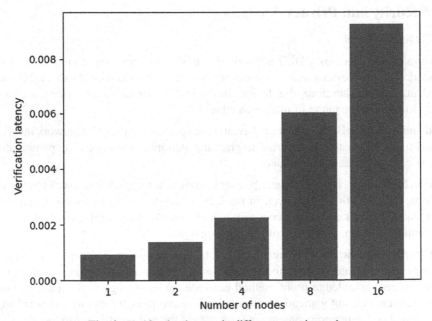

**Fig. 4.** Verification latency in different quantity nodes

The traditional zero-knowledge proof needs multiple rounds of interaction, which will produce a lot of communication cost in the process of user access. We compared the authentication speed of zero knowledge proof under different number of blockchain nodes, as shown in Fig. 4. The verification latency of the algorithm increases with the number of nodes. The fastest verification time was $9.298*10^{-4}$ s on a single node, and the highest verification latency was $92.45*10^{-4}$ s on 16 nodes. Experiments show that the number of blockchain nodes affects the efficiency of verification.

## 7  Conclusion

Our research focuses on privacy concerns related to identity in D2D networks. We have developed a new identity authentication method using blockchain smart contracts, which has guaranteed the efficiency and automation of the authentication process. This method provides a secure identity certificate to both the user and the relay node through the operator node, ensuring the security of the authentication system.

At the same time, an identity privacy protection method based on zero-knowledge proof protocol is proposed to meet the need of operators to protect user privacy. The privacy protection method based on zero-knowledge proof is used to protect the user's identity and privacy security in NCA. The user obtains the identity certificate and related key from the carrier, and only needs to pass the related key to the relay node for authentication. When the relay node authenticates the user's identity, it verifies the user's identity by reading the data on the blockchain and using the relevant key sent by the user. Through this zero knowledge proof method, the relay node realizes the user identity authentication without obtaining any user node identity information.

A technique for safeguarding user identities has been proposed to address the privacy concerns of operators. This technique is based on a zero knowledge proof protocol, which prevents any identity or privacy leakage in D2D network. To authenticate their identity, users must obtain an identity certificate and key from the carrier and provide the relevant key to the relay node. The relay node then verifies the user's identity by accessing the blockchain data, without obtaining any user node identity information. With this zero-knowledge proof method, user privacy is protected without compromising identity verification.

## References

1. Bi, Q.: P-RAN: a distributed solution for cellular systems in high frequency bands. IEEE Netw. **36**(4), 86–91 (2022)
2. Song, J., Harn, P.-W., Sakai, K., Sun, M.-T., Ku, W.-S.: An RFID zero-knowledge authentication protocol based on quadratic residues. IEEE Internet Things J. **9**(14), 12813–12824 (2022)
3. Khernane, N., Potop-Butucaru, M., Chaudet, C.: BANZKP: a secure authentication scheme using zero knowledge proof for WBANs. arXiv (2016). Accessed 08 Feb 2023
4. Rasheed, A.A., Mahapatra, R.N., Hamza-Lup, F.G.: Adaptive group-based zero knowledge proof-authentication protocol in vehicular Ad Hoc networks. IEEE Trans. Intell. Transp. Syst. **21**(2), 867–881 (2020)
5. Sun, Y., Cao, J., Ma, M., Zhang, Y., Li, H., Niu, B.: EAP-DDBA: efficient anonymity proximity device discovery and batch authentication mechanism for massive D2D communication devices in 3GPP 5G HetNet. IEEE Trans. Dependable Secure Comput. **19**(1), 370–387 (2022)

6. Liu, Y., et al.: A blockchain-based decentralized, fair and authenticated information sharing scheme in zero trust internet-of-things. IEEE Trans. Comput. **72**(2), 501–512 (2023)
7. Li, K.-C., Shi, R.-H.: A flexible and efficient privacy-preserving range query scheme for blockchain-enhanced IoT. IEEE Internet Things J. **10**(1), 720–733 (2023)
8. Goyat, R., et al.: Blockchain-based data storage with privacy and authentication in internet of things. IEEE Internet Things J. **9**(16), 14203–14215 (2022)
9. Chen, Y., Wang, Y., Zhang, Z.: Blockchain privacy protection based on zero-knowledge proof. J. Comput. Res. Dev. **56**(7), 1865–1872 (2019)
10. Liu, J., Zhang, Z., Wang, Y.: A hierarchical privacy protection scheme for blockchain based on zero-knowledge proof. J. Comput. Sci. Technol. **35**(2), 449–456 (2020)
11. Zhang, X., Wang, Y., Liu, L.: Privacy protection of smart contracts based on zero-knowledge proof. J. Comput. Eng. Appl. **57**(15), 196–203 (2021)

# A Privacy-Preserving Searchable Encryption Scheme for Data Protection in Smart Grid

Xun Zhang[1], Hong Zhao[1], Lei Di[1], and Zeng-Hui Yang[2]([✉])

[1] State Grid Gansu Electric Power Research Institute, Lanzhou 730070, China
[2] Information Security Center, State Key Laboratory of Networking and Switching Technology, Beijing University of Posts and Telecommunications, Beijing 100876, China
yangzh@bupt.edu.cn

**Abstract.** As smart grids continue to evolve at a rapid pace, huge amounts of power data have been collected by various devices in different locations. These enormous power data are of significant research and application value in machine learning. In the meanwhile, the privacy of training data is also becoming urgent to be addressed in the smart grid. Searchable encryption (SE) is an essential solution for protecting data privacy, which can achieve retrieval of ciphertext. In this paper, we present a blockchain-based searchable encryption scheme with fine-grained access control in smart grids. This scheme is designed to enable secure search and access to power data. Firstly, we track data uploads and searches using blockchain technology, which grants data access with transparency and traceability. In addition, in our scheme, the blockchain network stores the encrypted index, which protects the index from malicious tampering. Secondly, our proposed scheme can achieve retrieval of ciphertext while supporting fine-grained access control. Furthermore, Our scheme preserves the privacy of data. Users' attribute information and keyword information are protected in the data access process. Finally, there is an analysis of the security and advantages of our scheme. Experimental results and comparisons with other schemes show that our scheme is both cost-effective and practice-oriented.

**Keywords:** Privacy protection · Searchable encryption · Fine-grained access control · Blockchain · Smart grid

## 1 Introduction

Due to the smart grid [1] having developed rapidly, huge amounts of power data have been collected [2]. The increasing power data have meaningful worthiness for researching and applying [3,4]. Especially with the emergence of artificial intelligence, machine learning has significantly facilitated human life. For example, in the smart grid system, a trained model can help the staff to predict the electricity consumption habits of users. While machine learning has significant value, it also makes serious security and privacy risks. As a result, there is growing concern about data privacy. [5–7]. And how to solve power data storage and usage is an issue that cannot be ignored. The deployment of encryption technology is a useful measure to protect power data from disclosure, but it also brings challenges to the retrieval of data. The traditional retrieval method requires users to decrypt the data before performing the search operation, which not only consumes a lot of computing resources but also has a low data retrieval efficiency. Moreover, letting cloud servers decrypt data will lead to the exposure of plaintext to the server, which threatens the privacy of users. Therefore, in the scenario of having a large amount of sensitive data like smart grids, traditional retrieval methods can hardly satisfy the performance and security requirements. Fortunately, we can use searchable encryption (SE)[8] to address this issue. SE gives users the ability to perform keyword searches over encrypted data. This protects the privacy and improves retrieval efficiency. Thus, it is meaningful to apply SE in smart grids because it can keep electric data private and simultaneously achieve the search for ciphertext.

SE was introduced for the first time by Song et al. [8]. The scheme enables a single keyword retrieval over ciphertext, but the drawback is not efficient. Then, Boneh et al. [9] introduced public key encryption with keyword search (PEKS). Since then, SE technology has been comprehensively expanded. There has been a tendency to combine SE and attribute-based encryption (ABE) [10] to design new keyword search algorithms. ABE is regarded as advanced technology that can be used to control data access. Bethencourt et al. [11] introduced the ciphertext-policy attribute-based encryption (CP-ABE) scheme first, which is considered very appropriate for supporting fine-grained permissions. Later, Cheung et al. [12] presented a CP-ABE scheme with "AND gate" constructions, which makes the scheme support more comprehensive access policies. After that, to enhance the privacy of attributes, Nishide et al. [13] first introduced a semi-hidden access strategy CP-ABE scheme.

Due to the continuous development of smart grids, people are becoming more and more worried about the security of their power data. There are some SE schemes proposed and applied to smart grid environments in recent years. Li et al. [14] proposed a searchable symmetric encryption scheme that can safeguard the confidentiality of data generated in smart grids. Eltayieba et al. [15] presented a scheme using an ABE and applied it to the smart grid scenario. Zhang et al. [16] presented a scheme with multi-keyword retrieval in the context of smart grids. Wang et al. [17] also presented multi-keyword searchable public key encryption in the context of smart grids. In addition, there are also some SE schemes [18–21] that are being applied in other areas, including the Internet of Things, smart health.

In this article, we present a blockchain-based searchable encryption system with fine-grained access control in smart grids. This presented system has more comprehensive advantages than traditional SE schemes. In addition, we utilize blockchain technology [22] to track data upload and search operations. This scheme enables keyword-based search encryption, supports attribute-based fine-grained access control, and protects user privacy. Specifically, we optimize the scheme [23], which makes the size of system parameters smaller and the time of algorithms shorter. In addition, our improvements make the scheme appropriate for the smart grid scenario. This article focuses on the topic of privacy protection in the field of cyber security. Our proposed access control scheme can prevent unauthorized access to data by users. To be specific, the core contributions we have made are as follows.

1) We utilize blockchain technology to grant data access with transparency and traceability. In addition, in our scheme, the encrypted index will be stored in the blockchain network, which protects the index from malicious tampering.
2) Our proposed scheme allows ciphertext to be retrieved while supporting fine-grained access permissions. Furthermore, this designed scheme preserves the privacy of data. Users' attribute information and keyword information are protected in the data access process.
3) There is an analysis of the security and advantages of the designed scheme. Experimental results and comparisons with other schemes show that our scheme is both cost-effective and practice-oriented.

The remaining sections are organised as follows. In Sect. 2, we introduce the preliminary knowledge required for this paper. In Sect. 3, we give a brief overview of our scheme. And then, in Sect. 4, we introduce the detailed construction of the scheme. In Sect. 5, we provide a comprehensive analysis of the scheme. And finally, in Sect. 6, we conclude the scheme.

## 2 Preliminaries

### 2.1 Bilinear Pairing

Let $\mathbb{G}$ and $\mathbb{G}_T$ are cyclic groups of order $p$, a bilinear map is a map $e : \mathbb{G} \times \mathbb{G} \to \mathbb{G}_T$ that has the following characteristics:

1) Bilinear: $\forall x, y \in \mathbb{G}, m, n \in \mathbb{Z}_p, e(x^m, y^n) - e(x, y)^{mn}$.
2) Non-degenerate: $\exists f \in \mathbb{G}$ such that $e(f, f) \neq 1$.
3) Computable: For all $f \in \mathbb{G}$, $e(f, f)$ can be calculated.

### 2.2 Access Architecture

In this paper, the designed scheme is based on "AND gates" access architecture. We use $n$ to indicate the attribute number in the system universe. We utilize $\{A_1, A_2, ..., A_n\}$ to indicate the set of attribute names. Furthermore, we use $\{v_1, v_2, ..., v_n\}$ to indicate attribute values in the access architecture and attributes. Note that unlike the structure used in scheme [23], for each attribute $A_i$, each user has a unique attribute value $v_i$ in our scheme. In scheme [23], each attribute $A_i$ has a valueset $V_i = \{v_{i,1}, v_{i,2}, ..., v_{i,m}\}$. The set contains available values for attribute $i$ and $m$ is the size of set $V_i$. Therefore, in this structure, for each attribute $A_i$, each user has multiple attribute values. However, this structure is not appropriate for smart grids and will influence the efficiency of algorithms.

## 2.3   Searchable Encryption

Searchable encryption gives users the ability to perform keyword search operations on encrypted data. Generally, there are many kinds of searchable encryption algorithms, but we can roughly summarize the searchable encryption process into the following four steps.

1) Data Encryption. Users encrypt data locally by using public keys and keywords. And then, encrypted documents will be uploaded to cloud servers.
2) Trapdoor Generation. Users generate trapdoors for the search keywords by using secret keys. Note that trapdoors is not able to give out any details about keywords.
3) Data Retrieval. Cloud servers take trapdoors as input and perform the retrieval algorithm. Later, cloud servers will return all the encrypted documents containing the keywords corresponding to trapdoors.
4) Data Decryption. Users are allowed to decrypt encrypted documents by utilizing secret keys and obtain results.

## 3   System Overview

We present a scheme that combines blockchain and SE in smart grids. The proposed scheme can protect data privacy effectively in a smart grid environment. In our scheme, data owners are allowed to encrypt keywords associated with the ciphertext using access architecture. Users can only retrieve ciphertext successfully if attributes meet the access architecture. By utilizing the matching between attributes and access policies, we can support fine-grained permissions in smart grids. Furthermore, we employ blockchain technology to keep track of power data upload and retrieval. In this section, we introduce the architecture and the designed scheme definition. After that, we introduce the flow of the whole system.

### 3.1   System Architecture of the Scheme

In this subsection, we describe the architecture of the designed blockchain-based searchable encryption system. As shown in Fig. 1, we have five units, namely data owner (DO), data user (DU), authority center (AC), blockchain (BC), and cloud server (CS). As a next step, we describe the responsibilities of each unit in detail.

- **DO**: DO is responsible for encrypting and uploading data, which includes encrypted power data documents and encrypted keyword indexes. After performing the encryption operation, DO uploads encrypted data to cloud servers and uploads encrypted indexes to the blockchain network.
- **DU**: DU is allowed to retrieve encrypted power data. With the help of AC, DU can obtain a search trapdoor. By uploading the trapdoor to BC, DU can get the ID of the encrypted document they want to search, and then get the ciphertext from CS by using this ID. Finally, DU performs the decryption operation to get the original unencrypted data.

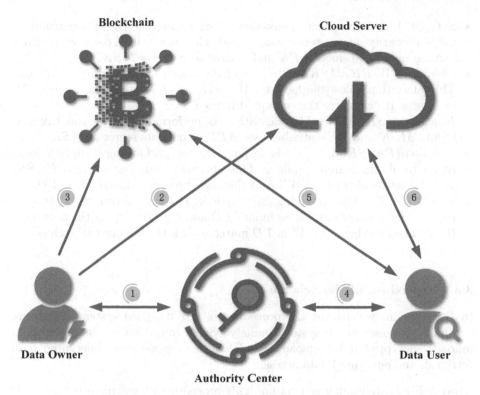

**Fig. 1.** System architecture of the proposed scheme.

- **AC**: AC is responsible for system initialization and key generation. AC can assist DO in generating public parameters and encrypting keyword indexes. In addition, AC can also generate secret keys by using attributes of DU. Note that in our scheme, AC is required to be fully trusted.
- **BC**: BC is responsible for recording the encrypted indexes that are uploaded by DO. In addition, any data upload and search will also be recorded on the blockchain network, which can prohibit some malicious repudiation. In addition, BC also executes the retrieval and returns the document ID to DU.
- **CS**: CS is mainly responsible for the storage of encrypted documents. Since it is too expensive to store encrypted data on BC, in this scheme, a large amount of encrypted data will be stored on CS. Du can use the file ID to get the expected encrypted data on CS.

## 3.2 Definition of the Scheme

In this subsection, we describe algorithms of the presented scheme, including *Setup*, *IndexEnc*, *KeyGen*, *TrapGen*, and *Test*.

- *Setup*$(1^\lambda)$: AC has the responsibility of executing the *Setup* algorithm. It takes a security parameter $\lambda$ as input and calculates system parameters which include public parameters $PK$ and master secret keys $MSK$.
- *IndexEnc*$(W, PK, MSK, AP)$: *IndexEnc* algorithm is performed by AC. This algorithm takes the keyword $W$, $PK$, $MSK$, and access policies $AP$ as inputs. It calculates the encrypted index $CT$.
- *KeyGen*$(MSK, ATT)$: This algorithm is performed by AC and takes as inputs $MSK$ and DU's attribute set $ATT$. It outputs secret keys $SK$.
- *TrapGen*$(PK, SK, W')$: DU is responsible for performing *TrapGen* algorithm to obtain search trapdoor. This algorithm takes as inputs $PK$, $SK$, and the retrieval keyword $W'$. And this algorithm returns trapdoor $TD$.
- *Test*$(TD, CT)$: After receiving the trapdoor, BC will execute the *Test* algorithm. This algorithm takes as inputs $TD$ and $CT$. It outputs the document ID if the search keyword $W'$ in $TD$ matches with the keyword $W$ included in $CT$.

### 3.3    Procedure of the Scheme

In this section, we present the procedure of our designed system. There are six main processes in our system, namely system initialization, encrypted data upload, encrypted index upload, search trapdoor generation, encrypted data retrieval, and encrypted data access.

**Step 1** Encrypted index generation. This procedure is performed between DO and AC. AC first initializes the system, including initializing the attribute space and generating system parameters. AC can generate $PK$, and $MSK$ by executing the *Setup* algorithm. After that, DO uploads the keyword associated with document $D$ to AC for encryption, and AC runs the *IndexEnc* algorithm and outputs the encryption result to DO.

**Step 2** Encrypted data upload. DO encrypts the power data document $M$ and specifies a document ID $ID_M$ for the document. After, DO uploads those encrypted file with the corresponding ID to cloud servers. The utilization of ID can uniquely identify the encrypted document and facilitate subsequent data access.

**Step 3** Encrypted index upload. DO constructs the encrypted index after getting the encrypted keywords. The encrypted index includes the encrypted keywords and corresponding document IDs. After that, DO uploads the encrypted index to BC. Using blockchain technology to store the encrypted index can protect it from malicious tampering.

**Step 4** Search trapdoor generation. In the procedure, DU sends his attribute set to AC. And AC first performs the *KeyGen* algorithm to get secret keys. After that, it returns $SK$ to DU. DU continues to execute the *TrapGen* algorithm to get the search trapdoor.

**Step 5** Encrypted data retrieval. DU can activate the search task by sending the search trapdoor to BC which executes the *Test* algorithm to match the keywords in the trapdoor with the encrypted index. BC will return the successfully matched results to DU. Consequently, after this process, DU will get the document ID of the desired search.

**Step 6** Encrypted data access. In the last step, DU will use the document ID obtained from BC to get the encrypted document. DU will send the document ID to CS which will return the corresponding encrypted data to DU according to the ID. DU will decrypt the obtained encrypted data to get the original unencrypted data. Until now, the whole flow of the scheme is finished.

## 4    Construction of the Proposed Scheme

Our designed scheme is made up of five algorithms: *Setup, IndexEnc, KeyGen, TrapGen*, and *Test*. Next, we will discuss these algorithms in detail.

$Setup(1^\lambda)$: The *Setup* algorithm takes a security parameter $\lambda$ as input and generates system parameters which will be used in the following algorithms. Let $\mathbb{G}$ and $\mathbb{G}_T$ be multiplicative cyclic groups of prime order $p$, where $p$ is determined by $\lambda$. And $e : \mathbb{G} \times \mathbb{G} \to \mathbb{G}_T$ be a bilinear map. To begin with, it chooses two generators $g_1, g_2 \in \mathbb{G}$. Then, it selects $\alpha, \beta, \gamma, t \in Z_p^*$, and computes $e(g_1, g_2)^\alpha$, $g_1^{\alpha/\gamma}, g_2^{\alpha/\beta}, g_2^\gamma$, and $g_2^t$. It also selects two hash functions: $H_1 : \{0,1\}^* \to \mathbb{G}$, $H_2 : \{0,1\}^* \to Z_p^*$. Next, $MSK$ is expressed as

$$MSK = \{\alpha, \beta, \gamma, t\}.$$

And $PK$ are published as

$$PK = \{g_1, g_2, e(g_1, g_2)^\alpha, g_1^{\alpha/\gamma}, g_2^{\alpha/\beta}, g_2^\gamma, g_2^t\}.$$

$IndexEnc(W, PK, MSK, AP)$: We takes as inputs the keywords $W$, $PK$, $MSK$, and access policies $AP$. To begin with, DO chooses a secret number $s \in Z_p^*$ randomly. And then, for $i \in [1, n-1]$, DO also selects $s_i \in Z_p^*$ randomly and computes $s_n = s - \sum_{i=0}^{n-1} s_i$. Next, for $i \in [1, n]$, it selects $a_i \in Z_p^*$ randomly and computes $f(x_i) = a_i(x_i - H_2(i||v_i)) + s_i$, where $v_i$ indicates the value of attribute $i$ in access policies $AP$. And we denote $a_{i0} = s_i - a_i H_2(i||v_i)$. Then, DO selects a random number $\mu \in Z_p^*$ and calculates $H_1(W)^{a_i\mu}$, where $1 \le i \le n$. After that, DO sends calculation results to AC.

AC continues to calculate $\prod_{i=1}^n H_1(W)^{a_i t\mu/\gamma} = H_1(W)^{\sum_{i=1}^n a_i t\mu/\gamma}$. And then, AC returns the encryption result to DO. Due to the presence of the random number $\mu$, the values during the communication of DO and AC will be protected from privacy disclosure. Finally, for keyword $W$, DO computes the encrypted keyword as

$$C_W = g_1^{\frac{(s-\sum_{i=1}^n a_{i0})\alpha}{\gamma}} \cdot H_1(W)^{\frac{\sum_{i=1}^n a_i t}{\gamma}},$$

$$C = g_2^{\frac{\alpha \sum_{i=1}^n a_i}{\beta}}, C_i = g_2^{a_i t}, i \in [1, n].$$

The encrypted index $CT$ are published as

$$CT = \{C_W, C, \{C_i\}_{1 \le i \le n}\}.$$

$KeyGen(MSK, ATT)$: This designed algorithm takes as inputs $MSK$ and DU's attributes $ATT$. There are $n$ attributes in $ATT$, and we refer to $v_i$ as the value of the $ith$ attribute. This algorithm first chooses $r \in Z_p^*$ randomly. And then it computes

$$D_0 = g_1^{r\beta}, D_i = g_1^{(H_2(i||v_i)+r)\frac{\alpha}{t}}, i \in [1, n].$$

The $SK$ are published as $SK = \{D_i\}_{0 \leq i \leq n}$.

$TrapGen(PK, SK, W')$: This algorithm takes $PK$, $SK$ and $W'$ as inputs. It selects a random number $\psi \in Z_p^*$. And using $SK$ to calculate

$$D_0' = g_1^{r\beta\psi}, D_1' = g_2^{\gamma\psi},$$

$$\tilde{D}_i = g_1^{(H_2(i||v_i)+r)\frac{\alpha\psi}{t}} \cdot H_1(W')^{\psi}, i \in [1, n].$$

The search trapdoor $TD$ are published as

$$TD = \{D_0', D_1', \{\tilde{D}_i\}_{1 \leq i \leq n}\}.$$

$Test(TD, CT)$: After receiving a trapdoor $TD$, BC performs $Test$ algorithm to find a keyword match. This algorithm takes $TD$ and $CT$ as inputs. And computing $R = R_1/R_2$, where $R_1 = \prod_{i=1}^n e(C_i, \tilde{D}_i)$ and $R_2 = e(C, D_0')$. Then, it computes $e(C_W, D_1')$. If the result of $R$ is equal to value of $e(C_W, D_1')$, it indicates that keywords inside the ciphertext match keywords inside the trapdoor. Then this algorithm returns document ID of the encrypted data. Otherwise, it outputs $\perp$ to denote that the keyword match failed.

Next, we will check the correctness of the scheme mentioned above. We first compute $R$,

$$\begin{aligned}
R &= R_1/R_2 \\
&= \frac{\prod_{i=1}^n e(C_i, \tilde{D}_i)}{e(C, D_0')} \\
&= \frac{\prod_{i=1}^n e(g_2^{a_i t}, g_1^{(H_2(i||v_i)+r)\frac{\alpha\psi}{t}} \cdot H_1(W')^{\psi})}{e(g_2^{\frac{\alpha \sum_{i=1}^n a_i}{\beta}}, g_1^{r\beta\psi})} \\
&= \frac{e(g_1, g_2)^{r\alpha\psi \sum_{i=1}^n a_i} \cdot e(g_1, g_2)^{\alpha\psi \sum_{i=1}^n (s_i - a_{i0})} \cdot e(H_1(W'), g_2)^{\psi t \sum_{i=1}^n a_i}}{e(g_1, g_2)^{r\alpha\psi \sum_{i=1}^n a_i}} \\
&= e(g_1, g_2)^{\alpha\psi \sum_{i=1}^n (s_i - a_{i0})} \cdot e(H_1(W'), g_2)^{\psi t \sum_{i=1}^n a_i}
\end{aligned} \tag{1}$$

Next, we compute $e(C_W, D_1')$,

$$\begin{aligned}
e(C_W, D_1') &= e(g_1^{\frac{(s - \sum_{i=1}^n a_{i0})\alpha}{\gamma}} \cdot H_1(W)^{\frac{\sum_{i=1}^n a_i t}{\gamma}}, g_2^{\gamma\psi}) \\
&= e(g_1, g_2)^{\alpha\psi(s - \sum_{i=1}^n a_{i0})} \cdot e(H_1(W), g_2)^{\psi t \sum_{i=1}^n a_i} \\
&= e(g_1, g_2)^{\alpha\psi(\sum_{i=1}^n s_i - \sum_{i=1}^n a_{i0})} \cdot e(H_1(W), g_2)^{\psi t \sum_{i=1}^n a_i} \\
&= e(g_1, g_2)^{\alpha\psi \sum_{i=1}^n (s_i - a_{i0})} \cdot e(H_1(W), g_2)^{\psi t \sum_{i=1}^n a_i}
\end{aligned} \tag{2}$$

Therefore, Eq. 1 and Eq. 2 have the same calculation results, when $W'$ inside the trapdoor is the same as $W$ inside the encrypted index.

# 5 Analysis of the Proposed Scheme

## 5.1 Security Analysis

The security features of the above-designed scheme are analysed in this subsection. Firstly, the above-designed scheme can ensure the credibility, integrity, and immutability of electric power data. By utilizing blockchain technology, we can track data upload and search operations. All operations about data upload and access are recorded as transactions on the blockchain, and all nodes on the blockchain network have the right to review these records. In addition, encrypted indexs will be stored in the blockchain, which protects the index from malicious tampering. In our scheme, users must complete both attribute and search keyword matches before being allowed to access files. The attribute information is contained in secret keys, and search keywords are encrypted in search trapdoors. Secret keys will be given to users by AC in the protected channel and the search trapdoor will be sent to the blockchain network to perform ciphertext retrieval. Although the blockchain will record the trapdoor, the trapdoor does not reveal the user's search keyword information because of the cryptographic operation. Moreover, our proposed scheme can be proven secure against chosen-keyword attacks. There is detailed security proof in the scheme [23], thus security proofs of the algorithms will not be the main focus of our paper. And more information on security proofs can be found in [23].

## 5.2 Performance Analysis

In the smart grid scenario, people prefer systems that have less waiting time. Therefore, our scheme focuses on efficiency to make improvements. In this section, we first analyzed the time and space complexity of algorithms and compared our scheme with the scheme [23]. Then, we implement our design and analyse its performance.

Firstly, we analyzed the space complexity of algorithms in Table 1, where $n$ means the number of attribute fields, $m$ indicates the size of each attribute field $i$, $|\mathbb{G}|$ and $|\mathbb{G}_T|$ denote the size of $\mathbb{G}$ and $\mathbb{G}_T$ respectively. Next, we analyzed the time complexity of algorithms in Table 2, where $E, E_T, M, P$ denotes the time of exponentiation calculation in $\mathbb{G}$, the time of exponentiation calculation in $\mathbb{G}_T$, the time of multiplication calculation in $\mathbb{G}$, and the time of pairing calculation respectively. By analyzing the designed algorithms, we can see that the time and space complexity of the above algorithms is less than scheme [23]. Since there will be multi-attribute values for each attribute field in Scheme [23], the algorithms have a high complexity. Our improvements reduce the time and space required to implement the algorithms. In addition, in the *Setup* algorithm, the parameter size of our scheme is constant.

Then, we execute the designed scheme on a computer that has a 64-bit Windows 11 system with 2.50 GHz 11th Gen Intel(R) Core(TM) i7-11700 CPU and 32.0 GB RAM. We utilize the Java Pairing-based Cryptography Library (JPBC) of version 2.0.0 [24]. In the experiments, we use type A parings. And

**Table 1.** Space Complexity Analysis.

| Schemes | Setup | IndexGen | KeyGen | TrapGen |
|---------|-------|----------|--------|---------|
| [23] | $(n \cdot m + 5)|\mathbb{G}| + |\mathbb{G}_T|$ | $(n \cdot m + 1)|\mathbb{G}|$ | $(n \cdot m + 1)|\mathbb{G}|$ | $(n \cdot m + 2)|\mathbb{G}|$ |
| Ours | $6|\mathbb{G}| + |\mathbb{G}_T|$ | $(n + 1)|\mathbb{G}|$ | $(n + 1)|\mathbb{G}|$ | $(n + 2)|\mathbb{G}|$ |

**Table 2.** Time Complexity Analysis.

| Schemes | Setup | IndexGen | KeyGen | TrapGen | Test |
|---------|-------|----------|--------|---------|------|
| [23] | $(n \cdot m + 3)E + E_T + P$ | $(n \cdot m + 3)E + M$ | $(n \cdot m + 1)E$ | $(2n \cdot m + 2)E + n \cdot mM$ | $(n \cdot m + 1)M + (n \cdot m + 2)P$ |
| Ours | $4E + E_T + P$ | $(n + 3)E + M$ | $(n + 1)E$ | $(2n + 2)E + nM$ | $(n + 1)M + (n + 2)P$ |

for each result, we repeat it 20 times and use the averages to judge the experimental results. As illustrated in Fig. 2, we can observe that the execution time is increasing with the attribute number. But note that the time of the *Setup* algorithm keeps steady. And the *Setup* and *TrapGen* algorithms are the least time-consuming and most time-consuming algorithms, respectively. This result is also consistent with the time complexity analysis of algorithms. In addition, as shown in Fig. 3, we compare the size of $PK$. In the scheme [23], the size of $PK$ increases with the number of attributes. Note that to show the changes in public parameters more conveniently, we set the size of each attribute field $m$ to 5 in the scheme [23]. As a comparison, the size of public parameters in scheme remains constant. This indicates that our scheme is also more suitable for some environments with a large number of attributes.

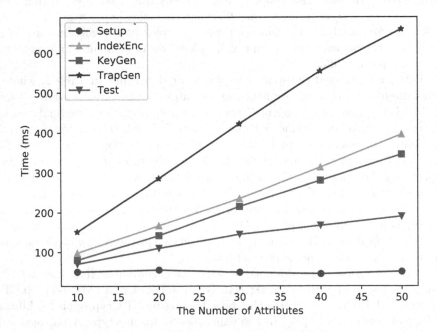

**Fig. 2.** The implementation time of algorithms.

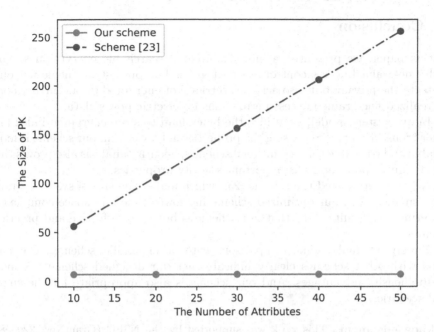

**Fig. 3.** The size of the public parameters.

## 5.3 Comprehensive Comparisons

In this subsection, we make extensive comparisons with other similar schemes. We compare the schemes in five main aspects, namely keyword search, privacy protection, blockchain, access structure, and mathematical structure. scheme [19] is a policy-hiding attribute-based scheme. However, this scheme cannot support keyword search and is not very efficient because it is constructed based on the composite group. Scheme [20] is a scheme that relies on lattice cryptography, which is considered an important technique to resist quantum attacks. And this scheme also uses blockchain to store indexes. However, it does not consider attribute hiding. Scheme [25] is a CP-ABE scheme with keyword search. This scheme is based on the LSSS which makes the scheme support more expressive access policies. Nevertheless, the scheme does not provide access architecture hidden. Our designed scheme is adapted from scheme [23]. Our Optimizations make the scheme more efficient and practical. Detailed comparisons are given in Table 3.

**Table 3.** Comparisons of Schemes.

| Schemes | Keyword Search | Privacy Protection | Blockchain | Access Structure | Mathematical Structure |
|---------|----------------|--------------------|------------|------------------|------------------------|
| [19] | ✗ | ✓ | ✗ | LSSS | Composite group |
| [20] | ✓ | ✗ | ✓ | AND gates | Lattice |
| [25] | ✓ | ✗ | ✗ | LSSS | Prime group |
| [23] | ✓ | ✓ | ✗ | AND gates | Prime group |
| Ours | ✓ | ✓ | ✓ | AND gates | Prime group |

# 6 Conclusion

In this paper, we presented a blockchain-based searchable encryption scheme with fine-grained access control in smart grids. The presented scheme not only protects the privacy but also achieves retrieval of encrypted data. Furthermore, we realized fine-grained access permissions for electric power data.

In our system model, we utilize the blockchain to store encrypted index and other transaction records. Using the properties of blockchain, our scheme is more reliable and traceable than traditional schemes. Security analysis also proves that the design is sound and has important security properties.

Next, we optimized the scheme [23], which makes the size of system parameters smaller. And our optimized scheme has lower time and space complexity. Experimental results show that our scheme is both cost-effective and practice-oriented.

Finally, we make some comparisons with other similar schemes. Comparisons with other schemes clearly indicate that our designed scheme has more comprehensive advantages. And our scheme is also appropriate for the smart grid scenario.

**Acknowledgement.** This work was supported by the NSFC (GrantNos.72293583, 62271070,61962009), and the Project (GrantNo.JCKY2021208B036).

# References

1. Massoud Amin, S., Wollenberg, B.: Toward a smart grid: power delivery for the 21st century. IEEE Power Energ. Mag. **3**(5), 34–41 (2005). https://doi.org/10.1109/MPAE.2005.1507024
2. Cai, D., et al.: Electric power big data and its applications. In: Muthuramalingam, T., Kayal, P., Ahmed, K. (eds.) Proceedings of the 2016 International Conference on Energy, Power and Electrical Engineering. AER-Advances in Engineering Research, vol. 56, pp. 181–184 (2016), International Conference on Energy, Power and Electrical Engineering (EPEE), Shenzhen, Peoples R China, OCT 30–31, 2016
3. Zhang, P., Xu, Y., Luo, F., Dong, Z.Y.: Power big data: new assets of electric power utilities. J. Energy Eng. **145**(3) (2019). https://doi.org/10.1061/(ASCE)EY.1943-7897.0000604
4. Xiaosheng, P., Diyuan, D., Shijie, C., Jinyu, W., Zhaohui, L., Lin, N.: Key technologies of electric power big data and its application prospects in smart grid. Proc. Chin. Soc. Electr. Eng. **35**, 503–511 (2015)
5. Pham, C.T., Mnsson, D.: A study on realistic energy storage systems for the privacy of smart meter readings of residential users. IEEE Access **7**, 150262–150270 (2019). https://doi.org/10.1109/Access.2019.2946027
6. Nguyen, T., Wang, S., Alhazmi, M., Nazemi, M., Estebsari, A., Dehghanian, P.: Electric power grid resilience to cyber adversaries: state of the art. IEEE Access **8**, 87592–87608 (2020). https://doi.org/10.1109/ACCESS.2020.2993233
7. Lu, R., Zhu, H., Liu, X., Liu, J.K., Shao, J.: Toward efficient and privacy-preserving computing in big data era. IEEE Netw. **28**(4), 46–50 (2014). https://doi.org/10.1109/MNET.2014.6863131

8. Song, D.X., Wagner, D., Perrig, A.: Practical techniques for searches on encrypted data. In: Proceeding 2000 IEEE Symposium on Security and Privacy. S&P 2000, pp. 44–55 (2000). https://doi.org/10.1109/SECPRI.2000.848445
9. Boneh, D., Di Crescenzo, G., Ostrovsky, R., Persiano, G.: Public key encryption with keyword search. In: Cachin, C., Camenisch, J.L. (eds.) EUROCRYPT 2004. LNCS, vol. 3027, pp. 506–522. Springer, Heidelberg (2004). https://doi.org/10.1007/978-3-540-24676-3_30
10. Sahai, A., Waters, B.: Fuzzy identity-based encryption. In: Cramer, R. (ed.) EURO-CRYPT 2005. LNCS, vol. 3494, pp. 457–473. Springer, Heidelberg (2005). https://doi.org/10.1007/11426639_27
11. Bethencourt, J., Sahai, A., Waters, B.: Ciphertext-policy attribute-based encryption. In: 2007 IEEE Symposium on Security and Privacy, Proceedings, pp. 321+. IEEE Symposium on Security and Privacy, IEEE Comp Soc TCSP (2007). https://doi.org/10.1109/sp.2007.11 IEEE Symposium on Security and Privacy (S&P 2007), Berkeley, CA, 20-23 (2007)
12. Cheung, L., Newport, C.: Provably secure ciphertext policy ABE. In: DiVimercati, S., Syverson, P., Evans, D. (eds.) CCS'07: Proceedings of the 14th ACM Conference on Computer and Communications Security, pp. 456–465. ACM SIGSAC (2007), 14th ACM Conference on Computer and Communication Security, Alexandria, VA, OCT 29-NOV 02 (2007)
13. Nishide, T., Yoneyama, K., Ohta, K.: Attribute-based encryption with partially hidden encryptor-specified access structures. In: Bellovin, S., Keromytis, A., Gennaro, R., Yung, M. (eds.) Applied Cryptography And Network Security, Proceedings. Lecture Notes In Computer Science, vol. 5037, pp. 111–129 (2008), 5th International Conference on Applied Cryptography and Network Security, Columbia University, New York, NY, JUN 03–06 2008
14. Li, J., Niu, X., Sun, J.S.: A practical searchable symmetric encryption scheme for smart grid data. In: ICC 2019–2019 IEEE International Conference On Communications (ICC). IEEE International Conference on Communications, IEEE; China Mobile; Huawei; ZTE; Qualcomm; Oppo; Natl Instruments (2019), IEEE International Conference on Communications (IEEE ICC), Shanghai, Peoples R China, MAY 20–24 (2019)
15. Eltayieb, N., Elhabob, R., Hassan, A., Li, F.: An efficient attribute-based online/offline searchable encryption and its application in cloud-based reliable smart grid. J. Syst. Architect 98, 165–172 (Sep 2019) .https://doi.org/10.1016/j.sysarc.2019.07.005
16. Zhang, D., Fan, Q., Qiao, H., Luo, M.: A public-key encryption with multi-keyword search scheme for cloud-based smart grids. In: 2021 IEEE Conference On Dependable And Secure Computing (DSC). IEEE (2021). https://doi.org/10.1109/DSC49826.2021.9346254,IEEE Conference on Dependable and Secure Computing (DSC), Aizuwakamatsu, JAPAN, JAN 30-FEB 02 2021
17. Wang, D., Wu, P., Li, B., Du, H., Luo, M.: Multi-keyword searchable encryption for smart grid edge computing. Electr. Power Syst. Res. 212. (NOV 2022) https://doi.org/10.1016/j.epsr.2022.108223
18. Miao, Y., Liu, X., Choo, K.K.R., Deng, R.H., Wu, H., Li, H.: Fair and dynamic data sharing framework in cloud-assisted internet of everything. IEEE Internet Things J. 6(4), 7201–7212 (AUG 2019). https://doi.org/10.1109/JIOT.2019.2915123
19. Zhang, Y., Zheng, D., Deng, R.H.: Security and privacy in smart health: efficient policy-hiding attribute-based access control. IEEE Internet Things J. 5(3, SI), 2130–2145 (JUN 2018). https://doi.org/10.1109/JIOT.2018.2825289

20. Li, C., et al.: Efficient medical big data management with keyword-searchable encryption in healthchain. IEEE Syst. J. **16**(4), 5521–5532 (Dec 2022).https://doi.org/10.1109/JSYST.2022.3173538

21. Xu, G., et al.: A searchable encryption scheme based on lattice for log systems in blockchain. CMC-Comput. Mater. CONTINUA **72**(3), 5429–5441 (2022). https://doi.org/10.32604/cmc.2022.028562

22. Nakamoto, S.: Bitcoin: a peer-to-peer electronic cash system. Decentralized business review, p. 21260 (2008)

23. Chaudhari, P., Das, M.L.: Privacy preserving searchable encryption with fine-grained access control. IEEE Trans. Cloud Comput. **9**(2), 753–762 (2021). https://doi.org/10.1109/TCC.2019.2892116

24. Caro, A.D., Iovino, V.: jPBC: java pairing based cryptography. Comput. Commun. (2011)

25. Ge, C., Susilo, W., Liu, Z., Xia, J., Szalachowski, P., Fang, L.: Secure keyword search and data sharing mechanism for cloud computing. IEEE Trans. Dependable Secure Comput. **18**(6), 2787–2800 (2021). https://doi.org/10.1109/TDSC.2020.2963978

# Data-Driven Distributed Autonomous Architecture for 6G Networks

Pengyu Li[1(✉)], Xinyu Chen[2], Zhenqiang Sun[1], Yanxia Xing[1], Jianfeng Zhou[2], and Wanpeng Fan[2]

[1] 6G Research Center, China Telecom Research Institute, Beijing, China
`lipengyu@chinatelecom.cn`
[2] Wireless Product Operation, ZTE Corporation, Nanjing, China

**Abstract.** Currently, the industry has put forward new scenarios, including ISAC (Integrated sensing and communication), computing and network coordination, and ubiquitous intelligence, for potential application to 6G networks in the future. Addressing the systematic and organic integration of data, computing, and communication becomes a primary concern. Meanwhile, the commercialization of 5G networks on a large scale has gradually revealed some new issues, such as network reliability, flexibility, rapid deployment, and service customization, all of which can benefit from further improvement. In this paper, we propose a new intelligent and simplified architecture for 6G networks. We detail its functionalities, key features and procedures, and illustrate how to implement a new service in the future based on this architecture. This preliminarily verifies the capability of the architecture to enhance the existing networks while meeting future service requirements.

**Keywords:** 6G · network simplicity · data-driven · user-centric · network AI · distributed autonomy

## 1 Introduction

With the commercial scale of the 5G networks, the industry has systematically carried out 6G technology research, 6G network architecture is the foundation of future network deployment, is related to the 6G network capability and service level to people's production and life, and directly reflects the commercial value of 6G networks. The design of the 6G network architecture should be based on service requirements and problem solving.

In terms of service requirements, the 2C (To Customer) market is predominantly influenced by the prospective demand for new applications like immersive XR (Extended Reality), holographic communications, and space-air-ground integration, with a focus on enhancing the functionalities and performance of 6G networks for communication, sensing and computing integration. The 2B (To Business) market, on the other hand, primarily caters to the differentiated demands of numerous industries on networks. The 5G networks have been standardized and defined regarding capability exposure. Furthermore, operators have also developed a capability exposure platform and conducted

H. Jin et al. (Eds.): IAIC 2023, CCIS 2059, pp. 149–163, 2024.
https://doi.org/10.1007/978-981-97-1280-9_12

trials on existing networks [1]. For 6G network research, the focus is on improving flexibility and robustness, enhancing customization capability of 2B, and providing network infrastructure guarantee services for the development of digital economy. To meet the requirements of new services, the 6G networks will incorporate AI/ML (Artificial intelligence and machine learning) services, computing services, ISAC services, in addition to connectivity services, providing assistance for AIaaS (AI as a Service) [2], CPaaS (Computing Power as a Service) [3], SaaS (Sensing as a Service) [3], as well as distributed intelligent native architecture integrating AI/ML capabilities [4], etc.

The commercialization of the 5G network is empowering all areas of social production and living. However, urgent issues have been encountered in the course of its practice.

- Need to reduce network deployment time and difficulty: the complexity of the core network architecture is hindered by various types of NFs (Network Functions) and logical interfaces, resulting in increased interactions and dependencies between the NFs, accompanied by the workload of extensive interoperability testing. Additionally, network deployment, scaling, operations, and maintenance pose challenges due to coupling issues. This predicament is even more compounded in the aftermath of malfunctions, as it becomes difficult to locate faults, troubleshoot, recover efficiently, and satisfy the demands of 6G network for speedy, adaptable, and flexible customization and deployment.
- Need to enhance network deployment flexibility: In the future space-air-ground scenario, the networks may be deployed on various high-speed moving carriers such as satellites, high-speed trains, airplanes, and high-altitude balloons. Therefore, the network design must fulfil the demands of dynamics, mobility, and flexible deployment.
- Need to address network robustness: the present network uses centralized networking. A small number of NFs and interface or link failures between NFs may result in the widespread propagation of failures. Therefore, it is imperative to introduce new mechanisms to enhance the self-healing capability of the network to improve its resilience.
- Data management and transmission efficiency needs to be improved: data is dispersed among various NFs, leading to redundancy and difficulty in management. With the emergence of new technologies such as AI and ISAC, meeting the real-time transmission of a vast amount of data using SBI is challenging. Therefore, there is a need for further decoupling and aggregation of data as well as for high volume data transfer requirements.
- The intelligent capabilities of the network need to be enhanced: The current AI functions in 5G networks utilize centralized collection and analysis of data pertaining to AI/ML [5], leading to significant communication overheads when dealing with large-scale networking, and is insufficient to fulfil requirements for efficient service provision as well as low latency. Besides, user-based AI services tend to be complex and inflexible. And there is still a lack of support for distributed learning and inference. Enhancements are required in both architectural and network functionality to support this aspect.

Therefore, continuous optimization of network architecture is necessary for 6G networks to address existing network challenges and meet new service demands. The paper presents the new architecture solution, with Sect. 2 detailing the functional architecture and key features. Section 3 explores the features and advantages of the new architecture through the general registration procedure example. In Sect. 4, a new service scenario supported by the new architecture is illustrated. The paper concludes with a brief summary.

## 2　Data-Driven Distributed Autonomous Architecture (DDAA)

### 2.1　Network Functional Architecture of DDAA

The upcoming 6G networks need to meet the requirements of future new services, at the same time, there is an urgent need to solve the current pain points such as complicated network architecture, lack of network adaptability to services, and insufficient agility. The design of 6G network's functional architecture requires optimizing and enhancing the Service Based Architecture (SBA) while improving the flexibility and elasticity of network services. Furthermore, it aims to enhance service efficiency, robustness, and reliability via network simplification, service restructuring, and other measures. The scope of SBA will be extended to the wireless access side, aiming to achieve end-to-end service-based interfaces in mobile networks. Furthermore, focus should be given to meeting the requirements of efficient transmission of large amounts of data in future scenarios.

To achieve our aim, we are proposing an architectural solution known as DDAA (Data-driven Distributed Autonomous Architecture). Its network functional architecture is illustrated in Fig. 1 and comprises of five streamlined network units, each equipped with customizable, on-demand microservices that can be tailored to meet specific needs. The network units efficiently interact with each other through a dual-bus structure. Traditional signaling messages and a small amount of data are transmitted through SBI (Service Based Interface), while bulk data is transmitted through the newly added DCI (Data Channel Interface), a feature that enables fast and effective data access and processing.

- NCU (Network Control Unit): NCU offers network control services that concentrate on restructuring services and interfaces. NCU incorporates and enhances the fundamental features of 5G networks, including access management, mobility management, session management, subscription management, authentication, and policy control. The services and interfaces are organized or restructured based on their functionality, deployment, capability exposure, and other needs to tackle the challenges of complex architecture and interoperability issues. It offers access adaptation functionality, which is highly flexible and supports various heterogeneous accesses, including 4G/5G, Wi-Fi, satellite, and fixed networks, additionally, it offers generalized and unified service capabilities. It adds an extension functionality that facilitates the adaptable and dynamic deployment of new or customized services, such as ISAC, CNC (Computing and Network Coordination), and others. Additionally, it introduces the DUE (digital twin User Equipment), representing the PUE (Physical UE) digitally, consequently boosting user satisfaction through enhanced personalized service experience enabled by stronger network AI capabilities.

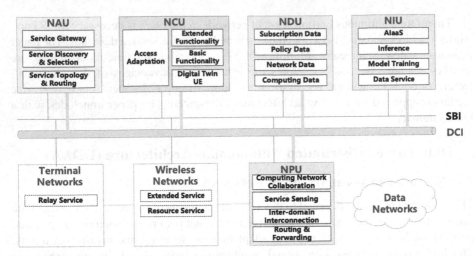

**Fig. 1.** Network functional architecture of DDAA

- NDU (Network Data Unit): NDU offers a variety of data services that enable data and network services to be further separated, while also providing enhanced data transmission mechanisms that work together to create a data plane. The system facilitates a logically unified and physically distributed management and access model for data. Along with traditional subscription and policy data, newly added network and computing data serve as fundamental data to enable CNC, AI/ML and other services. An independent data channel is introduced as another shared bus between internal services in the networks to collaborate with NDU to form an efficiently organized data plane, which supports the storage, processing, and transmission of large amounts of heterogeneous data between services within the networks and between UEs and services, etc., to meet the data requirements of new scenarios such as ISAC and AI.

- NPU (Network Packet Unit): NPU offers user-plane services with an emphasis on advancing new capabilities brought about by distributed autonomy and CNC. In addition to inheriting the traditional user-plane routing and forwarding of terminal service data and policy enforcement, the NPU adds autonomous domain/subnet routing and service continuity. It also introduces user-plane programmability, enabling flexible definition of user-plane processing logic. The NPU supports and enhances service-based interfaces, satisfying user-plane service discovery, control, and capability exposure. Additionally, it enhances service sensing and CNC capabilities.

- NIU (Network Intelligence Unit): the network intelligence services center on the evolution towards distributed and native intelligence as ubiquitous and fundamental services for 6G networks. NIU serves as the basic unit of network intelligence services, offering various AI/ML capabilities including data provision, training, and inference. It enhances network intelligence through the internal application of AI/ML technologies and enables flexible and customized AI services externally (AIaaS). NIU enables multi-node coordination across various domains such as terminals, wireless networks, core networks, applications, and network management systems. This coordination occurs at different levels, including within the service, within and across the autonomous domain/subnet through a hybrid approach that employs both standalone and service built-in units to form a distributed and end-to-end intelligent plane.

- NAU (Network Assistance Unit): NAU offers network support services that provide the necessary capabilities for distributed, open networks. The capabilities include service gateway for mapping and translation needed for offering cross-domain capabilities, service discovery and selection to manage services within and between autonomous domains/subnets and across different network domains (core, wireless, applications, etc.), network topology and routing for making topology connectivity decisions and routing configurations, etc. NAUs can function either as an internal service within the corresponding NUs (Network Units) or as an external standalone service.

Furthermore, with the advancement of ProSe (Proximity Services) and other services, the self-organizing network capability of terminals, alongside necessary policy control and assurance, can been enhanced, and gradually forming terminal networks, which can adapt to the industrial chain's rhythm and support capabilities including relay services and establishing connections with the core networks through SBI. After achieving the separation of the control plane and user plane of the Centralized Unit (CU), the wireless networks can potentially support resource services, AI/ML, ISAC, and other extended services through SBI on the premise of meeting service metrics.

## 2.2 Key Features of DDAA

The design of the DDAA functional architecture represents six key features.

- Network Simplification

The introduction of SBA enhances service management flexibility within the fundamental framework of the 5G network. However, certain aspects of the previous network remain intact, such as the point-to-point connection mode and dependencies between NFs in the control plane. Additionally, due to the requirements of different service interfaces, the transmission of user data and NF processing status compels frequent interactions between NFs, leading to complicated procedures.

The DDAA architecture implements the restructuring and aggregation of the fundamental NFs within the control plane. It integrates the functionalities of AMF (Access and Mobility Management Function), SMF (Session Management Function), NSSF (Network Slice Selection Function), and the logical processing of PCF (Policy Control function) and UDM (Unified Data Management) within a single NCU. Additionally, it carries out a fresh restructuring and design of the initial service-based interfaces and provides support for on-demand customization to attain network simplicity and elasticity, minimize deployment complexity, and improve deployment flexibility. NCU provides a unified multi-access service, supporting heterogeneous access and compatibility with existing 4G/5G functions through front-end adapters and unified processing at the back end. This reduces the complexity of network planning, design, and operation, enabling simplified access.

- Dual Bus Architecture

The SBI used in 5G networks is based on the HTTP protocol. According to the 3GPP standard, the maximum size of the JSON body of any HTTP request/response shall not exceed 16 million octets before compression is applied [6]. However, for new scenarios such as AI, ISAC, etc., relying solely on SBI to deliver data will fail to meet the requirements of rate and latency. Even if packet-splitting is utilized for transmission, it will considerably reduce the data transmission efficiency.

The DDAA architecture designs a dedicated data access interface (DCI) between each network unit. This interface collaborates with the SBI interface to facilitate a high volume of block data, SQL (Structured Query Language) data, log data, and other heterogeneous data transmissions. This is made possible through the incorporation of a new flexible evaluation and asynchronous processing mechanism, ensuring the system meets the requirements of highly efficient data access.

- Data-driven

In 5G networks, although the 3GPP specification states that data NFs like UDR (Unified Data Repository) or UDSF (Unstructured Data Storage Function) can centrally manage user data, application data, etc. [7], in reality, due to technical issues and interoperability challenges, the implementation is not satisfactory.

The DDAA architecture restructures data to prioritize user services using NDUs, enabling efficient use of professional database technology to simplify data processing logic. Additionally, it guarantees consistent services for data security, management, and access. It strengthens the stateless design of logical functions, which further separates dynamic data within NFs to improve network robustness. Furthermore, it adds terminal data, network data, computing data, and other relevant elements. To meet the demands of scenarios with intensive computing resources, data analysis, and aggregation like AI and ISAC, it is beneficial to aggregate user-related data to support user-based analysis and maximize data value.

- Digital twin UE

The networks currently follow a process paradigm, centered on connection functionality, which establishes data connections for UEs by invoking services among NFs. However, this approach does not support the management of the user dimension, making it more difficult to enhance user experience intuitively.

In the DDAA architecture, NCU has developed a processing mechanism focused on the user, incorporating DUE. This changes the interaction between UE and networks from a process paradigm to a declarative paradigm. DUE receives tasks from PUE, processes them on the network side and returns the results. This methodology reduces the frequency of interactions between UEs and networks, while utilizing network and AI/ML capabilities to accomplish more complex tasks. The DUE mechanism improves user-centric perception and service capabilities while keeping the terminal lightweight, resulting in a more powerful and intelligent device.

- Native network AI

Although the current 5G networks have incorporated AI service NFs as NWDAF (Network Data Analytics Function), there are several issues to be addressed. The relatively centralized collection of data may have an adverse impact on data utilization efficiency or result in network resource consumption. The NFs themselves lack AI capability, and the absence of a collaboration mechanism among multiple AI nodes is unfavorable for resource utilization, which may affect AI service efficiency. The user AI service mechanism and the user data collection process are intricate and inflexible. A significant volume of artificial intelligence data is inadequately transmitted through the SBI approach.

The DDAA architecture involves collaboration between NIUs and other NUs to achieve distributed intelligent native architecture. It has four main advantages. First, proposing "AI for user" capability, supporting scheduling of data, computing resource, and AI service capability around a single user based on the previously mentioned DUE, and implementing user-level AI services. Second, through collaborating standalone NIUs and built-in NIUs within NCUs, the proposed system enables cross-layer and cross-domain/subnet AI services in scale and ultra-distributed network environments. Additionally, it presents a general architecture to support various distributed AI/ML learning methods, such as federated learning [8], swarm learning [9], MARL (Multi-Agent Reinforcement Learning) [10], and split learning [11]. Third, the introduction of DCI enables the streamlined interaction of vast amounts of heterogeneous data, including data/models, and enhances the efficacy of both model training and inference. Finally, by utilizing the NDU to aggregate AI service data, in combination with terminal data, network data, and computing data, alongside knowledge, models and other AI data, network AI driven by both data and knowledge can be readily achieved.

- Distributed autonomy

Existing networks are primarily connected by NFs, while a few 2B customized networks with highly specialized demands are deployed in a lightweight and physically isolated manner. Furthermore, the relatively centralized network deployment approach and the interdependence among internal NFs render network failures prone to spreading (such as signaling storms) in case of any failure in individual NFs or interfaces or links between NFs.

The DDAA architecture offers a versatile and dynamic distributed autonomy capability, allowing for autonomous interconnections between domains/subnets through registration and discovery mechanisms. This results in extensive coverage for users and enhances network performance in terms of bandwidth and latency reduction. Moreover, the network displays significant robustness and resilience, preventing any potential failure from spreading outside a specific range to other autonomous networks.

The comparison of the network architecture between 5G and 6G DDAA is presented in Table 1.

**Table 1.** Comparison between 5G network architecture and 6G DDAA

|  | 5G network architecture | 6G DDAA |
|---|---|---|
| Network deployment time and difficulty | • Numerous NFs and interactions, strong dependencies<br>• Difficult interoperability<br>• Complex deployment procedures and extended deployment times | • Restructuring and aggregating similar NFs<br>• Unified mechanism for multi-network access<br>• Fewer interoperable interfaces to lower deployment difficulty and time |
| Network deployment flexibility | • Less likely to meet the demands of high-speed mobile scenarios like space-air-ground integration | • Distributed autonomy facilitates dynamic discovery and autonomous interconnections, providing proximity to users as needed, and meeting mobility and low-latency requirements<br>• Network simplification with on-demand customization support |
| Network robustness | • Relatively centralized network, susceptible to widespread failures<br>• Dispersed data, difficult recovery from disaster | • Distributed autonomy restricts the extent of failures, effectively containing the spread<br>• Stateless design enhances robustness |
| Data management and operational efficiency | • Dispersed data, redundant and difficult to manage<br>• Insufficient information regarding network and computing resources, not conducive to new scenarios | • Aggregate data and incorporate network resources and computing data<br>• Streamline data processing logic to facilitate data-driven |
| Data transmission efficiency | • The size of SBI data packets is restricted to ensure the efficient transmission of considerable data volumes | • Dual-bus architecture with added data channels facilitates efficient data transmission and high throughput |
| Network AI | • Centralized methods for collecting and analyzing AI/ML data, challenges to latency and service efficiency<br>• User-based AI services are complex and inflexible<br>• Lack of support for distributed learning and inference | • Native intelligence network architecture with distributed collaboration to enhance network intelligence<br>• "AI for user" capability to enhance user intelligence<br>• General architecture to support various distributed learning and inference methods |

## 3 DDAA Procedure Design Exploration

The paper aims to explore the features and benefits of the DDAA architecture through a typical general registration procedure representative example.

The UE registration procedure in 5G networks is exhaustively delineated in the 3GPP specification [12]. Unnecessary optional steps, exclusive to generic scenarios, have been omitted from the later part of the procedure for ease of analysis. It is suggested that these steps be further evaluated according to specific situations when available commercially. Additionally, Sect. 2 expounds on the optimization design for wireless and terminal aspects in the DDAA architecture. However, the focus of this section lies in analyzing the core network procedure, thus the steps related to UE and RAN will not be included in the procedure. After processing, Fig. 2 shows the standard 5G general registration procedure. For a detailed description of the procedure, refer to the specification as it will not be elaborated further in this paper.

The DDAA architecture facilitates the reorganization and consolidation of current NFs by means of NCUs. This includes the integration of UDM, PCF, and AUSF into home NCUs, the incorporation of new AMFs and SMFs into new NCUs, and the addition of old AMFs into old NCUs. Moreover, NDUs are introduced for the transmission of significant amounts of data. It should be noted that the DDAA architecture introduces a distributed autonomous domain/subnet design. In this design, the home NCU, new NCU, and old NCU generally belong to separate autonomous domains/subnets and can be interconnected using the NAU. However, for ease of comparing with the 5G procedure, the autonomous domains/subnets will not be depicted in the DDAA procedure, and the interconnection between different autonomous domains/subnets corresponds to the connection between internal NCUs reached via the NAU.

The general registration procedure of DDAA architecture will improve the steps marked in blue in Fig. 2, where the main modifications are also marked in blue, as shown in Fig. 3, and the optimized key points include:

1. Reduce the steps of home type NF selection: since the home NCU contains UDM and AUSF, the steps of "AUSF selection" and "UDM selection" are reduced to "Home NCU Discovery" in one step, see step 3 in the procedure.
2. Optimize the transmission and processing of user information: since the PCF and UDM are incorporated into the home NCU, the original multiple interactions between the new AMF, the UDM, and the PCF are simplified to a single interaction between the new NCU and the home NCU, i.e., step 6a "Data Acquisition & Policy Association Establishment/Modification", which includes new NCU registration, subscriber data acquisition and update subscription, AM policy association establishment/modification and policy acquisition, and UE policy association establishment and policy acquisition.
3. New efficient data transmission capability for large amount of data: due to the significant reduction of the interaction between the new NCU and the home NCU, as mentioned above, multiple information will be transmitted in one step (Step 6a), and therefore may lead to the need of large amount of data transmission. The NDU will, on the one hand, simplify and sort out the related information delivered in the original standard 5G procedure, e.g., removing redundancy, optimizing the data structure,

etc., and, on the other hand, considering factors such as changes in data type and data size, the NDU may choose to transmit the information to the new NCU through data channel, see step 6b in the procedure. In addition, as more scenarios and capabilities are supported in the future, the amount of UE context-related data may also increase significantly, and thus data channel may also be selected for transmission, see step 2b in the procedure.

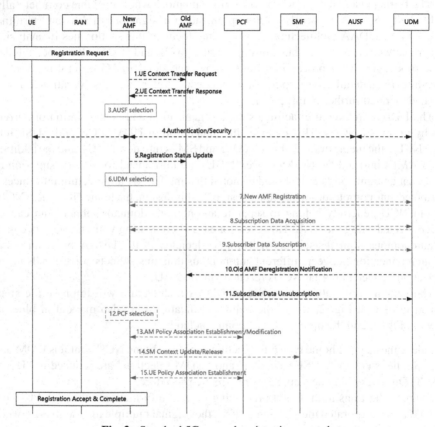

**Fig. 2.** Standard 5G general registration procedure

Furthermore, to enhance the robustness of the network and facilitate the data management and utilization, DDAA has also explored the stateless design of the networks. The initial idea is to optimize the management and delivery mechanism of dynamic data, such as UE context, by further separating the dynamic data, and aggregating it into home NCUs and NDUs as well, as with static data; moreover, the currently served NCUs no longer obtain information from the previously served NCUs but only interact with the home NCUs, in order to reduce the dependency among multiple NCUs. The general registration procedure for the Stateless DDAA architecture is shown in Fig. 4. The improvement points include:

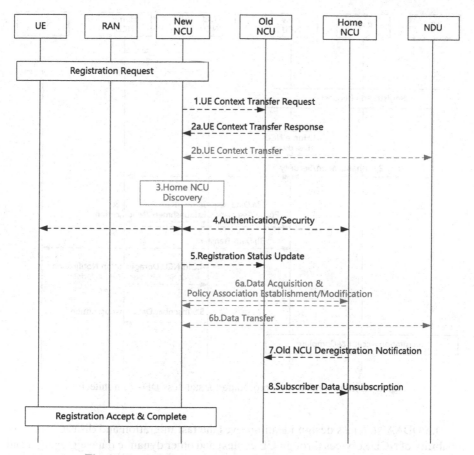

**Fig. 3.** General registration procedure for DDAA architecture

1. Optimize the UE context delivery mechanism: the UE context data is changed from being delivered in the new NCU registration procedure to being delivered to the home NCUs and NDUs by the old NCU in the process of providing services to the UE, i.e., before the UE registers with the new NCU, the old NCU has already updated the UE context to the NDU in time, refer to the step of "UE Context Transfer" marked in blue in the procedure. When the UE registers with the new NCU and the new NCU interacts with the home NCU, the delivery of the UE context is added in step 3a, "Data Acquisition & Policy Association Establishment/Modification".
2. Simplify the "Registration Status Update" step: since the UE context delivery mechanism has been optimized, the home NCU can obtain the UE context from the unified NDU, so it is possible to send the "Registration Status Update" by the home NCU to the old NCU to avoid the interaction between the new NCU and the old NCU, i.e., step 5 of the "Registration Status Update" from the new NCU to the old NCU in Fig. 3 can be changed to include an update of the registration status when the home NCU sends a de-registration notification to the old NCU, as shown in step 4 of Fig. 4.

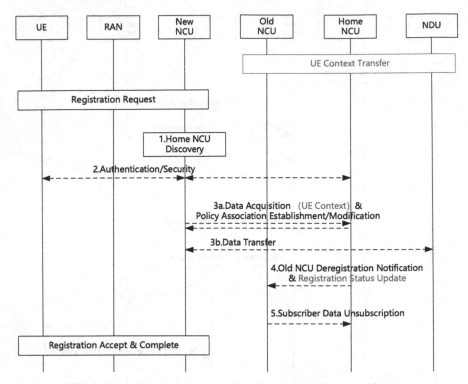

**Fig. 4.** General registration procedure for stateless DDAA architecture

The DDAA stateless design idea improves the fast migration and disaster recovery capability of NCU services through UE context and other dynamic data aggregation and optimization processing methods, and further reduces the dependency between NCUs. However, issues such as the delivery timing of UE contexts and other dynamic data need to be further explored, which should ensure the timeliness of data updates without affecting the service efficiency.

The above analysis of the standard 5G general registration procedure, the general registration procedure of the DDAA architecture, and the general registration procedure of the stateless DDAA architecture shows that the optimization of DDAA through NF aggregation and restructuring and the processing mechanism does bring about simplification of the procedures. The procedures comprise 15, 8 and 5 steps in sequence, with the DDAA architecture decreasing the number by close to 50% versus the existing 5G architecture. The stateless DDAA architecture then further shrinks it to 1/3. A schematic comparison of the procedures' step count is presented in Fig. 5. The simplification of the procedure facilitates the aggregation and structural optimization of the delivered information, as well as the transmission of extensive data. In conclusion, the comparative analysis of the registration procedures indicates the benefits of the DDAA architecture in terms of simplifying procedures and enabling large data transfers.

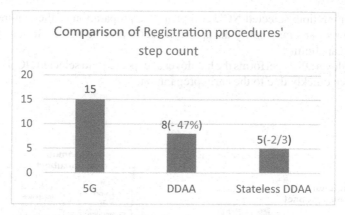

**Fig. 5.** Comparison of Registration procedures' step count

# 4 An Example of DDAA-Supported Applications

Immersive XR, ultra-high-definition video, unmanned aerial vehicle inspection, and V2X (Vehicle-to-Everything) communication are just a few of the applications anticipated to become mainstream in the future. These applications will place strict demands on network bandwidth, latency, reliability, as well as audio and video rendering quality. Furthermore, they must maintain a continuous service whilst on the move.

The DDAA architecture is designed to meet the service continuity requirements of high-performance applications, by integrating computing, network and intelligence, and providing a seamless user experience during service relocation. An example procedure is presented below, which combines V2X and edge computing. The DDAA architecture facilitates the timely and preemptive selection of local autonomous domain/subnet services and computing resource deployment to provide high-quality, seamless service for the rapidly moving vehicle (acting as a UE). Figure 6 illustrates the six key steps involved.

1. Trajectory prediction: the NIU performs the required AI/ML analysis by collecting DUE data, such as for UE mobility attributes and other data, to initiate mobile trajectory prediction for medium and high-speed type UEs.
2. Computing and network information acquisition: the current NCU obtains information from the NAU about the service capabilities and computing resources of NCUs in the neighboring edge autonomous domains/subnets to determine the list of suitable NCUs according to the expected UE mobile location given by the NIU analysis when the user is predicted to handover soon.
3. NCU selection: the current NCU may establish contact with suitable neighboring NCUs to obtain the current real-time status of service requirements, computing resources, etc., and finally select an NCU.
4. Establishing UE replica data: NCU passes data such as UE context and UE characteristics to the selected NCU to assist it in establishing DUE replica in advance.

5. Service migration: selected NCU completes the preparation of the required services and resources and may notify the applications for preloading of relevant content and required capabilities.
6. PUE handover: PUE performs the handover as expected and selected NCU can provide the service quickly due to the early preparation.

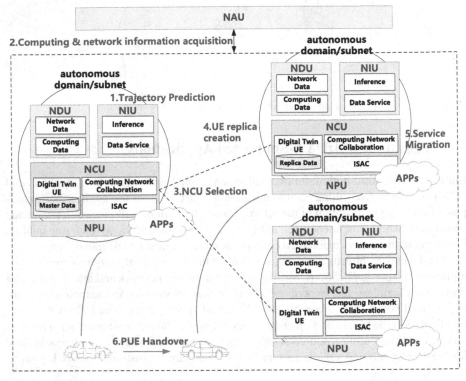

**Fig. 6.** Procedure example for a scenario combining V2X and edge computing

## 5  Conclusion

This paper outlines initial considerations for a new architecture for 6G networks, featuring an elaborate exposition of the DDAA architecture as well as its functionalities. The new architecture recommends optimal resolutions concerning network simplification, mass data transmission, data-driven, user-centric, native intelligence, and distributed autonomy.

The evolution of 6G network architecture is driven by services and technologies, with services serving as the primary driver and technologies as the enabler. The industry has identified several typical use cases and emerging technologies. While the use cases are still being developed and evaluated, they need further exploring to establish convincing service requirements. The new technologies have the potential to enhance network capabilities, but it is crucial that they meet the necessary requirements of new services. Therefore, the network architecture should result from the matching and collaboration of new services and technologies that require further research and validation.

**Acknowledgment.** This work was supported by the National Key R&D Program of China (No. 2020YFB1806700).

# References

1. Li,P., Xing, Y.: Capability exposure vitalizes 5G network. In: 2021 International Wireless Communications and Mobile Computing (IWCMC), Harbin City, China, pp. 874-878 (2021).https://doi.org/10.1109/IWCMC51323.2021.9498666
2. NGMN. 6G Use Cases and Analysis[R] (2022)
3. Xing, Y., Li, P., Li, J.: Discussion on 6G network architecture based on evolution. In: 2022 International Conference on Information Processing and Network Provisioning (ICIPNP), Beijing, China, pp. 20–23 (2022). https://doi.org/10.1109/ICIPNP57450.2022.00011
4. Li, P., Xing, Y., Li, W.: Distributed AI-native architecture for 6G networks. In: 2022 International Conference on Information Processing and Network Provisioning (ICIPNP), Beijing, China, pp. 57–62 (2022). https://doi.org/10.1109/ICIPNP57450.2022.00019
5. 3GPP TS 23.288, "Architecture enhancements for 5G System (5GS) to support network data analytics services (Release 17)", v17.7.0 (2022)
6. 3GPP TS 29.501, "Principles and Guidelines for Services Definition; Stage 3 (Release 18)", v18.3.0 (2023)
7. 3GPP TS 23.501, "System architecture for the 5G System (5GS); Stage 2 (Release 18)", v18.3.0 (2023)
8. Yang, Z., Chen, M., Wong, K.-K., Poor, H.V., Cui, S.: Federated learning for 6G: applications, challenges, and opportunities. Engineering **8**, 33–41 (2022)
9. Warnat-Herresthal, S., et al.: Swarm Learning for decentralized and confidential clinical machine learning. Nature **594**, 265–270 (2021)
10. Feriani,A., Hossain, E.: Single and multi-agent deep reinforcement learning for AI-enabled wireless networks: a tutorial. In: IEEE Communications Surveys & Tutorials, vol. 23, no. 2, pp. 1226–1252, Secondquarter 2021. https://doi.org/10.1109/COMST.2021.3063822
11. Gupta, O., Raskar, R.: Distributed learning of deep neural network over multiple agents. J. Netw. Comput. Appl. **116**, 1–8 (2018)
12. 3GPP TS 23.502, "Procedures for the 5G System (5GS); Stage 2 (Release 18)", v18.3.0 (2023)

# Decoupled Knowledge Distillation in Data-Free Federated Learning

Xueqi Sha(iD), Yongli Wang(✉)(iD), and Ting Fang(iD)

Nanjing University of Science and Technology, NanJing 210000, China
yongliwang@njust.edu.cn

**Abstract.** As the digital age advances, enormous mobile devices are interconnected with the Internet, resulting in a huge amount of distributed data. In response to this demand, federated learning (FL) arises. FL permits clients to collaborate on training a network model under the coordination of a central server by merely transmitting model parameters instead of private data with the central server, maintaining the decentralization and dispersion of the training data. However, distributed model training brings out the challenge of system heterogeneity and data heterogeneity. To mitigate these issues, numerous studies have employed methods such as regularized local loss, Bayesian approaches, meta-learning, and multi-task learning, which inevitably incur higher overhead. Consequently, achieving a trade-off between model accuracy and communication overhead has become a focus of FL algorithm research. This paper proposes a new method, Federated Learning via Decoupled Knowledge Distillation and Generative Model(FedDKDGen). We conduct massive experiments to explore the impact of statistical heterogeneity, communication rounds, and active-user ratio on model performance, respectively. Experimental results demonstrate that FedDKDGen performs admirably on EMNIST and MNIST, better exploiting the high-level semantics of the logits layer. As a result, the model accuracy advantage becomes increasingly pronounced as the data heterogeneity grows. Moreover, the model accuracy steadily outperforms the FL algorithm compared to, accounting for the lower communication cost due to the faster convergence rate. This allows FedDKDGen to achieve the same model accuracy as other algorithms with fewer communication rounds.

**Keywords:** Federated learning · Decoupled knowledge distillation · statistical heterogeneity

## 1 Introduction

With the development of computer networks, billions of edge devices are accessible to the Internet, which are typically outfitted with high-frequency-use sensors(e.g. cameras, microphones, GPS) that have access to ample data [1,3] suitable for training Neural Network Models and, in turn, improving the client experience. However, such data are usually highly sensitive or in huge quantities, and

© The Author(s), under exclusive license to Springer Nature Singapore Pte Ltd. 2024
H. Jin et al. (Eds.): IAIC 2023, CCIS 2059, pp. 164–177, 2024.
https://doi.org/10.1007/978-981-97-1280-9_13

the employment of traditional training methods would pose numerous difficulties and risks. Especially in the healthcare and financial industries, different customer organizations and geographic distribution have resulted in multiple data islands, and this highly confidential data cannot be stored centrally for obvious reasons. In order to better address the data privacy issue in model training, federated learning (FL) arises in response to demand.

FL permits clients to collaboratively train a network model under the coordination of a central server by transmitting model parameters only with the central server instead of private data, keeping the training data decentralized and dispersed. Therefore, the server can analyze and learn from multiple data owners [2] under the premise of overcoming privacy challenges. In addition, clients can achieve better modeling performance compared to isolated training.

The process of FL is divided into the following steps:(1) the client trains the global model using privacy data; (2) the central server aggregates the model parameters of clients to derive a new global model. Classical federated learning algorithms (e.g. FedAvg [1], FedProx [10]) follow this step, whereas such models encounter two challenges: (1) the system heterogeneity leads to the global model being unsuitable for a subset of clients owing to different clients featuring varying computing capabilities, network latency, input-output performance; (2) this statistical heterogeneity leaves shared global models difficult to generalize to each client [4,5] because cross-device and cross-island data are non-IID essentially. Consequently, simple model aggregation progresses with great difficulty in personalized federated learning scenarios with data heterogeneity.

Several studies have exploited regularized local loss [10,11], Bayesian [18], meta-learning [6–8], multi-task learning [9,26], migration learning [12], and other PFL algorithms to mitigate statistical heterogeneity, which achieved promising generalization performance by using local feature data to regulate global model parameters.

The balancing problem of model accuracy and calculation communication overhead has always been the focus of research in FL algorithms. [13–15]adopt the idea of weight pruning to alleviate the communication and computation expense, while [3,4,16,24] speed up the FL process by reducing the communication parameters. Knowledge distillation [17]is also a practical solution to reduce calculation communication overhead. [17], as the pioneering work of KD, firstly proposed the idea of knowledge distillation: extract useful information from the huge, massively growing, formidable teacher model to build a lightweight, easy-to-deploy, compact student model, and try to maintain the accuracy of the teacher model. This technique is relevant to the global and local models of FL, which can reduce resource consumption to overcome the challenge of communication computing power [7].

However, knowledge distillation typically relies on a proxy dataset [19,20, 31,35], which has significant impacts on the model accuracy and generalization performance due to the size, feature distribution, and quality of the public shared dataset. Besides, there also exists a substantial risk of privacy leakage. Therefore [21,22]proposed data-free federated knowledge distillation algorithms to solve the problem of proxy datasets. However, some methods use logits-based KD algorithms [17], which require more communication times to balance the accuracy, the others use the intermediate-feature-based KD algorithms [28–30], which has a higher computational cost.

It is apparent that the optimization of model accuracy and computational communication costs are interdependent, as improving accuracy inevitably leads to higher computational overhead or more communications, whereas reducing costs is accompanied by more or less loss of accuracy. Knowledge distillation, a powerful tool for lightweight neural networks, suffers from potential privacy vulnerabilities of shared datasets.

In this work, we propose Federated Learning via Generative model and Decoupled Knowledge Distillation(FedDKDGen), which incorporates Decoupled Knowledge Distillation(DKD) [23] to support federated learning [22]. The proposed method optimizes data-free federated knowledge distillation and substitutes traditional KD with more flexible DKD, thereby achieving superior model accuracy. Along with the increasing difficulty and heterogeneity of the dataset, the accuracy advantage of FeDKDGen becomes increasingly apparent and converges faster, achieving equivalent model accuracy as the other algorithms only through fewer communication rounds, which reduces the communication cost to a certain extent.

## 2  Related Work

The classical FL algorithm yields the global model by aggregating the client model parameters. FedAvg [1] uses a weighted average for model aggregation, which is simple and low-cost but has many accuracy shortcomings. FedProx [10] builds on the FedAvg and addresses the system and data heterogeneity, two improvements are provided to the iterative update strategy, allowing variable local epochs and applying a proximal term to the empirical loss. LG-FedAvg [25] trains smaller global models by sharing compact local representations of each client instead of all model parameters, decreasing communication costs but without optimizing communication efficiency. LotteryFL [3] uses the lottery hypothesis to learn lottery networks as subnetworks, with fewer communication parameters based on the compact size of lottery networks. FedSkel [4] applies a pruning strategy to identify the backbone network of the model and updates merely the essential parts of the model to achieve communication-efficient FL. FedGB [24]based on FedSkel, implements the dynamic selection of the backbone network and interacts merely with the backbone information of local models for the global model aggregation. FedLP [15]adopts a further iterative pruning mechanism in the local model training and global model updating, in addition, proposes two pruning schemes against IID and non-IID scenarios for mitigating resource consumption and preventing model attacks.

Lightweight neural networks can also be accomplished through knowledge distillation. Hinton [17] uses logits-based distillation, which employs softened softmax transformations to allow the STUDENT model to learn more soft targets, i.e., the similarity information of the incorrect category, rather than just the ground truth. Inspired by Hinton, [27]proposed Born-Again Neural Networks, with the dual goal of matching training labels and teacher model outputs, to "born again" the student model, resulting in better performance than the teacher model. [29] observing that similar semantic inputs will induce similar activation patterns, hence guides the training of the student network by retaining the similarity knowledge extracted by the teacher, so that the student network will no longer simply emulate the teacher model outputs but capture semantic patterns similar to the teacher. CRD [28] exploits comparative learning for the intermediate layer feature distillation to capture the relevance and higher-order output dependencies of structured feature knowledge. DKD [23] redirects attention to the logits layer of the output results by rephrasing Hinton's formula [17] into target class knowledge distillation (TCKD) and non-target class knowledge distillation (NCKD) and decoupled them, improving the flexibility of logits distillation and obtaining similar or even better results than feature-based distillation methods.

## 3   Design of FedDKDGen

### 3.1   Overview

Knowledge distillation is a powerful tool for lightweight neural networks, but there remains the security risk of sharing datasets, and more significantly, the two mainstream distillation methods have non-negligible shortcomings. Conventional distillation based on logits utilizes the last layer output to compute the KL-Divergence, which is computationally computed but not accurate enough. Knowledge distillation from deep features of intermediate layers exhibits superior performance on various computational vision tasks but also triggers greater computational cost and storage space.

In this paper, we propose FedDKDGen as a solution to these issues, as shown in Fig. 1. FedDKDGen eliminates the need for each client to provide a portion of data to constitute a proxy dataset, thus reducing the risk of privacy leakage. Bypassing the increase of data heterogeneity, the advantage of FedDKDGen accuracy becomes increasingly obvious. Firstly, FedDKDGen solves the security risk problem of shared datasets and improves the generalization performance of the model with a lightweight generator as the global model, thereby reducing communication overhead. Secondly, the logits-based DKD gives a fuller expression to the semantic value of the logits layer, saving more computation and storage costs than KD based on the intermediate feature layer, while maintaining comparable training accuracy. Lastly, FedDKDGen demonstrates a faster convergence speed and could achieve the same model accuracy as other algorithms [1,10,22] with fewer communication rounds and decreased communication costs.

### 3.2   Federated Learning via Generative Model

FedDKDGen performs knowledge extraction on the server and knowledge distillation on the clients. server merely demands the distribution of client data,

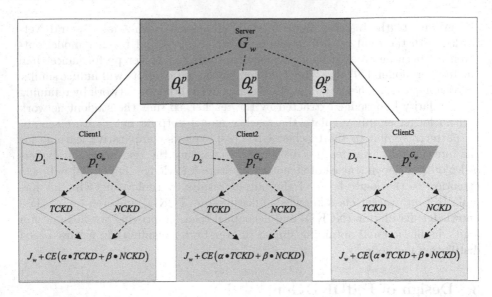

**Fig. 1.** Overview of FedDKDGen: The server learns a generator $G_W$ to combine data from several local clients without accessing their data. The generator is distributed to local users, whose knowledge is decoupled and distilled to user models to modify their interpretations of a desirable feature distribution.

rather than the privacy data. The server derives an empirical estimate of the global data based on the distribution of client sample labels. To further minimize the high-dimensional computational cost and preserve data privacy, this loss is mapped to a latent space, where the computation of the loss function is performed. [22].

$A$ denotes the data over an instance space, $B$ denotes the data over an output space, and $C$ denotes the data over a latent space. There exists a labeling function: $f = A \rightarrow B$ extracts the feature distribution of the original data, completing the mapping from the instance space to the output space. There exists a labeling distribution of the client sample data without the privacy data in $B$. The model parameter $\theta$ can be expressed as two components $\theta = \left[\theta^f, \theta^p\right]$ for feature extraction and prediction respectively. The $Loss\left(\theta\right)$ denotes the loss of the model parameter $\theta$.

The server learns a Maximum Likelihood Estimation (MLE) conditional distribution based on the global labeling distribution: $L^* = B \rightarrow A$, which generates samples that match the global labeling distribution:

$$L^* = \arg\max_{L:B\rightarrow A} E_{B\sim P(B)} E_{A\sim L(A|B)} \left[\log P\left(B|A\right)\right] \tag{1}$$

where $P\left(B\right)$ is the ground truth over the output space, which can be estimated by the labeling distribution shared by clients; $P\left(B|A\right)$ is the posterior distribution of the output space, which can be estimated by ensemble wisdom from user models.

$$\hat{p}(B) \propto \sum E_{A \sim P(\hat{D}_k)}\left[I\left(f\left(A\right) = B\right)\right]$$

$$\log \hat{p}(B|A) \propto \frac{1}{K}\sum_{i=1}^{K} \log P\left(B|A; \theta_k\right) \tag{2}$$

To decrease the overhead of high-dimensional operations and the risk of user data leakage, the data $B$ over the output space is mapped to the $C$ over the latent space by an induced distribution: $G^* = B \to C$. The data over the latent space still follows the global labeling distribution and has more compact features with fewer feature dimensions.

$$G^* = \underset{G:B \to C}{\arg\max} E_{B \sim \hat{P}(B)} E_{C \sim G(C|B)}\left[\sum_{i=1}^{K} \log P\left(B|C; \theta_k\right)\right] \tag{3}$$

FedDKDGen is intended to recover a latent space by learning a conditional generator $G_w$ to perform knowledge extraction which is optimized on the server side with the following optimization objective:

$$\underset{w}{\min} J\left(w\right) = E_{B \sim \hat{P}(B)} E_{C \sim G_w(C|B)}\left[l\left(\theta\left(\frac{1}{K}\sum_{i=1}^{K} g\left(C; \theta_k\right)\right), B\right)\right] \tag{4}$$

where $\theta$ is the sigmod function and $g$ is the output of the logits layer. The generator $G_w$ is continuously optimized by $w \leftarrow w - \alpha \nabla_w J(w)$. After broadcasting the lightweight generator $G_w$ to the client, FedDKDGen performs knowledge distillation on the clients. Each client model can be sampled from $G_w$ to acquire an augmented representation over the feature space. The optimization objective of the client model $\theta\left(k\right)$ is to maximize its probability of producing the desired prediction for the augmented sample:

$$\underset{\theta(k)}{\min} J\left(w\right) = \hat{L}_k(\theta_k) + E_{B \sim \hat{P}(B)} E_{C \sim G_w(C|B)} \times \left[l\left(h\left(C; \theta_k\right), B\right)\right] \tag{5}$$

where $\hat{L}_k(\theta_k)$ denotes the empirical loss of the client localized data.

So far, the server broadcasts a lightweight generator, predict layer of the global model $\theta$, global label distribution $P(B)$ to the clients, which construct enhanced data samples, updates the local label distribution and local model parameters by knowledge distillation, as well as returns the label distribution and model parameters to the server. By interactively learning a lightweight generator that primarily depends on the prediction rule of local models, this cyclic interaction has made it possible to distill information without using any data.

### 3.3 Decoupled Knowledge Distillation

FedDKDGen uses the DKD for knowledge distillation on the client side. The classic logits distillation [17] introduces a distillation temperature $T$ that amplifies the dark knowledge contained inside the soft targets of the teacher model as well as giving the student model both the similarity probability of the ground truth and a rich amount of non-correct category similarity information [23]. Assuming that there are a total of $C$ classes in the training sample, the probability that a sample is recognized as each class can be denoted as $q = [q_1, q_2, ..., q_g, ..., q_C]$, where g is the ground truth class. It can be determined by using a soften softmax function:

$$q_j = \frac{\exp(z_j/T)}{\sum_{k=1}^{C} \exp(z_k/T)} \tag{6}$$

where $z_j$ is the logits layer output probability that the sample data belongs to the $j$th class and $T$ is the distillation temperature.

Given that $d = [q_g, q_{\backslash g}]$ represents the probability of the target class and non-target classes respectively, and that $\hat{q}_j$ is the distributional probability of each category in the non-target classes, it could be derived:

$$q_g = \frac{\exp(z_g/T)}{\sum_{k=1}^{C} \exp(z_k/T)}, q_{\backslash g} = \frac{\sum_{k=1, k \neq g}^{C} \exp(z_k/T)}{\sum_{k=1}^{C} \exp(z_k/T)}, \hat{q}_j = \frac{exp(z_j/T)}{\sum_{k=1, k \neq g}^{C} \exp(z_k/T)} = \frac{q_j}{q_{\backslash g}} \tag{7}$$

Supposing that $TE$ represents the teacher model and $ST$ represents the student model, traditional KD uses the Cross-Entropy loss function as the optimization objective, so the KL-Divergence of the difference between the teacher model and the student model can be expressed a

$$Loss_{KD} = q_g^{TE} \log\left(\frac{q_g^{TE}}{q_g^{ST}}\right) + \sum_{j=1, j \neq g}^{C} \left(q_j^{TE} \log\left(\frac{q_j^{TE}}{q_j^{ST}}\right)\right) \tag{8}$$

According to (7) it can be seen that $q_j = \hat{q}_j \times q_{\backslash g}$, thus we can reformulate the KL-Divergence of the traditional teacher model with the student model [23]:

$$Loss_{KD} = q_g^{TE} \log\left(\frac{q_g^{TE}}{q_g^{ST}}\right) + q_{\backslash g} \times \log\left(\frac{q_{\backslash g}^{TE}}{q_{\backslash g}^{ST}}\right) \sum_{j=1, j \neq g}^{C} \hat{q}_j^{TE}$$

$$+ \left(q_{\backslash g} \sum_{j=1, j \neq g}^{C} \hat{q}_j^{TE} \log\left(\frac{\hat{q}_j^{TE}}{\hat{q}_j^{ST}}\right)\right) \tag{9}$$

$$= \left(KL(q^{TE} \| q^{ST})\right) + (1 - q_g)\left(KL(\hat{q}^{TE} \| \hat{q}^{ST})\right)$$

$(KL(q^{TE} \| q^{ST}))$ denotes the difference in the probability distributions between the teacher model and the student model for the target class, referred to as target class knowledge distillation (TCKD), $(KL(\hat{q}^{TE} \| q^{ST}))$ denotes the difference in the loss function between the two models for the non-target class, referred to as non-target class knowledge distillation (NCKD). [23] pointed out that TCKD is more effective for difficult datasets, and NCKD contains considerably more dark knowledge, which is crucial for the success of logits-based distillation, but its ability is suppressed by the coefficient $(1 - q_g)$ in the classical KD algorithm. The loss formulation of KD can thus be reformulated as decoupled TCKD and NCKD and assigned decoupled coefficients $\alpha$ and $\beta$ respectively:

$$Loss_{KD} = \alpha TCKD + \beta NCKD \tag{10}$$

Expanding $\alpha$ and $\beta$ permits DKD to transfer more "difficult" knowledge and utilize the high-level semantics of logits-based distillation to a greater extent, thereby extracting richer dark knowledge.

# 4  Experiments

In this section, we compare the performance of our algorithm with other federated learning algorithms on two datasets, MNIST [33] and EMNIST [34].

**Table 1.** Top-1 test accuracy under different data heterogeneity on MNIST. The smaller the $\gamma$, the higher the data heterogeneity. Distillation temperature $T$, TCKD coefficient $\alpha$, and NCKD coefficient $\beta$ were transformed for the proposed FedDKDGen.

| Methods | Data Heterogenetiy | | | | |
|---|---|---|---|---|---|
| | 0.1 | 0.5 | 1.0 | 5.0 | 10.0 |
| FedAvg | 0.91181 | 0.93965 | 0.94162 | 0.94343 | 0.94475 |
| FedProx | 0.90880 | 0.93629 | 0.93896 | 0.94152 | 0.94071 |
| FedDistill | 0.70244 | 0.78121 | 0.82567 | 0.87657 | 0.88495 |
| FedGen | 0.95817 | 0.97276 | 0.97224 | 0.96596 | 0.97424 |
| FedDKDGen0.5_1_2 | 0.93725 | 0.96082 | 0.95955 | 0.96081 | 0.96141 |
| FedDKDGen0.5_1_3 | 0.92708 | 0.95854 | 0.96057 | 0.96071 | 0.96242 |
| FedDKDGen0.5_2_3 | 0.92821 | 0.96103 | 0.96211 | 0.96485 | 0.96424 |
| FedDKDGen0.5_3_2 | 0.94517 | 0.96212 | 0.96129 | 0.96525 | 0.96586 |
| FedDKDGen1_1_2 | 0.94536 | 0.96310 | 0.96487 | 0.96495 | 0.96576 |
| FedDKDGen2_1_2 | 0.91803 | 0.94627 | 0.94736 | 0.94980 | 0.95020 |

MNIST and EMNIST are both handwritten character recognition datasets for image tasks. MNIST is a reduced version of NIST, containing only digits, with a total of 60,000 training images and 10,000 test images. EMNIST is a more challenging Benchmark developed on the basis of MNIST divided into six categories: ByClass, ByMerge, Balanced, Letters, Digits, and MNIST. In our experiments, EMNIST merely consists of letters with 37 classes, containing 88,800 training images and 14,800 test images. Dirichlet distribution is used to create non-IID data distribution [32], the smaller $\gamma$ represents the higher statistical heterogeneity and the larger difference among local models.

We use Convolutional Neural Network (CNN) as the classifier baseline. We contrast proposed algorithm FedGen with FedAvg [1], FedProx [10], FedDistill [21], and FedGen [22]. In order to provide a comprehensive comparison of different algorithms, this paper conducts comparative experiments on data statistical heterogeneity, communication round, and active-user ratio.

## 4.1  Impact of Data Heterogeneity on Model Accuracy

The parameters for a complete federated learning process are as follows: there are 20 user nodes with active-user ratio=0.5. The server interacts with each client for 200 communication rounds with each step setting batch size = 32, and 10 clients are randomized to participate in the training process. Each client

**Table 2.** Top-1 test accuracy under different data heterogeneity on EMNIST. The smaller the $\gamma$, the higher the data heterogeneity. Distillation temperature $T$, TCKD coefficient $\alpha$, and NCKD coefficient $\beta$ were transformed for the proposed FedDKDGen.

| Methods | Data Heterogenetiy | | | | |
|---|---|---|---|---|---|
| | 0.01 | 0.1 | 0.5 | 1.0 | 5.0 |
| FedAvg | 0.59367 | 0.68898 | 0.76443 | 0.77838 | 0.78715 |
| FedProx | 0.57167 | 0.68755 | 0.75699 | 0.76934 | 0.77755 |
| FedDistill | 0.51400 | 0.60388 | 0.63972 | 0.67941 | 0.70650 |
| FedGen | 0.68733 | 0.77449 | 0.82513 | 0.84186 | 0.84455 |
| FedDKDGen0.5_1_2 | 0.72933 | 0.80806 | 0.85011 | 0.85766 | 0.85925 |
| FedDKDGen0.5_1_3 | 0.72800 | 0.80378 | 0.84710 | 0.85486 | 0.85750 |
| FedDKDGen0.5_2_3 | 0.72167 | 0.80480 | 0.84955 | 0.85782 | 0.85940 |
| FedDKDGen0.5_3_2 | 0.72267 | 0.80582 | 0.85091 | 0.85978 | 0.86225 |
| FedDKDGen1_1_2 | 0.70867 | 0.78571 | 0.82602 | 0.83404 | 0.83535 |
| FedDKDGen2_1_2 | 0.66600 | 0.74714 | 0.77767 | 0.78367 | 0.78195 |

receives a global model 20 times for local training, and the model is trained with a learning rate of 0.98 as the decay coefficient. It should be clarified that FedDKDGen0.5_1_2 is a variation of FedDKDGen with specific parameters $T = 0.5, \alpha = 1, \beta = 2$.

As shown in Fig. 2,Tables 1 and 2, in terms of MNIST, with the increase of data heterogeneity (the decrease of $\gamma$), the accuracy of all five models decreases, but the extent of the accuracy decrease for MNIST is much lower than that for EMNIST. This is due to the fact that EMNIST letters contain 37 classes, resulting in the dataset being more difficult and more sensitive to changes in $\gamma$. From Tables 1 and 2, it can be observed that for both datasets, FedDKDGen achieves excellent results, and its accuracy is steadily superior to that of FedAvg, FedProx, and FedDistill, and is comparable to that of FedGen. For EMNIST, the advantage of FedDKDGen is more pronounced. Across different levels of statistical heterogeneity, FedDKDGen achieves higher accuracy than FedGen. Especially for $\gamma = 0.01$, the accuracy of the model exceeds FedGen by 4%.

As illustrated in Figs. 2(b) and 2(d), we adjust the distillation temperature $T$, TCKD coefficient $\alpha$ and NCKD coefficient $\beta$ for FedDKDGen. We observe that the results are improved when the distillation temperature is less than 1, because the generative-based federated knowledge distillation algorithm, for the sake of privacy security and computational convenience, maps the labels of the data multiple times, which inevitably loses part of ground truth. A distillation temperature of less than 1 is well to recover this knowledge and achieve a better distillation result. Additionally, for the two coefficients $\alpha$ and $\beta$, the best result is obtained when $\alpha = 3, \beta = 2$. TCKD contains more hard knowledge, transferring "difficult" knowledge and NCKD contains more dark knowledge. However, in the scenario of this paper, the model needs more ground truth, so $\alpha > \beta$ can achieve the best performance.

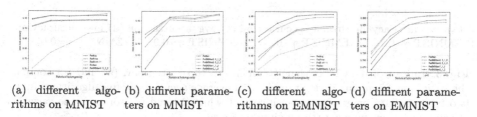

(a) different algo-rithms on MNIST  (b) diffirent parameters on MNIST  (c) different algo-rithms on EMNIST  (d) diffirent parameters on EMNIST

**Fig. 2.** Impact of data heterogeneity on accuracy. (a) and (c) compare the heterogeneity-accuracy variation plots of five algorithms: FedAvg, FedProx, FedDistill, FedGen, and FedDKDGen, on MNIST and EMNIST respectively. (b) and (d) alter the FedDKDGen parameters to compare with FedGen. The distillation temperature $T$ exposes a significant impact on test accuracy.

## 4.2 Impact of Communication Round on Model Accuracy

The data heterogeneity of EMNIST is set as $\gamma=0.01$ and $\gamma=0.1$, the parameters of FedDKDGen are set as $T = 0.5, \alpha = 3, \beta = 2$, for the first 20,30,...,200 rounds of communication are studied for the mean and maximum values of model test accuracy.

**Table 3.** Average accuracy and top-1 accuracy under different communication rounds on EMNIST with data heterogeneity $\gamma = 0.01$. FedDKDGen is parameterized with $T = 0.5, \alpha = 3, \beta = 2$. FedDKDGen outperforms other algorithms in terms of both average accuracy and top-1 accuracy at each sampled communication round.

| Methods | | rounds on EMNIST where $\gamma=0.01$ | | | | | | |
|---------|-----|---------|---------|---------|---------|---------|---------|---------|
| | | 20 | 50 | 80 | 110 | 130 | 170 | 200 |
| FedAvg | avg | 0.18718 | 0.28841 | 0.35101 | 0.39043 | 0.41561 | 0.43626 | 0.45248 |
| | max | 0.33133 | 0.43100 | 0.51000 | 0.53867 | 0.55467 | 0.58333 | 0.59367 |
| FedProx | avg | 0.13525 | 0.26015 | 0.32045 | 0.36139 | 0.39272 | 0.41429 | 0.43122 |
| | max | 0.26333 | 0.40633 | 0.47067 | 0.52433 | 0.55367 | 0.55367 | 0.57167 |
| FedDistill | avg | 0.39465 | 0.44053 | 0.45778 | 0.46865 | 0.47596 | 0.48180 | 0.48629 |
| | max | 0.45300 | 0.48033 | 0.49367 | 0.50100 | 0.50633 | 0.51033 | 0.51400 |
| FedGen | avg | 0.20867 | 0.32686 | 0.39700 | 0.44684 | 0.48596 | 0.51422 | 0.53618 |
| | max | 0.34133 | 0.48367 | 0.55300 | 0.63267 | 0.66933 | 0.68000 | 0.68733 |
| FedDKDGen | avg | 0.21002 | 0.33287 | 0.42093 | 0.47905 | 0.51999 | 0.54972 | 0.57288 |
| | max | 0.33000 | 0.50733 | 0.63167 | 0.68467 | 0.70167 | 0.70933 | 0.72933 |

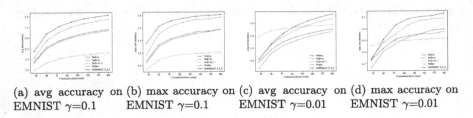

(a) avg accuracy on (b) max accuracy on (c) avg accuracy on (d) max accuracy on
EMNIST $\gamma$=0.1      EMNIST $\gamma$=0.1      EMNIST $\gamma$=0.01      EMNIST $\gamma$=0.01

**Fig. 3.** Impact of data heterogeneity on model top-1 accuracy. (a) and (c) compare the variation of communication rounds with average accuracy for five algorithms under two types of data heterogeneity on EMNIST while (b) and (d) compare the variation of communication rounds with maximum accuracy.

From Fig. 3 and Table 3, it can be discovered that the average accuracy of each algorithm reveals a steady and consistent growth with an increase in the number of communication rounds. However, the rate of increase gradually slows down over time. For the relatively low-difficulty datasets, the improvement of the model accuracy encounters a bottleneck, e.g. on EMNIST with $\gamma = 0.1$, the improvement in model accuracy is minimal between rounds 100 and 200, with the average and maximal accuracy almost identical. FedDKDGen demonstrates exceptional performance in terms of both average accuracy and maximum accuracy at each communication round interval. It achieves comparable accuracy to other algorithms with fewer communication rounds, resulting in faster convergence and reduced communication costs.

## 4.3   Impact of Active-User Ratio on Model Accuracy

Set the heterogeneity of EMNIST data as $\gamma$=0.2 and $\gamma$=0.5 and the parameters of FedDKDGen as $T = 0.5, \alpha = 3, \beta = 2$, total user =30. We adjusted the proportion of active users aiming to study the test accuracy of 200 rounds of communication.

As shown in Fig. 4, the increase of active-user ratio can generally improve the accuracy of the model, but this improvement is somewhat marginal, especially for the FedAvg and FedProx algorithms. The positive correlation between the active-user ratio and model test accuracy is not apparent. By contrast, for the FedGen and FedDKDGen algorithms, an increase in active-user ratio leads to an improvement in model accuracy.

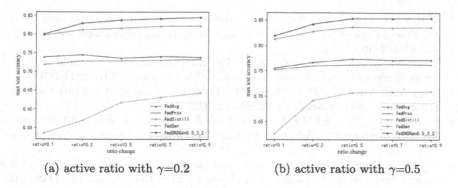

(a) active ratio with $\gamma$=0.2          (b) active ratio with $\gamma$=0.5

**Fig. 4.** Impact of active-user ratio on top-1 accuracy on EMNIST with data heterogeneity$\gamma = 0.2$ and $\gamma = 0.5$.

## 5    Conclusion

Our experiments show that TCKD and NCKD decoupled logits distillation algorithm maximally exploits the semantic value of the logits layer. DKD can bring higher quality federated learning models, enabling FedDKDGen to reach the model accuracy of FedAvg or FedProx in fewer communication rounds, thus reducing communication overhead to some extent. The advantages of FedDKD-Gen become more pronounced when the labeling of the dataset is more complicated and the differences in data distribution are greater. Although FedDKDGen has excellent performance in terms of model accuracy as well as the client-side computational cost of logits-based KD is lower than KD based on the intermediate feature layer, the generator-based training of the server model still has a high computational overhead. FedDKDGen has not achieved a significant relief in computational cost yet. This is an interesting direction for the future, which can try to reduce the number of parameters for computation and communication by model heterogeneity [36] and other methods.

## References

1. McMahan, B., Moore, E., Ramage, D., Hampson, S., y Arcas, B.A.: Communication-efficient learning of deep networks from decentralized data. In: Artificial intelligence and statistics, pp. 1273–1282, pp. 1273–1282. PMLR (2017)
2. Kairouz, P., et al.: Advances and open problems in federated learning. Found. Trends® Mach. Learn. **14**(1–2), 1–210 (2021)
3. Li, A., et al.: LotteryFL: empower edge intelligence with personalized and communication-efficient federated learning. In: IEEE/ACM Symposium on Edge Computing (SEC), pp. 68–79. IEE 2021 (2021)
4. Luo, J., Yang, J., Ye, X., Guo, X., Zhao, W.: FedSkel: efficient federated learning on heterogeneous systems with skeleton gradients update. In: Proceedings of the 30th ACM International Conference on Information & Knowledge Management, pp. 3283–3287 (2021)

5. Tan, Y., et al.: FedProto: federated prototype learning across heterogeneous clients. In: Proceedings of the AAAI Conference on Artificial Intelligence, vol. 36, no. 8, pp. 8432–8440 (2022)
6. Chen, F., Luo, M., Dong, Z., Li, Z., He, X.: Federated meta-learning with fast convergence and efficient communication. arXiv preprint: arXiv:1802.07876 (2018)
7. Tan, A.Z., Yu, H., Cui, L., Yang, Q.: Towards personalized federated learning. IEEE Trans. Neural Netw. Learn. Syst. (2022)
8. Khodak, M., Balcan, M.F.F., Talwalkar, A.S.: Adaptive gradient-based meta-learning methods. In: Advances in Neural Information Processing Systems, vol. 32 (2019)
9. Huang, Z.A., et al.: Federated multi-task learning for joint diagnosis of multiple mental disorders on MRI scans. IEEE Trans. Biomed. Eng. **70**(4), 1137–1149 (2022)
10. Li,, T., Sahu, A.K., Zaheer, M., Sanjabi, M., Talwalkar, A., Smith, V.: Federated optimization in heterogeneous networks. Proc. Mach. Learn. Syst. **2**, 429–450 (2020)
11. Wu, C., Wu, F., Qi, T., Huang, Y., Xie, X.: FedCL: federated contrastive learning for privacy-preserving recommendation. arXiv preprint: arXiv:2204.09850 (2022)
12. Wang, K., Mathews, R., Kiddon, C., Eichner, H., Beaufays, F., Ramage, D.: Federated evaluation of on-device personalization. arXiv preprint: arXiv:1910.10252 (2019)
13. Anwar, S., Hwang, K., Sung, W.: Structured pruning of deep convolutional neural networks. ACM J. Emerg. Technol. Comput. Syst. (JETC) **13**(3), 1–18 (2017)
14. Yu, F., Qin, Z., Chen, X.: Distilling critical paths in convolutional neural networks. arXiv preprint: arXiv:1811.02643 (2018)
15. Shi, Y., et al.: Efficient federated learning with enhanced privacy via lottery ticket pruning in edge computing. arXiv preprint: arXiv:2305.01387 (2023)
16. Shi, S.,et al.: A distributed synchronous SGD algorithm with global top-k sparsification for low bandwidth networks. In: 2019 IEEE 39th International Conference on Distributed Computing Systems (ICDCS), pp. 2238–2247. IEEE (2019)
17. Hinton, G., Vinyals, O., Dean, J.: Distilling the knowledge in a neural network. arXiv preprint: arXiv:1503.02531 (2015)
18. Corinzia, L., Beuret, A., Buhmann, J.M.: Variational federated multi-task learning. arXiv preprint: arXiv:1906.06268 (2019)
19. Li, D., Wang, J.: FedMD: heterogenous federated learning via model distillation. arXiv preprint: arXiv:1910.03581 (2019)
20. Sattler, F., Korjakow, T., Rischke, R., Samek, W.: FedAUX: leveraging unlabeled auxiliary data in federated learning. IEEE Trans. Neural Netw. Learn. Syst. (2021)
21. Jeong, E., Oh, S., Kim, H., Park, J., Bennis, M., Kim, S.L.: Communication-efficient on-device machine learning: federated distillation and augmentation under Non-IID private data (2018)
22. Zhu, Z., Hong, J., Zhou, J.: Data-free knowledge distillation for heterogeneous federated learning. In: International Conference on Machine Learning, pp. 12878–12889. PMLR (2021)
23. Zhao, B., Cui, Q., Song, R., Qiu, Y., Liang, J.: Decoupled knowledge distillation. In: Proceedings of the IEEE/CVF Conference on Computer Vision and Pattern Recognition, pp. 11953–11962 (2022)
24. Yang, Z., Sun, Q.: A dynamic global backbone updating for communication-efficient personalised federated learning. Connect. Sci. **34**(1), 2240–2264 (2022)
25. Liang, P.P., et al.: Think locally, act globally: federated learning with local and global representations (2020)

26. Wang, X., Fu, L., Zhang, Y., Wang, Y., Li, Z.: MMatch: semi-supervised discriminative representation learning for multi-view classification. IEEE Trans. Circuits Syst. Video Technol. **32**(9), 6425–6436 (2022)
27. Furlanello, T., Lipton, Z., Tschannen, M., Itti, L., Anandkumar, A.: Born again neural networks. In: International Conference on Machine Learning, pp. 1607–1616. PMLR (2018)
28. Tian, Y., Krishnan, D., Isola, P.: Contrastive representation distillation. arXiv preprint: arXiv:1910.10699 (2019)
29. Tung, F., Mori, G.: Similarity-preserving knowledge distillation. In: Proceedings of the IEEE/CVF International Conference on Computer Vision, pp. 1365–1374 (2019)
30. Romero, A., Ballas, N., Kahou, S. E., Chassang, A., Gatta, C., Bengio, Y.: FitNets: hints for thin deep nets (2015)
31. Seo, H., Park, J., Oh, S., Bennis, M., Kim, S.L.: 16 federated knowledge distillation. Mach. Learn. Wirel. Commun., 457 (2022)
32. Hsu, T.M.H., Qi, H., Brown, M.: Measuring the effects of non-identical data distribution for federated visual classification. arXiv preprint: arXiv:1909.06335 (2019)
33. LeCun, Y., et al.: MNIST handwritten digit database, vol. 7, no. 23, p. 6 (2010). URL: https://yann.lecun.com/exdb/mnist
34. Cohen, G., Afshar, S., Tapson, J., Van Schaik, A.: EMNIST: extending MNIST to handwritten letters. In: International Joint Conference on Neural Networks (IJCNN), pp. 2921–2926. IEEE (2017)
35. Shi, N., Lai, F., Kontar, R.A., Chowdhury, M.: Fed-ensemble: improving generalization through model ensembling in federated learning. arXiv preprint: arXiv:2107.10663 (2021)
36. Diao, E., Ding, J., Tarokh, V.: HeteroFL: computation and communication efficient federated learning for heterogeneous clients. arXiv preprint: arXiv:2010.01264 (2020)

# An Improved SSD Seal Detection Algorithm Based on Channel-Spatial Attention

Xiangchao Shao[✉], Yin Deng, and Yanji Zhou

Dongguan Power Supply Bureau of Guangdong Power Grid Corp., Dongguan 523000, Guangdong, China
728264172@qq.com

**Abstract.** In response to the challenges in seal detection due to small seal targets, faint imprints, and text occlusions, which result in low detection accuracy and difficulties in detection, an improved SSD (Single Shot multi-boxes Detectors) seal detection algorithm based on channel-spatial attention is proposed, achieving fast and accurate detection of document seal images. Firstly, the Channel-Spatial Attention Module (CSAM) is introduced to separately obtain channel attention and spatial attention features from multi-scale features. The Attention Fusion Layer (AFL) is used to integrate context information of features from different scales; combined with the feature fusion layer, it enhances the expression of features to be detected. Secondly, Shuffle Net V2 is introduced as the backbone network to replace the VGG network, effectively improving the detection speed of the model. Additionally, this paper constructs a seal detection dataset specifically for seal detection and proposes a random occlusion seal image augmentation strategy. Experiments show that compared to the original algorithms, the improved algorithm increased detection accuracy by 5.2% points and improved the inference speed by 14 FPS, which essentially meets the practical needs of the project.

**Keywords:** Multi-scale · Channel-Spatial Attention Module · Shuffle Net V2 · SSD · Augmented strategy

## 1 Introduction

With the continuous development of computer technology, the form of documents has gradually transitioned from physical paper documents to electronic documents [1]. Seals play a crucial role in modern economic society; they are a key component of documents, symbolizing their authority and validity. Therefore, how to effectively verify the seals within documents has become of utmost importance [2].

In the early stages, researchers attempted to employ knowledge-based approaches to process target images, designing a series of image processing methods based on features such as color, shape, and edge operators for the extraction and detection of seal target regions. For instance, BAO et al. [3] utilized RGB color space features to detect seals in traditional Chinese paintings, assuming that seals are red; however, in reality, seals may change color due to oxidation, limiting the performance of this method. Additionally,

Liu Zhihui et al. [4] proposed a shape-based elliptical seal automatic positioning method, but this method involves complex sampling calculations and is only applicable to the detection of elliptical or circular seals. Kang Yaqi et al. [5] combined the Canny operator with the SN color space for adaptive edge detection of seals; nevertheless, this method is computationally complex and heavily relies on the processing results of the SN color space, making it less robust. Unlike traditional target image detection algorithms, seal image detection often deals with smaller and more subtle target variations.

Addressing the two primary challenges faced in seal image detection: **1.** Seal images exhibit diverse shapes, often accompanied by overlapping text in practical applications, leading to issues like text occlusion and adhesion. **2.** Constrained by factors such as ink and pressure, archives commonly contain incomplete seal graphics and faint imprints.

In order to swiftly and accurately detect these targets, this paper proposes a seal target detection approach based on channel-spatial attention, effectively integrating multi-scale contextual information and spatial-channel attention features for target detection. This paper contributes in the following ways:

1. The paper constructs a seal detection dataset and introduces a random synthetic augmentation strategy based on seal and text images, beneficial for enhancing the diversity of seal data and improving model generalization performance.
2. The paper presents a multi-scale channel-spatial attention module, incorporating an attention feature fusion layer, which combines contextual semantic information, effectively enhancing the model's feature representation capability.
3. The paper introduces an improved SSD seal detection algorithm based on channel-spatial attention, coupled with the lightweight backbone network Shuffle Net V2 [6], achieving rapid and accurate detection.

## 2  Related Work

With the rise of deep learning, neural networks have a strong adaptive learning ability and generalization due to adaptive parameter adjustments. There has been extensive development in object detection algorithms based on deep learning, which can be divided into two-stage and one-stage object detection algorithms according to the detection scenario.

Two-stage detection refers to detection algorithms that include a candidate region proposal phase and an object classification phase. These algorithms often have high accuracy. Lu Haitao and others combined Faster RCNN [7, 8] with data augmentation to effectively detect Manchu ancient books and seals. Kaiming He [9] proposed R-FCN, using position-sensitive score maps to realize information perception of target areas, effectively improving the speed of object detection. However, although the position-sensitive score map avoids many candidate areas, it inevitably increases the perception channel dimension. Zeng et al. [10] proposed Light-head RCNN, which significantly reduces its perception channel to increase the speed of the model, but there is a slight decrease in accuracy. Nevertheless, the two-stage object detection scheme is limited by the selection of candidate regions, which greatly reduces the inference speed of the model, making it unsuitable for actual deployment and rapid detection.

To accelerate the inference speed, Redmon introduced YOLO [11], a one-stage object detection method that uses regression to predict target areas. Due to the lack of anchoring and constraints on the predicted area coordinates, the YOLO algorithm has a low detection accuracy. Wei Liu combined the idea of regression prediction with the Anchor multi-box detection strategy to propose the SSD [12] detection algorithm. To speed up detection, Yan Zhenzhen [13] combined the structure of Mobile Net V2 [14] with SSD, using a lightweight backbone network to detect official documents. One-stage object detection algorithms have struck a balance between detection accuracy and speed. However, one-stage algorithms are insensitive to small object detection and are prone to omissions and misjudgments in cases of occlusion. CY Fu et al. [15] proposed DSSD, which adds context information to the original SSD object detection, effectively avoiding the lack of effective combination between multi-layer features, but due to the hourglass structure design and transposed convolution, it results in high computational complexity. Although the DSSD structure has certain improvements for small objects, it also adds more inference time. Yang Chaochen et al. [16] introduced channel dimension weighted information based on the densely connected DSSD algorithm to further improve small object detection.

With the widespread application of the attention mechanism in the field of natural language processing [17], researchers have tried to apply the attention mechanism to the visual image field to strengthen the focus on target feature information and suppress background information. Jaderberg [18] proposed the use of a spatial domain attention network to extract attention features under the feature space to strengthen model expression. H Jie and others introduced SENet (Squeeze and Excitation Network) [19], which uses channel dimension attention to extract features based on the idea of compression and expansion. Later, Woo [20] proposed a convolution-based attention block feature extraction method that combines spatial domain and channel domain attention mechanisms. However, the local bias induction of convolution lacks global information features. Zhang Rui et al. [21] proposed the use of various attention mechanisms combined with the FPN (Feature Pyramid Network) [22] algorithm to improve the detection of distant nighttime vehicles. The design of the FPN structure increases the coupling degree of the model, and obtaining multiple attention mechanisms increases computational complexity. Zeng Tianhao et al. [23] introduced a foreground object attention module combined with the YOLO algorithm to improve traffic sign recognition, and Chen Dehai et al. [24] proposed a feature attention module to enhance small object detection.

## 3  Method

This section will provide a detailed explanation of the improved SSD algorithm. It mainly includes parts such as the SSD algorithm model, channel-space attention module, Shuffle Net V2 network, and optimization objectives.

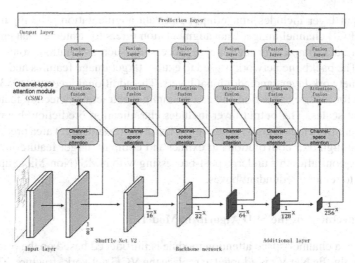

**Fig. 1.** Improved SSD algorithm model structure diagram.

## 3.1 SSD Algorithm Model

The SSD object detection algorithm belongs to a one-stage object detection algorithm, which consists of three parts: the input layer, the backbone network, and the output layer. Its specific structure is shown in Fig. 2.

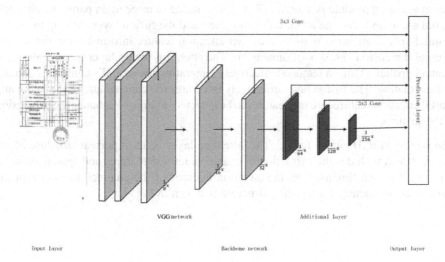

**Fig. 2.** SSD algorithm structure diagram.

The input layer includes image input and data augmentation. The input image is usually an RGB channel image. Data augmentation refers to strategies that enhance the diversity of data samples through image processing, such as rotation, contrast adjustment, etc. The backbone network is used to extract target image features and consists of VGG [25] network layers and additional layers. The additional layers use 3x3 convolution blocks for feature extraction and downsampling, extracting detection features from six different scales. The output layer includes classification prediction, bounding box regression prediction, and post-processing. Classification and target area predictions are achieved through 3x3 convolutions to extract and compute target feature information. Predicted regions after this undergo post-processing with NMS (Non-Max Suppression) operations to remove redundant boxes.

## 3.2  Improvement of the SSD Algorithm Model

In this paper, a channel-space attention module is introduced based on the original SSD model, and Shuffle Net V2 is adopted to replace the VGG network structure. The overall structure is shown in Fig. 1. It consists of four parts: the input layer, the backbone network, the channel-space attention module, and the output layer. The backbone network includes the Shuffle Net V2 module and the additional layer. The channel-space attention module consists of the channel-space attention layer, the attention fusion layer, and the fusion layer. This section will provide a detailed introduction to the improved channel-space attention module and the Shuffle Net V2 network structure.

**Channel-Spatial Attention Module (CSAM).** The basic structure of the channel-spatial attention module is shown in Fig. 1. It consists of three main parts: the channel-spatial attention layer, the attention fusion layer, and the fusion layer. Among them, the channel-spatial attention is used to extract attention feature information in the feature channel dimension and spatial dimension. The attention fusion layer fuses multi-scale channel-spatial attention features, effectively integrating the model's contextual semantic information. The fusion layer effectively integrates the channel attention feature and spatial attention feature into the features to be detected, aiming to enhance the expression of the features.

*Channel-Spatial Attention Layer.* This attention layer is an important structure of the model, which includes the extraction of channel attention features and spatial attention features. Through this module, the attention features of the channel and space of the features are extracted. The specific structure is shown in Fig. 3.

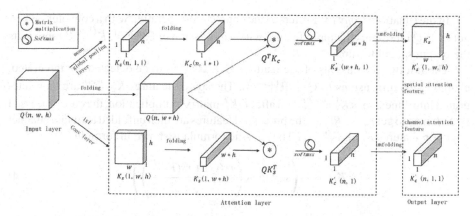

**Fig. 3.** Channel-Space Attention Feature Structure Diagram.

**Spatial Attention Feature:**
The features extracted from the backbone network are processed through global average pooling to obtain channel-based feature information, as shown in formula (1).

$$K_c = GA(Q) \tag{1}$$

In the equation: $K_c \in \mathbb{R}^{n \times 1 \times 1}$ represents the channel feature, $Q \in \mathbb{R}^{(n,w,h)}$ represents the input feature, $n$ stands for feature channels, $w$ is the feature width, $h$ denotes the feature height; $GA$ signifies the global averaging operation.

Features Q are integrated in the last two dimensions, with the dimensions changing to $Q \in \mathbb{R}^{(n,wh)}$. After transposition, the feature $Q^T \in \mathbb{R}^{(wh,n)}$ is obtained. The channel feature is then integrated dimensionally into $K_c \in \mathbb{R}^{(n,1)}$. Multiplying the $Q^T \in \mathbb{R}^{(wh,n)}$ and features $K_c \in \mathbb{R}^{(n,1)}$, followed by the *Soft* max function, produces the feature $K_{s'} \in \mathbb{R}^{(w*h,1)}$. The feature $K_{s'}$ is then dimensionally split and refolded to obtain the spatial attention feature. The specific information conversion formula is shown as formula (2).

$$K_{s'} = rs\left( soft \max\left( \frac{rs(Q)^T \times rs(K_c)}{\sqrt{d}} \right) \right) \tag{2}$$

In the equation: $Q \in \mathbb{R}^{(n,w,h)}$ represents the input feature; $rs$ stands for dimension transformation operation; $d$ indicates the number of elements and serves as a coefficient factor; $K_c \in \mathbb{R}^{(n,1,1)}$ represents the channel feature; $K_{s'} \in \mathbb{R}^{(1,w,h)}$ represents the spatial attention, and its dimensions align with the spatial dimensions of the information in the previous layer.

**Channel Attention Feature:**
The features extracted from the main network are processed with a $1 \times 1$ convolutional layer operation to obtain a feature map with a channel dimension of 1. The formula for this information transformation is shown in formula (3).

$$K_S^{(1,w,h)} = Conv1 \times 1\left( Q^{(n,w,h)} \right) \tag{3}$$

In the equation: $Conv 1 \times 1$ refers to the convolution operation with a convolution kernel of 1; $Q$ represents the input feature, with a dimension of $\mathbb{R}^{(n,w,h)}$, $K_s^{(1,w,h)}$ represents the output spatial feature。

The last two dimensions of the features $Q \in \mathbb{R}^{(n,w,h)}$ to be detected are integrated, changing the dimensions to $Q \in \mathbb{R}^{(n,w \times h)}$. The spatial features $K_s^{(1,w,h)}$ are then integrated into dimensions $K_s^{(1,w \times h)}$, and after $QK_s^T$ matrix multiplication, they are processed by $Soft$ max to get $K_{c'} \in R^{(n,1)}$; the processed features are then unfolded to obtain channel attention features $K_{c'} \in \mathbb{R}^{(n,1,1)}$. The specific formula is shown in formula (4).

$$K_{c'} = rs\left( soft \max \left( \frac{rs(Q) * rs(K_s)^T}{\sqrt{d}} \right) \right) \tag{4}$$

In the equation: $K_s \in \mathbb{R}^{(1,w,h)}$ represents the spatial feature; $rs(\cdot)$ represents the dimension integration and transformation operation; $Q \in \mathbb{R}^{(n,w,h)}$ represents the feature to be detected; $d$ indicates the number of elements, which is a coefficient factor; $K_{c'}$ represents the channel attention feature.

Therefore, every feature selected from the main network will obtain spatial attention features and channel attention features at the corresponding scale.

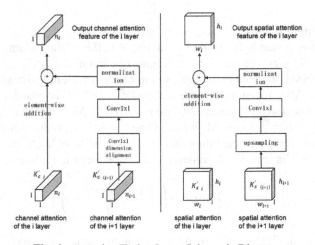

**Fig. 4.** Attention Fusion Layer Schematic Diagram.

*Attention Fusion Layer.* After passing through the channel-spatial attention layer, each resolution feature will obtain the channel-spatial attention layer for the corresponding dimension. In order to effectively integrate multi-scale contextual information for object detection, this section integrates multi-scale attention features through the attention fusion layer. The specific structure is shown in Fig. 4. Taking the channel-spatial attention feature $K_{ci'}$; $K_{si'}$ of the $i$ layer as an example, the channel attention feature $K_{c(i+1)'}$ and the spatial attention feature $K_{s(i+1)'}$ of the $i + 1$ layer are aligned in feature dimensions through a $1 \times 1$ convolutional layer and up-sampling operation, respectively, and then they are passed through another $1 \times 1$ convolution and normalization layer to achieve $K_{ci'}$; $K_{si'}$ consistent dimensions. The fusion of attention features is realized by element-wise addition. The specific expressions are shown in formula (5) and (6).

$$K'_{ci} = K'_{ci} \oplus Bn\left(Conv\left(Conv\left(K'_{c(i+1)}\right)\right)\right) \tag{5}$$

In the equation: $K_{c'(i+1)}$ represents the channel attention feature of the $i + 1$ layer; $Conv(\cdot)$ represents $1 \times 1$ convolution; $Bn(\cdot)$ stands for the normalization layer; $\oplus$ indicates element-wise addition fusion operation. $K_{c'i}$ represents the channel attention feature of the $i$ layer.

$$K'_{si} = K'_{si} \oplus Bn\left(Conv\left(up\left(K'_{s(i+1)}\right)\right)\right) \tag{6}$$

In the equation: $K'_{s(i+1)}$ represents the spatial attention feature of the $i + 1$ layer; $up(\cdot)$ stands for the up-sampling layer; represents $1 \times 1$ convolution; $Bn(\cdot)$ stands for the normalization layer; $\oplus$ indicates element-wise addition fusion operation. $K_{s'i}$ represents the spatial attention feature of the $i$ layer.

*Fusion Layer.* The channel attention features and spatial attention features integrated by the attention fusion layer carry contextual semantic information and attention feature information. The acquired channel-spatial attention information is further fused with the features extracted by the backbone network and used for object detection. The specific structure is shown in Fig. 5.

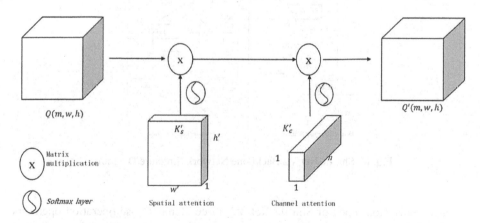

**Fig. 5.** Fusion Layer Structural Schematic.

The features $Q$ obtained from the backbone network are individually fused with the spatial attention $K_{s'}$ and channel attention $K_c^i$, which have been processed through a $1 \times 1$ convolution and *Soft* max layer, by element-wise multiplication along corresponding dimensions. The specific expression is as shown in formula (7).

$$Conv(\cdot)Q' = Q \otimes soft\max\left(\frac{K'_s}{\sqrt{d_s}}\right) \otimes soft\max\left(\frac{K'_c}{\sqrt{d_c}}\right) \tag{7}$$

In the equation: $Q \in \mathbb{R}^{(n,w,h)}$ represents the features from the backbone network; $\otimes$ represents element-wise multiplication; $K_s' \in \mathbb{R}^{(1,w,h)}$ represents spatial attention; $K_c' \in \mathbb{R}^{(n,1,1)}$ represents channel attention; $d_s, d_c$ represents the scaling factor, with a value of $d_s = w * h, d_c = n$; $Q' \in \mathbb{R}^{(n,w,h)}$ represents the fused features used for object detection.

**ShuffleNet v2structure.** This section introduces the structure of Shuffle Net V2 to replace the VGG network in the original SSD object detection algorithm. Its structure is relatively simple, and the inference speed is very fast. The overall structure of the model is shown in Chart 1, where the input image size is $224 \times 224$, containing downsampling images of 5 scenes.

The outputs after MaxPool, Stage2, Stage3, and Stage4 respectively correspond to the feature sizes of $1/8\times$, $1/16\times$, $1/32\times$, and $1/64\times$ in the SSD detection image. The layer named StageX contains two different types of operation blocks. Its specific structure is shown in Fig. 6. One is a feature extraction block (with a stride of 1), and the other is a downsampling block (with a stride of 2).

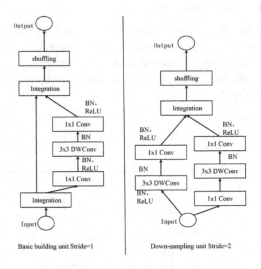

**Fig. 6.** Shuffle Net V2 Backbone Network Structure Diagram.

In the operation block of Shuffle Net V2, three element-wise operation operators are included: channel division, integration layer, and channel shuffling. In the feature extraction block (Stride = 1), the data is divided into two parts through channel division. One part of the features goes through the depthwise separable convolution layer and then concatenates with the other part of the features for channel integration. The order of the concatenated feature channels is then shuffled for output. In this operator, the size of the feature channel dimension remains unchanged at the input and output ends. In the downsampling block (Stride = 2), data passes through the depthwise separable convolution layer with Stride = 2 to reduce the feature size. It is then dimensionally integrated, and the output features are shuffled in their channels for output.

On one hand, the Shuffle Net V2 network effectively inherits the lightweight nature of the depthwise separable convolution operation of the Mobile Net network. On the other hand, through channel division processing, it further reduces the computation of the model. The overall model parameter size of Shuffle Net V2 is about 2.3M, while that of VGG16 is around 138M, greatly reducing the model's computation.

**Loss Function.** This paper uses L1 loss and cross-entropy loss to optimize the loss. During the objective training optimization phase of the model, the detection results output by the model are compared with the real la $L_{loc}$ bels. The classification loss is calculated using cross-entropy loss, while the localization loss is measured and calculated using L1 loss. The overall loss calculation is shown in formula (8).

$$L(x, c, p, g) = \frac{1}{N}(L_{cls}(x, c) + \lambda L_{loc}(x, p, g)) \tag{8}$$

where $L$ denotes the overall loss; $L_{cls}$ represents the classification confidence loss; represents the localization loss; $\lambda$ is the coefficient factor, typically taken as 1.0; $N$ denotes the number of matching prediction boxes to the true boxes; $c$ represents confidence, $p$ denotes the prediction box, $p$ denotes the true box, and $x$ denotes the matching coefficient.
The specific expression for the classification loss is shown in formula (9).

$$L_{cls}(x, c) = - \sum_{i \in Pos}^{N} X_{ij}^k \log\left(c_i^k\right) - \sum_{i \in Neg} \log\left(c_i^0\right) \tag{9}$$

where $x_{ij}^p$ denotes the matching situation of the $i$ prediction box to the $j$ true box being of the $k$ class; it is 1 if matched, otherwise 0. $c_i^k$ represents the prediction value of the $i$ prediction box being of the $k$ class. $c_i^0$ represents the prediction value of the $i$ prediction box being of the 0 class background.The specific expression for the localization loss is shown in formula (10).

$$L_{loc}(x, p, g) = \sum_{i \in Pos}^{N} \sum_{m \in \{cx, cy, w, h\}} X_{ij}^k smoth_{L1}\left(p_i^m - g_j^m\right) \tag{10}$$

where $cx$, $cy$, $w$, $h$ respectively represent the horizontal coordinate, vertical coordinate, width, and height of the prediction box. $p_i^m$ represents the $i$ prediction box, and $g_i^m$ represents the $j$ true box; $x_{ij}^k$ represents the matching situation of the $i$ prediction box with the $j$ true box being of the $k$ class. It is 1 if matched and 0 if not matched.

## 4 Experiments

### 4.1 Dataset and Augmentation Strategies

We collected 3,200 archive seal images by screening the network and work archive images. We used the Lableme annotation software to annotate, with the annotation format being the Pascal VOC text annotation format, thus constructing the seal detection dataset for this paper. At the same time, the dataset was randomly divided into training and test sets according to the 9:1 principle.

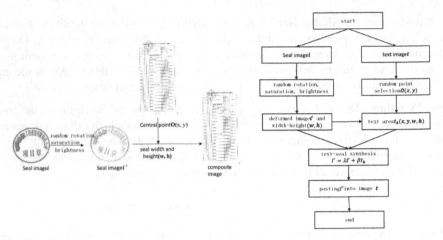

**Fig. 7.** Seal Text Image Synthesis Strategy Flowchart.

Through manual analysis, most of the seal images are relatively clear, with only a small portion of images showing issues like text overlap and blurry imprints. In order to enhance the detection of unclear imprints and overlapping text, and to enhance the generalization ability of the model, this paper introduces an augmentation strategy based on random synthesis of seal and text images. This is in addition to the common data augmentation strategies such as rotation, stitching, random noise, and contrast adjustment. The specific process is shown in Fig. 7.

Firstly, the obtained seal images $\beta = 1 - \lambda$ are adjusted through random rotation, noise, saturation, brightness, and other image processing techniques to get the deformed seal images $i\prime$ and their corresponding width and height $(w, h)$. Then, a point $O(x, y)$ is randomly selected from the text image $t$, serving as the center coordinate of the synthesis area. Combined with the width and height of the seal image $i\prime$, the target area $t_k(x, y, w, h)$ for the composite seal text image is determined. The seal image $i\prime$ and text image are overlaid and synthesized using a linear addition method, where the weighting follows a uniform distribution between $(0.35, 0.65)$.

Through this strategy, a large number of different types of seal text images are synthesized based on a few seal images. This significantly increases the number and diversity of training samples.

### 4.2 Experimental Analysis

The experiments in this paper were conducted under the Ubuntu 18.04 operating system. Other experimental environment configurations include: CPU: Intel(R) Core (TM) i7-9700K 3.60 GHz@ 4.90 GHz; GPU: Nvidia GeForce RTX2080Ti 11 GB; CUDA-Toolkit:10.2; Memory: 32 GB; Deep Learning Framework: Pytorch-1.9.

This paper tested various different models on the test set, and the results are shown in Table 2. The improved SSD object detection algorithm proposed in this paper achieved a detection accuracy of 96.3%, a 5.2% increase compared to the original SSD algorithm; the recall rate reached 96.7%, a 4.1% increase compared to the SSD algorithm. At the same time, in terms of GPU inference speed, the improved SSD object detection algorithm proposed in this paper achieved a speed of 62fps, about 14fps faster than SSD.

**Table 1.** Shuffle Net V2 Network Structure Table.

| layer | Output size | Convolution size | Step size | frequency | channel |
|---|---|---|---|---|---|
| Input layer | 224 × 224 | | | | 3 |
| Convolution 1 | 112 × 112 | 3 × 3 | 2 | 1 | 24 |
| MaxPool | 56 × 56 | 3 × 3 | 2 | 1 | 24 |
| Stage2 | 28 × 28 | | 2 | 1 | 116 |
| | 28 × 28 | | 1 | 3 | |
| Stage3 | 14 × 14 | | 2 | 1 | 232 |
| | 14 × 14 | | 1 | 7 | |
| Stage4 | 7 × 7 | | 2 | 1 | 464 |
| | 7 × 7 | | 1 | 3 | |
| Conv5 | 7 × 7 | 1 × 1 | 1 | 1 | 1024 |

**Table 2.** Performance of different models on the test dataset.

| method | Precision (%) | Recall (%) | FPS |
|---|---|---|---|
| Yolo | 72.4 | 82.3 | 40 |
| Faster RCNN | 94.8 | 95.2 | 10 |
| YoloV3 | 92.1 | 93.7 | 50 |
| SSD | 91.1 | 92.6 | 48 |
| Ours | 96.3 | 96.7 | 62 |

To further validate the impact of the attention module, context fusion strategy, and data augmentation strategy on model performance, this study designed a series of ablation experiments based on the improved SSD object detection algorithm. The results are shown in Table 3. Among them, the backbone network model of this model is Shuffle Net V2, and its accuracy has a certain decline compared to the original VGG network. The addition of a data augmentation strategy improved the model's detection accuracy by 1.7% points, and when the data augmentation strategy was added to the improved strategy, the model's performance improved by 0.5% points. Overall, the augmentation strategy has a certain effect, which is beneficial for enhancing the model's generalization ability. Adding the channel-space attention layer improved the model performance by

Table 3. Comparison Table of Ablation Experiment Data.

| Data augmentation | Channel-Spatial attention layer | Attention fusion layer | Precision (%) |
|---|---|---|---|
| - | - | - | 89.6 |
| ✔ | - | - | 91.3 |
| ✔ | ✔ | - | 95.2 |
| | ✔ | ✔ | 95.8 |
| ✔ | ✔ | ✔ | 96.3 |

3.9% points, and after adding the attention fusion layer, the model performance improved by 1.1% points.

### 4.3  Experimental Results Presentation

Some of the detection results in this paper are shown in Fig. 8. They encompass situations involving obscured seal text, complex backgrounds, rotated seals, and partial seals. The algorithms consistently exhibit good detection performance. The detection confidence is relatively higher compared to the SSD algorithm, as depicted in Fig. 8.

To address issues such as false positives related to QR codes, rectangular patterns, and the challenges posed by complex backgrounds and various intricate icons, the improved SSD algorithm demonstrates superior discrimination capabilities. This is specifically illustrated in Fig. 9.

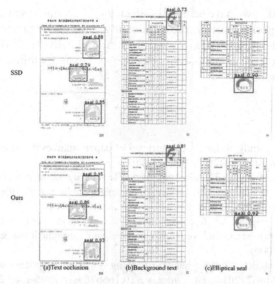

**Fig. 8.** Detection Comparison Chart between SSD and Improved SSD Algorithm.

**Fig. 9.** Comparison Chart of False Positives Testing between SSD and Improved SSD Algorithm.

## 5 Conclusion

This study is devoted to addressing key challenges in the field of seal detection, including difficulties in effectively detecting obscured seal text, overlapping seals, and seals with shallow imprints. Building upon the foundation of SSD object detection, we have introduced the Channel-Spatial Attention Module (CSAM) to enhance target detection performance. Furthermore, by replacing the traditional VGG detection model with the

Shuffle Net V2 model, we have achieved satisfactory detection results. However, overall, although our algorithm performs exceptionally well on GPU devices, its inference efficiency on edge devices remains relatively low. Additionally, the SSD object detection algorithm is constrained by multi-scale detection and image resolution, leaving ample room for optimization.

# References

1. Yi, L., Zou, B.: A study on intelligent retrieval of paper archives based on deep learning. Ji Dian Bing Chuan Arch. **2022**(06), 99–103 (2022)
2. Wang, S., Ma, D., Wan, Y.: The direction of public security management of seal industry in the information age - on "seal text service." J. People's Public Secur. Univ. China (Soc. Sci.) **35**(06), 104–116 (2019)
3. Bao, H., Xu, D., Feng, S.: An effective method to detect seal images from traditional Chinese paintings. In: 2009 international conference on wireless communications & signal processing, pp. 1–4. IEEE (2009)
4. Liu, Z.: Implementation of an automatic positioning system for elliptical seals. Internet Things Technol. **8**(04), 61–67 (2018). https://doi.org/10.16667/j.issn.2095-1302.2018.04.019
5. Yao, M., Mou, X., Chen, P., Zhao, M., Li, Z.: Research on detection, positioning, and recognition of seals in images. Inf. Technol. Inf. **2018**(12), 148–150 (2018)
6. Zhang, X., Zhou, X., Lin, M., Sun, J.: ShuffleNet: an extremely efficient convolutional neural network for mobile devices. In: Conference on Computer Vision and Pattern Recognition (CVPR), pp. 6848–6856 (2018)
7. Lu, H., Wu, L., Zhou, J., et al.: Manchu document seal detection based on Faster R-CNN and data augmentation. J. Dalian Minzu Univ. **20**(05), 455–459 (2018)
8. Ren, S., He, K., Girshick, R., Sun, J.: Faster R-CNN: towards real-time object detection with region proposal networks. IEEE Trans. Pattern Anal. Mach. Intell. **39**(6), 1137–1149 (2017). https://doi.org/10.1109/TPAMI.2016.2577031
9. Dai, J., Li, Y., He, K., et al.: R-FCN: object detection via region-based fully convolutional networks. In: Neural Information Processing Systems, pp. 379–387 (2016)
10. Li, Z., Peng, C., Yu, G., et al.: Light-head R-CNN: In Defense of Two-Stage Object Detector (2017). http://arxiv.org/abs/1711.07264
11. Redmon, J., Divvala, S., Girshick, R., Farhadi, A.: You only look once: unified, real-time object detection. In: 2016 IEEE Conference on Computer Vision and Pattern Recognition (CVPR), pp. 779–788 (2016)
12. Liu, Wei, et al.: SSD: Single shot multibox detector. In: Leibe, B., Matas, J., Sebe, N., Welling, M. (eds.) Computer Vision – ECCV 2016, pp. 21–37. Springer International Publishing, Cham (2016). https://doi.org/10.1007/978-3-319-46448-0_2
13. Yan, Z.: Application of SSD-MobileNet in the detection of official document archive seals. Chongqing Normal University (2020).https://doi.org/10.27672/d.cnki.gcsfc.2020.000596
14. Sandler, M., Howard, A., Zhu, M., et al.: MobileNetV2: inverted residuals and linear bottlenecks. In: Conference on Computer Vision and Pattern Recognition (CVPR), pp. 4376–5107 (2018)
15. Fu, C.Y., Liu, W., Ranga, A., et al.: DSSD: Deconvolutional Single Shot Detector (2017). http://arxiv.org/abs/1701.06659
16. Yang, C., Chen, J., Xing, K., et al.: Research on small object detection algorithm based on improved DSSD. Comput. Technol. Dev. **32**(06), 63–67 (2022)
17. Vaswani, A., Shazeer, N., Parmar, N., et al.: Attention is all you need. In: International Conference on Neural Information Processing Systems (NIPS 2017), pp. 6000–6010 (2017)

18. Jaderberg, M., Simonyan, K., Zisserman, A., et al.: Spatial transformer networks. In: International Conference on Neural Information Processing Systems, 2015:2017–2025
19. Hu, J., Shen L., Sun, G.: Squeeze-and-excitation networks. In: Conference on Computer Vision and Pattern Recognition (CVPR), pp. 7132–7141 (2018)
20. Woo, S., Park, J., Lee, J.-Y., Kweon, I.S.: CBAM: convolutional block attention module. In: Ferrari, V., Hebert, M., Sminchisescu, C., Weiss, Y. (eds.) Computer Vision – ECCV 2018, pp. 3–19. Springer International Publishing, Cham (2018). https://doi.org/10.1007/978-3-030-01234-2_1
21. Zhang, R., Gao, S., Zhao, X., et al.: Improved YOLOv5s-based algorithm for unmanned driving night vehicle target detection. In: Electronic Measurement Technology 2023, pp. 1–9 (2023). http://kns.cnki.net/kcms/detail/11.2175.TN.20230523.1113.002.html
22. Lin, T.Y., Dollar, P., Girshick, R., et al.: Feature pyramid networks for object detection. In: Proceedings of the IEEE Conference on Computer Vision and Pattern Recognition, pp. 2117–2125 (2017)
23. Zeng, T., Chen, L.: Small object detection based on traffic signs [J/OL]. Laser Magazine, pp. 1–8 (2023). http://kns.cnki.net/kcms/detail/50.1085.TN.20230517.1026.006.html
24. Yan, X., Jia, Y., Zhao, L., et al.: Seal recognition based on improved YOLOv5 model. Elect. Measur. Technol. 46(02), 169–174 (2023)
25. Simonyan, K., Zisserman, A.: Very deep convolutional networks for large-scale image recognition. Computer Science 2014 (2014)

# A Multiple Fire Zones Detection Method for UAVs Based on Improved Ant Colony Algorithm

Fanglin Xue, Peng Geng, Huizhen Hao[✉], Yujie He, and Haihua Liang

School of Information and Communication Engineering, Nanjing Institute of Technology, Nanjing 211167, China
haohuizhen@njit.edu.cn

**Abstract.** In response to the increasing severity and frequency of forest fires caused by dry climate and temperature rise, and the difficulty of identifying the fire initiation point in the early stage due to the large forest area, an effective method of utilizing unmanned aerial vehicles (UAVs) for early fire search and extinguishing is used in this article. This is a path planning problem in multi-agent multi-objective search in uncertain environments. This article proposes an improved ant colony algorithm based on the idea of competition cooperation. Based on the advantages of ant colony algorithm in path planning, exclusion attraction pheromones are added to the original algorithm, which enhances the detection area and fire extinguishing speed. On the NetLogo simulation platform, by comparing the convergence time and coverage area of the improved ant colony algorithm with the random walk algorithm and ant colony algorithm, it is found that the detection performance of the improved ant colony algorithm has significantly improved. The proposed UAVs based forest wildfire detection model based on improved ant colony algorithm is of great significance for the suppression of forest fires.

**Keywords:** Multi-objective search · UAVs · Ant colony algorithm · Competition cooperation

## 1 Introduction

Forest fires are sudden natural disasters that spread rapidly due to factors such as terrain and wind direction, making rescue operations difficult. Forest fires gain global attention and are considered a significant global environmental issue [1, 2]. At the present time global warming leads to an increase in temperatures, especially in dry and hot areas, resulting in an increase in the frequency and scale of forest fires. This causes severe environmental damage, as well as casualties and economic losses. According to statistics, over 200,000 forest fires occur worldwide on average each year, burning approximately 0.1% of the total forest area [3]. Recently, on August 8th local time, a wildfire broke out on Maui Island, Hawaii, USA. Maui Island is known as a vacation island. After the fire, houses and buildings are damaged, most of the buildings on the island are destroyed, many families are displaced, and many natural landscapes and historical sites

© The Author(s), under exclusive license to Springer Nature Singapore Pte Ltd. 2024
H. Jin et al. (Eds.): IAIC 2023, CCIS 2059, pp. 194–208, 2024.
https://doi.org/10.1007/978-981-97-1280-9_15

are threatened by wildfires. As of August 12th local time, the death toll from the Maui Island wildfire has reached 89, making it the deadliest wildfire in the United States for over a century.

The enormous impact of forest fires has attracted a great deal of attention, making firefighting methods particularly important. Common methods for extinguishing forest fires include human firefighters and helicopters [4]. Human firefighters usually participate in extinguishing fires in flat terrain areas, mainly extinguishing surface fires and creating fire zones. Helicopters are suitable for extinguishing fires in areas such as high mountains that are difficult for firefighters to reach. Water bombs released from the air are mainly used for rapidly spreading fires. From the data, it can be seen that the United States uses approximately 1000 forest firefighting aircraft annually, Canada has over 500 aircraft, and Russia has over 800 aircraft. However, these methods certain limitations: firefighters face difficulties in operating in toxic smoke, high temperatures, and flames caused by forest wildfires, which can jeopardize their safety [5]. When helicopters are used to transport firefighting resources, their effectiveness is greatly reduced due to impaired visibility caused by smoke and terrain obstacles during forest fires. In this case, small unmanned aerial vehicles (UAVs) become a more suitable option. The mobility and flexibility of drones in forest areas are excellent, allowing them to conduct precise reconnaissance and monitoring in complex terrain environments, providing critical information support for fire suppression operations. If drone swarms can quickly detect and determine the size of forest fires, and allocate firefighting resources more scientifically, they can extinguish forest fires faster.

In the field of using UAVs for fire fighting, an important research topic is: how to find multiple random fire zones in an uncertain environment? This problem belongs to the category of path planning in a multi-agent environment with multitarget. An important approach to researching this type of problem is to use metaheuristic algorithms to approximate the optimal solution, which is more practical. Many researchers have conducted research in this area. Zhang et al. started from the path planning task of unmanned aerial vehicle groups and classified, reviewed and statistically analyzed the related research [6]. Li et al. studied the high-efficiency coverage problem in unmanned aerial vehicles executing search, reconnaissance and other tasks [7]. In the collaborative search and attack task of multiple drones, Zhen et al. not only plans the task for UAVs, but also considers collision avoidance constraints between UAVs [8]. Rutuja Shivgan et al. develop a solution to the path planning problem of multiple UAVs with the objective of minimizing the total energy consumption of all UAVs. This research contribution is significant as it focuses on energy optimization for UAVs, which is an important consideration for prolonging the flight time and increasing the operational efficiency of these vehicles [9]. The above literature is based on the detection of static targets in a known environment and is not suitable for the dynamic characteristics of complex forest environments and the spreading of fire. This article proposes an improved ant colony algorithm based on an information pheromone rejection-attraction mechanism for uncertain dynamic environments. The algorithm releases rejection information pheromones at the search path and releases attraction information pheromones at the fire points, thus solving the problem of multi-UAV cooperative search. Compared with traditional algorithms, the proposed algorithm has a higher coverage rate and search efficiency.

## 2   Model Design

This article investigates the detection of forest fire areas using an improved ant colony algorithm-based UAV swarm. The overall research scenario includes the fire propagation model and the model for the UAV swarm to detect the forest fire area. The research scenarios primarily encompass the aspects shown in Table 1.

**Table 1.** Names and purposes of the main components in the research scenario

| Index | Name | Purpose |
|---|---|---|
| 1 | Simulation Area | The environment is a bounded and fixed-sized space, where unmanned aerial vehicles (UAVs) and fire points exhibit dynamic behavior |
| 2 | Drone | Unmanned aerial vehicles (UAVs) are objects capable of continuous movement within the space |
| 3 | Fire | The positions of fire points are random, which reflects a more realistic scenario |

Due to the focus of this research on forest fire area detection, the model design is based on the following assumptions: the unmanned aerial vehicles (UAVs) have unlimited battery power, and they carry an unlimited supply of pheromones for information dissemination. The damage to unmanned aerial vehicles (UAVs) caused by fires is negligible. The schematic diagram of the research framework is depicted in Fig. 1.

### 2.1   Fire Model

Common fire spread models typically include simple discrete models, diffusion models, and models based on computational fluid dynamics (CFD). In the simple discrete model, each tree or area is treated as an individual unit, simulating fire spread through propagation rules between adjacent units. In the diffusion model, fire spread is modeled as a thermal conduction process, usually described by diffusion equations or similar equations. More complex fire spread models can incorporate CFD and other methods to consider additional factors such as wind direction, wind speed, tree density, humidity, etc., aiming to accurately simulate fire propagation [10]. Fire spread is a complex and dynamic problem, specific models need to be determined based on the actual situation, data, and research objectives.

The fire spread model utilized in this study is based on the Fire model within the NetLogo model library [11]. Building upon this foundation [12], improvements were made. In a controlled environment, trees were randomly generated based on the density of the forest. In contrast to the basic Fire model, ignition did not start from a single edge but rather from the random selection of five tree locations, thereby introducing greater randomness to the ignition points. Additionally, the combustibility factor of trees was incorporated. In real forest wildfires, due to various factors such as rainfall, weather

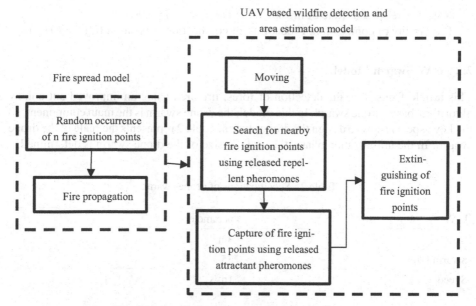

**Fig. 1.** Investigating the scene framework diagram

humidity, and tree characteristics, not all trees may be combustible, thus obstructing the unhindered spread of flames.

The rules for fire propagation are as follows: during the simulation, for each fire flame, it checks the four neighboring patches to determine if they are green patches (representing trees). If a neighboring block is a green patch, indicating the presence of a tree, it ignites the neighbor and generates a new fire flame. Subsequently, the original fire flame transitions into an ember, with the ember's color gradually darkening until it is eventually removed when it approaches black.

The combustion rule entails selecting a green patch (representing a tree) and transforming it into a fire flame, with its color set to red. The color of the ignited green patch is then changed to dark red, indicating that the tree has been burned down. The fire propagation model is depicted in Fig. 2.

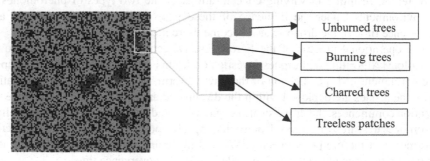

**Fig. 2.** The fire propagation model

Rate of fire spread: Rate of fire spread at position $(i, j) = F(i, j)$.
Probability of combustion: Probability of combustion at position $(i, j) = P(i, j)$.

## 2.2 UAV Swarm Model

This article focuses on the detection of forest fire area using an improved ant colony algorithm-based drone swarm. In this model, the drone swarm is the main component, so the key aspect is how to design the drone swarm. Table 2 represents the scale of the drone swarm. In the initialization phase, the scale or size of the drone swarm is determined.

Table 2. Specifications of drone swarm.

| UAV | parameters |
| --- | --- |
| Size | 3m |
| Swarm size | 35 |
| Speed | 1m/s |

The action module of the drone swarm operates at runtime, where each individual drone performs different actions based on the current environmental conditions. If a drone detects flames at its current position, it executes the operation to capture the fire point and releases attractive pheromones. Conversely, if there are no flames detected at the drone's position, it performs the operation to search for fire points and releases repellent pheromones. During the food capturing process, the drone eliminates the flames at its current position. When searching for fire spots, drones tend to areas with high attracting-pheromones. During the flight of the search target, they adjust their position based on the repulsive pheromones along the way and stay away from areas with high repulsive pheromones, thereby expanding the search area of the drone group.

The improved algorithm for unmanned aerial vehicle flight module is random walking and probability selection. The flight direction is adjusted based on repellent and attractive pheromones, which are used to update the environment by releasing pheromones. The volatilization coefficients of the repellent and attractive pheromones are different, resulting in varying concentrations of the two types of pheromones in the environment. Under the influence of these pheromones, the coverage rate and convergence time of the drone swarm become more ideal.

The operational process of the Forest Fire Detection Model using the improved Ant Colony Algorithm and drones investigated in this article begins with the setup of the environment and specification of the drone swarm. During the model initialization phase, the drones are deployed within the designated area of $(x < 4, y < 4)$. Once the operation commences, the drone swarm engages in random walking, releasing repellent pheromones to search for targets. Upon detecting a fire point, the drones release attractive pheromones at the fire point location. When a fire point is successfully captured, it is incrementally counted to facilitate the calculation of convergence time.

# 3  Improved Ant Colony Algorithm

Ant Colony Optimization (ACO) [13] is a swarm intelligence optimization algorithm proposed by Italian scholar Marco Dorigo, inspired by the collective foraging activities of ants. Ants use pheromones to communicate and choose pathways based on their pheromone concentration, forming a positive feedback mechanism. This algorithm has wide applications in fields such as combinatorial optimization problems and path planning problems [14, 15]. There are also many different studies on ant colony algorithm, some of which are combined with other meta heuristic algorithms [16]. In terms of optimization problems, in addition to research based on ant colony algorithm, there are also studies based on other algorithms such as genetic algorithm and BCO algorithm [17].

## 3.1  The Basic Procedure of Ant Colony Algorithm

The basic process of ant colony algorithm [18] is described as follows.

(1) Initialization: Set the initial values for the number of ants, pheromone concentration, problem space, and other relevant parameters.
(2) Ant search: Each ant selects a path based on pheromone levels and metaheuristic rules, gradually constructing the solution space.
(3) Update pheromones: After all ants have completed path selection, update the pheromone concentration based on the quality of the chosen paths.
(4) Evaluate paths: Compute the quality of each path using objective functions or problem-specific evaluation methods.
(5) Update best solution: Update the global or local best solution based on the path quality.
(6) Repeat iterations: Repeat steps (2)-(5) based on the complexity and requirements of the problem until termination conditions are met.
(7) Output results: Output the best solution or an approximate best solution as the algorithm's solution, for further analysis and application.

## 3.2  Innovation Based on Ant Colony Algorithm

The model studied in this article is a forest fire area detection model for UAV swarms based on an improved ant colony algorithm. This algorithm, called the Competition-Cooperation Ant (C-C Ant) algorithm, introduces two types of pheromones, namely repulsive pheromones and attractive pheromones, in addition to the basic ant colony algorithm. The UAV swarm releases repulsive pheromones when searching for targets, and releases attractive pheromones at the detected fire points. The volatility coefficients and increment values of the repulsive and attractive pheromones are different, resulting in different levels of these pheromones in the environment. By adjusting the direction of UAV flight based on the different content of pheromones in the neighborhood, the algorithm incorporates random walking during flight to expand the search range and avoid ineffective searches, thus addressing the problem of cooperative search in multi-UAV scenarios. Compared to traditional algorithms, this algorithm offers higher coverage and search efficiency.

At the initialization stage, pheromone is distributed on each grid block with an initial value of $\tau_0$. The transition probability for C-C Ant can be described as Eq. (1).

$$P(j) = \frac{e^{\alpha_j}}{\left(\sum (e^{\alpha_i}) + \sum \left(e^{\beta_i \cdot \gamma_i}\right)\right)} \tag{1}$$

In the Eq. (1), $P(j)$ represents the transition probability of target $j$, $\alpha_j$ represents the attractive pheromone of target $j$, $\alpha_i$ represents the attractive pheromone of target $i$, $\beta_i$ represents the repellent pheromone of target $i$, and $\gamma_i$ represents the distance from target $i$. This formula maps the concentration of pheromones on targets to the probability space, and it is used to determine the probability of selecting the next target based on the relative magnitude of the pheromones.

The updating method of pheromones in this algorithm differs to some extent from the traditional ant colony algorithm. The updating formula for pheromones is as Eq. (2).

$$\tau_j(t+1) = \rho\left(\tau_j(t) + \Delta\tau_j(t+1)\right) \tag{2}$$

In the Eq. (2), $\tau_j(t)$ and $\tau_j(t+1)$ represent the pheromone levels before and after updating within the grid block, while $\rho$ is the pheromone evaporation coefficient. $\Delta\tau_j(t+1)$ represents the pheromone increment within the grid block.

The above equation represents the pheromone updating formula. However, to account for the different roles of repellent and attractive pheromones in the environment, they have distinct evaporation coefficients and pheromone increments. This prevents the accumulation of pheromones, facilitating the movement of the swarm of unmanned aerial vehicles (UAVs).

The flowchart in Fig. 3 illustrates the process of the competitive and cooperative algorithm.

The improved algorithm for this study, C-C Ant algorithm, incorporates both repulsive and attractive pheromones based on the ant colony algorithm. Building upon the fire spread model, a certain amount of initial pheromone is introduced. During the random walk of the drone swarm, a decision process is undertaken. When nearby fire points are detected, the drones capture the fire points and release attractive pheromones. Conversely, if no fire points are detected in the vicinity, the drones perform a search for fire points and release repulsive pheromones. During the search for fire points, the direction is adjusted based on the neighboring attractive pheromones while avoiding the repulsive pheromones as much as possible. To verify the effectiveness of the improved algorithm, coverage ratio and convergence time are introduced. Comparative analysis with the ant colony algorithm and random walk algorithm demonstrates the superior effectiveness of the improved algorithm.

## 4 Simulation Evaluation

This article utilized NetLogo [11], a best-suited tool for simulating collective intelligence, as the simulation and agent-based modeling language. NetLogo is a widely used language and environment for multi-agent modeling, extensively employed in modeling and simulating complex systems. By coding in NetLogo, users can construct intricate

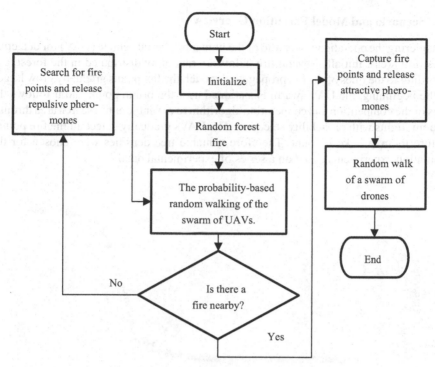

**Fig. 3.** Flowchart of the C-C Ant algorithm

models, simulate agent behaviors, and investigate interactions and system evolution. In order to simulate the scenario of a forest fire and the coordination mechanism of a drone swarm, a model was built using NetLogo, which enabled the simulation of the detection of wildfire areas by the drone swarm during a forest fire.

The simulation experiments in this article are conducted on a Lenovo Xiaoxin 14IIL2020 laptop equipped with an Intel® Core™ i5-1035G1 CPU @ 1.00 GHz, 16 GB of memory, and running on a 64-bit Windows 10 Home edition operating system. The experiments involved the proposed model for detecting wildfire areas using the improved ant colony algorithm implemented with a drone swarm.

In NetLogo, the view is a two-dimensional plane (X/Y) composed of individual patches with a size of 4. To ensure smooth execution of the model and avoid performance issues in NetLogo, the search area of the model is set to a size of 101 * 101, totaling 10,201 patches. Additionally, the positions of the trees are randomly distributed based on the specified tree density. In order to facilitate the smooth operation of the model, the drones in this simulation experiment are assumed to have sufficient battery power to cover the entire area. The convergence time of the algorithm and the coverage ratio of the drone swarm's detection area are used as criteria for comparing the efficiency of the model.

## 4.1  Scenario and Model Execution Overview

Considering the operational scenario of the unmanned aerial vehicle (UAV) for detecting forest fire areas: Initially, several fire points are randomly distributed in the forest, and according to the rules of the fire propagation model, the fire points gradually grow larger. At the beginning, the UAV swarm is stationed near the origin point, and it follows the rules of the competition and cooperation algorithm to cover the entire search area through random flights with probability selection. The UAVs gradually detect all the fire points, capture them, and keep count. Therefore, suitable tree densities were chosen for the simulation experiments based on a series of experimental data.

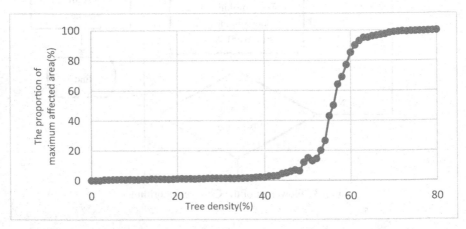

**Fig. 4.** The effect of tree density on fire area

The Fig. 4 above shows the average proportion of affected area after 20 experiments, under different tree densities. From the graph, it is evident that the affected area varies with different tree densities. However, within the range of [50%, 60%] tree density, the change in the affected area by fire is more pronounced. Therefore, for testing the UAV swarm's detection of forest fire areas, simulating a forest environment with a tree density within [50%, 60%] is considered more appropriate. The proportion of affected area is calculated using the Eq. (3).

$$Fa = \frac{Affected - Area}{10201} \tag{3}$$

In the Eq. (3), *Affected − Area* represents the area of the burned region.

## 4.2  Fire Model Under Different Tree Densities

For the sake of comprehensive research and to improve the realism of the simulations, fire models were executed for different scenarios under two tree density levels. Figures 5 and 6 depict the fire models with tree densities of 50% and 60% respectively. In these models, five ignition points were randomly selected at the same time, and the resulting changes in the affected area after tree combustion were observed. This approach was employed to ensure the thoroughness of the article and enhance the simulation's alignment with real-world conditions.

The simulation experiments conducted in this article demonstrate the successful operation of the fire model under investigation. It is noteworthy that the tree densities of [50%, 60%] exhibit significant effects, rendering them suitable for utilizing drone swarms in detecting forest fire areas.

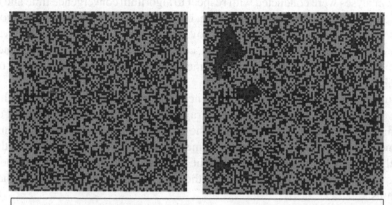

The above figures illustrate the propagation model of a fire with a tree density of 50%. The left image corresponds to ticks=2, while the right image corresponds to ticks=32.

**Fig. 5.** The fire models with tree densities of 50%

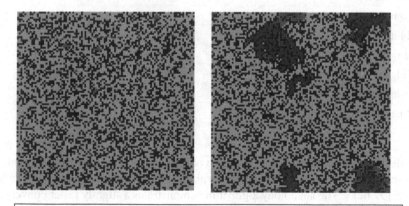

The above figures illustrate the propagation model of a fire with a tree density of 60%. The left image corresponds to ticks=2, while the right image corresponds to ticks=32.

**Fig. 6.** The fire models with tree densities of 60%

### 4.3   Multi-Drone Target Search Under Different Densities

During the modeling process, it was observed that a drone quantity of 35 is more suitable for the operation of the model. Excessive or insufficient drone numbers can negatively impact the functioning of the model and may even cause NetLogo to malfunction due to excessive model size or long computation time.

In order to investigate the efficiency of the proposed improvement algorithm (C-C Ant algorithm) in multi-agent multi-objective search under uncertain environments, comparative analyses were conducted with respect to algorithm convergence time and coverage ratio of the drone swarm detection area, comparing it with the basic Ant algorithm, random walk algorithm (RW), and the improved algorithm. To comprehensively validate the efficiency of the improvement algorithm, corresponding experiments were conducted regarding coverage ratio and convergence time for environments with tree density of 50% and 60%.

Figure 7 shows the algorithm convergence time in a forest environment with a tree density of 50%. The graph displays the average number of fire points for the ant colony algorithm, random walk algorithm, and the improved algorithm after 20 simulation experiments in the 50% tree density environment, confirming the effectiveness of the proposed algorithm in this article. To simplify the comparison process and validate the effectiveness of the proposed algorithm, we set the parameters listed in Table 3 for the three algorithms.

**Table 3.** General simulation parameters

| Search-area | 101 * 101 |
|---|---|
| Drones-number | 35 |
| Initial ignition point | 5 |
| Probability-of-spread | 4 |
| Attract-rate | 0.7 |
| Repelling-rate | 0.5 |
| Attract-increase | 0.1 |
| Repelling-rate | 0.2 |

As depicted in the above graph, it can be observed that in a forest environment with a tree density of 50%, the random walk algorithm exhibits the longest convergence time, reaching 197 ticks. The ant colony algorithm demonstrates a slightly faster convergence time of 183 ticks, whereas the C-C Ant algorithm exhibits the shortest convergence time of only 64 ticks. This indicates that the improved algorithm is effective in reducing the algorithm convergence time in the case of a 50% tree density.

Figure 8 illustrates the algorithm convergence time in a forest environment with a tree density of 60%. The graph displays the average number of fire points for the ant colony algorithm, random walk algorithm, and the improved algorithm after 20 simulation experiments in the 60% tree density environment. To validate the effectiveness of the

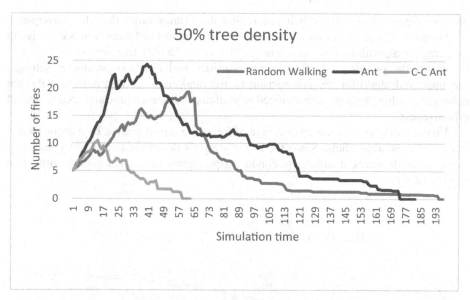

**Fig. 7.** Algorithm convergence time with 50% tree density

algorithm convergence time, the experimental parameters used are identical to those listed in Table 3.

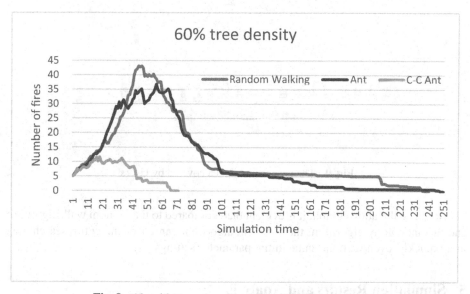

**Fig. 8.** Algorithm convergence time with 60% tree density

As evident from the above graph, in a forest environment with a tree density of 60%, both the random walk algorithm and the ant colony algorithm exhibit comparatively prolonged convergence times, being more than three times longer than the convergence time of the C-C Ant algorithm. This signifies that the improved algorithm is effective in reducing the algorithm convergence time in the case of a 60% tree density.

Under the premise of different tree densities and identical parameter settings, the improved algorithm, in comparison to the random walk algorithm and the ant colony algorithm, demonstrates reduced algorithm convergence time and greater overall effectiveness.

Figure 9 displays the model coverage of forest fire area detection by a drone swarm under different algorithms. Since the drone swarm's detection area is independent of varying tree densities, it suffices to conduct experiments in an environment with a tree density of 50%.

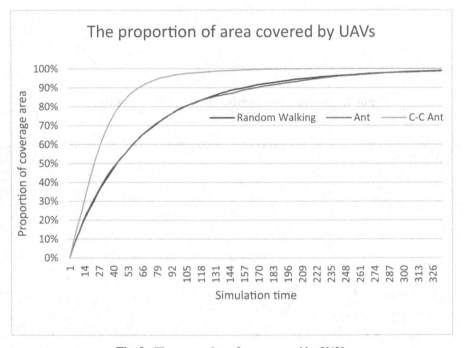

**Fig. 9.** The proportion of area covered by UAVs

Based on the above graph, it is evident that, compared to the random walk algorithm and the ant colony algorithm, the C-C Ant algorithm can cover the entire search area more quickly, even with the same initial parameter settings.

## 5   Simulation Results and Analysis

Based on the aforementioned simulation experiments, it can be concluded that, compared to the random walk algorithm and the ant colony algorithm under the same initial parameter settings, the effectiveness of the improved algorithm is apparent, whether in a tree density of 50% or 60%. This confirms the success of the competitive and cooperative algorithm studied in this article in terms of forest fire area detection by a drone swarm.

To validate the effectiveness of the algorithm, certain differentiations were made in the release of pheromones by the drone swarm for ease of comparison. In the establishment of global variables, attracting pheromones and repelling pheromones were respectively defined. Additionally, all the regions traversed by the drone swarm were colored according to the corresponding blocks, facilitating the analysis of the swarm's coverage. Killing the captured fire points by the drone swarm served the purpose of countability and prevented the further spread of the fire.

As mentioned earlier, the improved algorithm studied in this article is more effective compared to the random walk algorithm and the ant colony algorithm. It successfully achieves the objectives set forth in this article, as the drone swarm under the C-C Ant algorithm can detect forest fire areas in a shorter time.

## 6 Conclusion and Future Research Directions

Forest fires remain a hotspot topic globally, with a majority of research focusing on unmanned aerial systems for firefighting or wildfire studies [19]. In this article, the focus is on the detection of forest fire areas by a drone swarm. The proposed model can successfully achieve the objective in a simulated environment by releasing repelling pheromones during target search and attracting pheromones upon fire detection. The repelling and attracting pheromones have different volatilities, and random walk is employed during flight, using probability selection when determining direction, significantly enhancing the effectiveness of the drone swarm in detecting forest fire areas.

However, the present study did not consider the issue of energy consumption of the drone swarm and the firefighting aspects of fire suppression, which could be further explored in future research. Besides the detection of fire areas studied in this article, the evacuation of personnel during forest fires is also a topic that can be further investigated in future studies.

**Acknowledgments.** This work was supported by the National Natural Science Foundation of China (grant number 41972111) and the Second Tibetan Plateau Scientific Expedition and Research Program (STEP) (grant number 2019QZKK020604).

## References

1. McAlpine, R.S., Wotton, B.M.: The use of fractal dimension to improve wildland fire perimeter predictions. Can. J. For. Res. **23**(6), 1073–1077 (1993)
2. Martinez-de Dios, J.R., Arrue, B.C., Ollero, A., et al.: Computer vision techniques for forest fire perception. Image Vis. Comput. **26**(4), 550–562 (2008)
3. Singh, R., Gehlot, A., Akram, S.V., Thakur, A.K., Buddhi, D., Das, P.K.: Forest 4.0: digitalization of forest using the Internet of Things (IoT). J. King Saud Univ. Comput. Inf. Sc. **34**(8), 5587–5601 (2022). https://doi.org/10.1016/j.jksuci.2021.02.009
4. Akhloufi, M.A., Couturier, A., Castro, N.A.: Unmanned aerial vehicles for wildland fires: sensing, perception, cooperation and assistance. Drones **5**(1), 15 (2021)

5. Cummings, M.L., Clare, A., Hart, C.: The role of human-automation consensus in multiple unmanned vehicle scheduling. Hum. Factors **52**(1), 17–27 (2010)
6. Zhang, H., Xin, B., Dou, L., et al.: A review of cooperative path planning of an unmanned aerial vehicle group. Front. Inf. Technol. Elect. Eng. **21**(12), 1671–1694 (2020)
7. Li, J., Xiong, Y., She, J.: An improved ant colony optimization for path planning with multiple UAVs. In: 2021 IEEE International Conference on Mechatronics (ICM), pp. 1–5. IEEE (2021)
8. Zhen, Z., Xing, D., Gao, C.: Cooperative search-attack mission planning for multi-UAV based on intelligent self-organized algorithm. Aerosp. Sci. Technol. **76**, 402–411 (2018)
9. Shivgan, R., Dong, Z.: Energy-efficient drone coverage path planning using genetic algorithm. In: 2020 IEEE 21st International Conference on High Performance Switching and Routing (HPSR), pp. 1–6. IEEE (2020)
10. Dhall, A., Dhasade, A., Nalwade, A., et al.: A survey on systematic approaches in managing forest fires. Appl. Geogr. **121**, 102266 (2020)
11. Netlogo: Netlogo. Accessed September 2017 (2017). http://ccl.northwestern.edu/netlogo/
12. Tisue, S., Wilensky, U.: Center for connected learning and computer-based modeling northwestern University, Evanston, Illinois. In: NetLogo: A Simple Environment for Modeling Complexity, Citeseer (1999)
13. Dong, W., Zhou, M.C.: A supervised learning and control method to improve particle swarm optimization algorithms. IEEE Trans. Syst. Man Cybern. Syst. **47**(7), 1135–1148 (2016)
14. Bobkov, S.P., Astrakhantseva, I.A.: The use of multi-agent systems for modeling technological processes. J. Phys. Conf. Ser. **2001**(1), 012002 (2021)
15. Yue, W., Xi, Y., Guan, X.: A new searching approach using improved multi-ant colony scheme for multi-UAVs in unknown environments. IEEE Access **7**, 161094–161102 (2019)
16. Zhang, Y., Wang, S., Ji, G.: A rule-based model for bankruptcy prediction based on an improved genetic ant colony algorithm. Math. Probl. Eng. **2013**(pt.14), 1–10 (2013)
17. Zhang, Y., Wu, L.: Weights optimization of neural network via improved BCO approach. Progr. Electromagn. Res. **83**(5), 185–198 (2008)
18. Cekmez, U., Ozsiginan, M., Sahingoz, O.K.: Multi-UAV path planning with multi colony ant optimization. In: Abraham, A., Kr, P., Muhuri, A.K., Muda, N.G. (eds.) Intelligent Systems Design and Applications, pp. 407–417. Springer International Publishing, Cham (2018). https://doi.org/10.1007/978-3-319-76348-4_40
19. Kumar, A., Wu, S., Huang, Y., et al.: Mercury from wildfires: global emission inventories and sensitivity to 2000–2050 global change. Atmos. Environ. **173**, 6–15 (2018)

# Research on the Regulation and Network Security of Smart Tourism Public Information Service Platform

Kun-Shan Zhang[1]([✉]), Chiu-Mei Chen[1], and Wen-Yu Chang[2]

[1] Guangdong Province, University of Zhao Qing, Zhao Qing City 526061, China
963677272@qq.com
[2] Central Queensland University, Queensland, Australia

**Abstract.** The new economy represented by the digital economy has become a new engine of economic growth, and data as a core production factor has become a fundamental strategic resource. In order to play the role of digital governance, various regions have launched the construction of smart cities. By building smart cities, we can achieve a deep integration of informatization, industrialization, and urbanization, Based on the success model of D&M information system, this study takes tourists using Zhao Qing tourism service platform as the survey object, and discusses the satisfaction analysis that affects tourists' choice of tourism public service platform. Data were collected by questionnaire to verify the research model and hypothesis relationship. The results show that use attitude, self-efficacy, information quality, system quality and service quality have a positive and significant impact on tourists' satisfaction in choosing Zhao Qing tourism service platform. According to the research results, Based on the research results, propose suggestions for optimizing tourism service platforms and network security.

**Keywords:** Network Security · Wisdom of tourism · User characteristics · Quality characteristics · Information service platform

## 1 Introduction

The application of public information in government institutions has greatly increased the convenience of their work and significantly improved work efficiency. However, due to factors such as network security management system, management measures, and technology, their network security issues should also be given full attention. Network security management should avoid security risks to the maximum extent and improve the stability and reliability of network operation.

The networking of public information services can help alleviate the big city disease and improve the quality of urbanization. Smart cities achieve refined and dynamic management by focusing on the three major areas of public safety, public management, and public services, improving the effectiveness of urban management and improving the quality of people's livelihood in the city. The construction of smart cities mainly involves collecting basic information on management elements and continuous operation data in various fields of urban operation, digital sign data, urban IoT perception data, and City Information Modeling (CIM) data to achieve "one network unified management" of urban operation and "one network access" of government services.

Public information is based on urban information data, establishing a three-dimensional urban spatial model and an organic synthesis of urban information In terms of scope, it is an organic combination of GIS data for large scenes, BIM data for small scenes, and the Internet of Things Compared with traditional GIS-based digital cities, CIM refines the granularity of data to an electromechanical component and a door inside a single building in the city, upgrading the traditional static digital city into a perceptible, dynamic online, and virtual real interactive digital twin city, providing a data foundation for urban agile management and refinement.

In order to provide effective services and management for the operation and management of smart cities, it is inevitable to collect and process a large amount of personal information, such as tourists' use of information service platforms, social security payment information, personal employment and medical information, etc. The effectiveness of tourists' use of information service platforms will affect their evaluation and satisfaction with tourist destinations. Therefore, improving the information service system will greatly contribute to the development of the tourism industry and drive economic growth.

Leisure and entertainment activities represented by tourism have become an important part of people's social life. Tourism has gradually become a pillar industry of national economic development. Tourism and informatization are developing at a high speed. Smart tourism has realized the comprehensive improvement of tourism management, tourism service and tourism marketing.

Zhao Qing is a third tier city. Compared with other first tier cities, its scientific and technological information services are not very mature, which is more conducive to tap the development resources of tourism information services. Based on the development characteristics of first tier cities and foreign information service platforms and the research status of information services, this paper analyzes the problems and deficiencies of Zhao Qing public information service platform, and provides targeted suggestions for the future development of Zhao Qing smart tourism informatization.

## 2    Literature Discussion

This paper systematically summarizes the relevant literature of smart tourism public information service platform. Through sorting and comparison, it is found that domestic tourism public information service pays more attention to meeting the personalized needs of tourists, is committed to building the service purpose of "customer is God", improves the content of tourism public information, provides comprehensive services, and provides personalized information services under the rapid development of big data, so as to improve the level of tourism public information service, Increase the attraction of tourism destinations to improve customer satisfaction. Through the research, it is found that the information service platform plays a very important role in tourists' travel. Tourism activities have shown the trend of popularization, individual tourists and personalization. Tourists need timely, real and complete tourism information to ensure the safety, smoothness and comfort of the journey [7]. (Lu Shangyu, 2020) China's smart tourism information service platform has developed rapidly, but the scope of tourism information publicity is small, the information lacks integrity and authority, and the payment channel is unsafe, which does not play a real role in online transactions The tourism management information system is not perfect, and the tourism e-government lacks long-term strategic planning and implementation program.

(Dong Lei, 2019) the interaction of experience is weak, and the information is scattered and shared poorly. Although online travel websites are increasing, most of these information cannot be shared. The information content is large, the overall quantity is huge, the content is scattered, the value of the information provided is too low, and there is no guiding role, which is not conducive to tourists to obtain timely, accurate and diverse information [3].

Schmidt (2002) believes that a more convenient way of tourism is taking shape. The formation of this new way of tourism comes from the continuous changes in the collection, exchange and evaluation mechanism of tourism information with the emergence and development of new technologies such as Internet and location-based services [10]. In the development of smart tourism, Europe attaches great importance to the construction and application of infrastructure, and is committed to building an integrated market. As early as 2001, it began to implement the project of "creating user-friendly personalized mobile tourism services". At the level of public services, we will fully develop and apply telematics technology, establish a wireless data communication network covering Europe, and realize the functions of traffic management, navigation and electronic charging. Many regions also adopt QR code technology, which is connected with smart city information to provide tourism services for tourists. The intelligent tour guide software and tourism planning software developed by various companies are widely used to restore the original appearance of historic sites on the basis of global positioning system and identification software.

Ulrike gretzel believes that smart tourism is used to describe the increasing dependence of tourism destinations, their industries and tourists on emerging ICT forms.

## 2.1   Public Information Services

Tourism public information service is the core of the construction of tourism public service, which refers to timely transmitting the developed and processed information products to relevant tourists in a convenient way in order to meet the needs of tourists for tourism destination basic information, tourism product promotion information, tourism safety information, public environment and other related information services. Tourism public information service refers to meeting the service needs of tourists for tourism safety information, tourism product information and basic tourism information, and sending tourism information to tourism websites in a convenient form in time, so that tourists can have an understanding of the whole process of tourism. Zhang Xueting, Liu Fuying, Yan Wei and Lu Canhui (2021) believe that tourism public services refer to the general name of products and services with obvious public welfare and publicity provided by the government, the market or other social organizations to meet the common needs of tourists and stakeholders.

Therefore, from a practical point of view, strengthen the construction of Information infrastructure to improve the application of Big data infrastructure construction in tourism, promote the convergence and integration of data resources, build and improve the tourism information service platform system, deepen the application of Big data to create the construction of Big data innovation system, improve the timely and accurate guarantee of the network platform, so that tourists can travel in Leave Home Safe. The research results have great practical significance.

## 2.2    Smart Tourism and Public Information Services

Wen xuye (2019) believes that smart tourism is an important part of tourism information services, providing tourists with more tourism information and ensuring the timeliness and reliability of information transmission. Smart tourism also fully embodies the people-oriented idea. The provision of information services highlights the characteristics of humanization and ensures that smart tourism information services are provided to tourists at the first time, so as to effectively promote the development of China's tourism industry in the direction of intelligence and intelligence.

Ye Fei (2020) believes that smart tourism essentially pursues service, which further improves the new requirements for tourism public services. In the era of smart tourism, it is necessary to actively develop and build a public service system for tourists, and the tourism industry is facing great problems. The depth of integration between tourism and public services should be improved.

Chen Qi and Guo Dan (2020) believe that the essence of smart tourism is to improve tourism services, tourism experience and tourism management, rely on intelligent technology, including information technology and communication technology of tourism industry, make use of tourism resources and tourism value and competitiveness of tourism industry, and realize the modernization of tourism management and the expansion of industrial scale.

In the early 21st century, Canada's Gordon Phillips (2000) first proposed the definition of smart tourism, which refers to the planning, research, sales, and operation of travel goods through comprehensive and sustainable management methods. Schmidt (2002) pointed out that some more convenient and efficient travel methods are also emerging, and the emergence of these new travel methods is derived from the collection, exchange, and evaluation mechanisms of travel information. With the formation and development of new generation information technologies such as networks and location services, they are increasingly changing and forming. By utilizing remote information processing skills, building a wireless data communication network that covers the world can improve navigation, traffic management, and electronic toll collection. The application of QR code technology in multiple regions not only brings the common link of smart city information closer, but also creates tourism service demands for tourists. The widespread application of intelligent tour guide software and tourism planning software utilizes global positioning systems and recognition software to restore the original appearance of relics.

In practice, this study aims to verify the satisfaction of tourists in balancing the integration of smart tourism and public information service platforms with economic development.

## 2.3 User Satisfaction

In this study, the satisfaction of tourists with the use of Zhaoqing City's tourist public information content servers was used to evaluate whether tourists arc willing to choose and continue to use Zhaoqing City's tourist public information content servers. The key reasons for the success of tourist public information content servers are which characteristics tourists prefer to use, and whether they are willing to continue using tourist public information content servers. Therefore, this study refers to Xia Wen's (2019) evaluation scale for satisfaction in the study of users' willingness to continue using tourism virtual communities, and divides tourist satisfaction into two sub dimensions: willingness to choose and willingness to use.

## 2.4 Network Security is the Foundation of Digital Construction

Cybersecurity refers to a series of practices that protect enterprise institutions, their critical systems, and sensitive information from digital attacks by deploying personnel, policies, processes, and technologies. With the arrival of the digital age, data, networks, and digital technologies will be more widely applied in the digital economy, digital society, and the construction process of digital government. This means that the potential of data is further activated, network capabilities are further enhanced, and digital technologies, including artificial intelligence, big data, blockchain, cloud computing, etc., are further developed and innovated. Data, networks, and digital technologies are closely related to network security, and they play an important role in promoting and collaborating with network security. The development of data, networks, and digital technologies will put forward higher standards and requirements for network security capabilities, and the development and progress of network security capabilities will promote the wider and deeper application of data, networks, and digital technologies.

The 2035 Vision and Goals outline mentions the cultivation and growth of emerging digital industries such as artificial intelligence, big data, blockchain, cloud computing, and network security, the improvement of national cybersecurity laws and regulations, the strengthening of research and development of key technologies for cybersecurity, and the enhancement of the comprehensive competitiveness of the cybersecurity industry. The country places equal importance on cybersecurity with artificial intelligence, big data, blockchain, and cloud computing [16], indicating its important significance in the national new industrial strategy, It is an important component of digital strategy. With the support of a series of national policies and supporting measures, the development space of cybersecurity is infinitely vast, including technological research and development, product innovation, service upgrading, consulting business and other business formats of cybersecurity, which will flourish and usher in new opportunities and growth points. In the process of upgrading the government system and building a digital government, network security will be carried out together with the application and innovation of new government infrastructure, new architecture, and new service models [14]. Digital government and network security will promote and integrate each other. Without network security, digital government will be difficult to sustain. Establish matching risk management methods, processes, and standards for innovation and applications in new business formats, new services, cloud technology, big data, artificial intelligence, blockchain, and the Internet of Things. For industries that directly or directly reach consumers, such as government, finance, energy, electricity, internet, telecommunications, and service industries, continuously track legal and regulatory requirements, and further improve information technology risks, network security Data security and privacy protection capabilities to prevent various traditional and new network security risks [17].

# 3  Research Design and Methods

According to the research purpose and literature discussion, this study is based on the success model of D & M information system [2] (Delone wh 2003), takes the user characteristics and quality characteristics as the research dimension, and discusses the factors affecting the use intention through the degree of tourists' willingness to choose the tourism public information service platform.

## 3.1  Research Assumptions

According to the literature discussion, research framework and theoretical model of scholars, the research hypothesis is put forward:

Hypothesis 1: user characteristics have a positive impact on tourist use satisfaction.

H1–1: use attitude has a positive impact on tourists' use satisfaction.

H1–2: self efficacy has a positive impact on tourists' use satisfaction.

Hypothesis 2; Quality characteristics have a positive impact on tourist satisfaction.

H2–1: information quality has a positive impact on tourists' use satisfaction.

H2–2: system quality has a positive impact on tourists' use satisfaction.

H2–3: service quality has a positive impact on tourists' use satisfaction.

## 3.2  Research Object and Sampling Method

The respondents of this study are tourists who have used Zhao Qing tourism public information service platform, such as Zhao Qing tourism official website or Zhao Qing tourism official account. The convenient sampling method in the non probability sampling method is adopted. The electronic version of the questionnaire is made with the help of the questionnaire star and distributed on the wechat social network platform. A total of 230 questionnaires are distributed, including 33 respondents who have not used Zhao Qing tourism public information service platform, There are 197 respondents who have used Zhao Qing tourism public information service platform. After deducting 63 invalid questionnaires, there are 134 valid questionnaires.

# 4  Analysis of Research Results

## 4.1  Sample Basic Data

This paper collects the basic information of the respondents of a five-star hotel in Shekou, Shenzhen through the questionnaire. The basic information of the respondents in this questionnaire is as follows:

(1) Gender: the number of female consumers participating in the survey is significantly higher than that of male consumers. The proportion of men participating in the survey was 30.9% and women accounted for 69.1%.

(2) Age: most of the consumers who participated in the survey were under the age of 25, accounting for 72.4% of the total sample; People aged 25–40 account for 18.2% of the total number of samples, people aged 41–60 account for 7.8% of the total number of samples, and people over 61 account for only 0.6% of the total number of samples.

(3) Education level: in this survey, the proportion of respondents with education level at or below middle school is 9.9%, and high school is 13.8%. Undergraduate (including junior college) accounts for the largest proportion in the sample, accounting for 74%, master's accounting for 1.2% and doctor's accounting for 1.1%.

(4) Occupation: in this survey, the proportion of students in school is the largest, accounting for 68%, followed by those in the service industry, accounting for 11.6%, followed by those in the education industry and other occupations, accounting for 4.4%, those in the manufacturing industry, accounting for 9.9%, and those in government institutions, accounting for only 6.1% of the total sample.

(5) Monthly income: in this survey, the proportion of income less than 2000 yuan is the largest, accounting for 64.7% of the sample data, 13.2% of 2001–4999 yuan, 20.4% of 5000–9999 yuan and 1.7% of more than 10000 yuan.

## 4.2 Reliability and Validity Analysis

The overall reliability of this study is Cronbach's αThe coefficient is 0.971, and all substructures are consistent with the overall Cronbach's αThe values are greater than 0.8 and have a high reliability level, indicating that the reliability and stability of each item in the questionnaire have met the requirements of general academic research, indicating that the scale has a certain degree of credibility.

The KMO and Bartlett sphere test results of the overall factor test of this study are shown in the table below. KMO is 0.963, and Bartlett sphere test p value is significantly 0.001, indicating that the correlation between variables is high, which is suitable for extracting common factors.

In this study are between 0.594–0.73, less than 0.75 and Cronbach's α of individual dimensions, and most of the correlation coefficients are significant, indicating that the scale has a certain differential validity. After the above reliability After the evaluation of content validity, convergent validity and differential validity, on the whole, the internal and external quality of the questionnaire is good, and the sample data is suitable for the next statistical analysis.

## 4.3 Analysis of Verification Results

This study analyzes the impact between user characteristics, quality characteristics and use satisfaction. According to the above empirical analysis, the hypothesis test results are shown in Table1, and the exploration hypothesis path results are shown in Fig. 1.

216    K.-S. Zhang et al.

**Table 1.** Hypothesis test results

| Research hypothesis | | Standardized beta | T-value | p-value | Verification results |
|---|---|---|---|---|---|
| H1–1 | Use attitude has a positive impact on tourists' use satisfaction | 0.269 | 3.479 | 0.001 | Hypothesis established |
| H1–2 | Self-efficacy has a positive impact on tourists' use satisfaction | 0.574 | 7.419 | 0.000 | Hypothesis established |
| H2–1 | Information quality has a positive impact on tourists' use satisfaction | 0.179 | 2.769 | 0.006 | Hypothesis established |
| H2–2 | System quality has a positive impact on tourist use satisfaction | 0.274 | 3.951 | 0.000 | Hypothesis established |
| H2–3 | Service quality has a positive impact on tourists' use satisfaction | 0.497 | 8.215 | 0.000 | Hypothesis established |

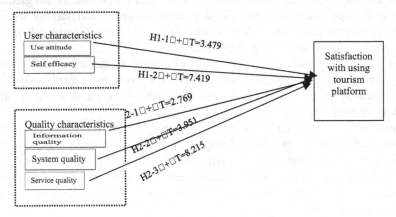

**Fig. 1.** Hypothetical path results

# 5 Conclusions and Suggestions

## 5.1 Effects of Use Attitude and Self-Efficacy on Use Intention

According to the test of linear regression analysis, it can be concluded that tourists' use attitude and self-efficacy have a significant positive impact on tourists' use satisfaction. The main factors affecting tourists' satisfaction with the use of Zhao Qing tourism public information service platform are that it can provide tourists with new functions of the platform, make tourists feel the happiness of the journey in the use process, enable tourists to quickly complete reservation services in the use process, improve tourists' decision-making efficiency, be easy to operate and meet tourists' needs.

## 5.2 Impact of Quality Characteristics on Use Intention

According to the linear regression analysis, it can be concluded that tourists' information quality, system quality and service quality have a significant positive impact on tourists' use satisfaction. A platform that can help tourists achieve their goals and show good tourism product information, ensure the privacy and safety of tourists, ensure that there is no jam in the use of tourists, understand the actual needs of tourists and provide personalized services is more popular with tourists.

## 5.3 Suggestions

According to the above empirical research results and conclusions, this study puts forward the following suggestions for Zhao Qing tourism public information service platform.

*(1) Tourism public information service platform should provide personalized and customized services.*

personalized route recommendations are provided for tourists' playing preferences, so that tourists can get the best travel path and the best playing experience in the shortest time. At present, the popular online services such as live broadcast marketing, VR live experience and providing intelligent navigation function ensure the freshness of tourists' platform use and gradually and efficiently meet the needs of tourists.

*(2) Grasp the market opportunities brought by the external environment for the tourism public information service platform.*

The external environment such as tourism policies, network technology, and economic development has prompted more tourists to choose tourism platforms to travel to Zhao Qing. We will deepen the promotion of mass tourism, actively develop smart tourism, improve tourism public facilities, improve the quality of tourism services, and enhance the quality of urban development.

**Fund Project:** The humanities and Social Sciences project of Zhao Qing University "applied economic management" Research on the development impact and regulation of smart tourism public information service platform "to serve the local economic and social development plan project.

218     K.-S. Zhang et al.

# References

1. Chu, L., Wu, J., Zhu, W., Zhang, W., Xiang, H.: Comparative analysis and innovative construction ideas of cycling tourism information sharing platform. J. Hubei Univ. Econ. (Human. Soc. Sci. Edn.) **18**(10), 48–52 (2021)
2. Delone, W.H., Mclean, E.R.: The Delone and Mclean model of information system success: a ten-year update. J. Manag. Inf. Syst. **19**, 9–30 (2003)
3. Dong, L.: Design and Implementation of Intelligent Scenic Spot Information Management Service System. Zhejiang University of technology, Hangzhou (2019)
4. Duan, J.: Construction of digital tourism information platform in Zhao Qing City. Elect. Technol. Softw. Eng. **08**, 143–144 (2019)
5. Huang, Y.: Research on the Impact Mechanism of Reservation Tourism App Users' Continuous Use Behavior. Fujian Agriculture and Forestry University, Fuzhou (2019)
6. Jinlong: Research on the construction of smart tourism application model system under big data. Ship Voc. Educ. **8**(05), 69–71 + 77(2020)
7. Lu, S.: Research on the Construction of Urban Smart Tourism Evaluation System from the Perspective of Public Service. South China University of technology, Guangzhou (2020)
8. National travel agency statistical survey report: Market management department of the Ministry of culture and tourism. Release Date: 2022/05/10, Index Number: 357a11–04–2022–0001 (2021)
9. New Trends in Cybersecurity: Interpreting Cybersecurity in the 14th Five Year Plan of the People's Republic of China through the Two Sessions, Outline of the 14th Five Year Plan for National Economic and Social Development and the Long-Range Goals for 2035, Deloitte China, 10 May 2022
10. Schmidt-Belz, B., Makelainen, M., Nick, A., Poslad, S.: Intelligent brokering of tourism services for mobile users. In: Wöber, K.W., Frew, A.J., Hitz, M. (eds.) Information and Communication Technologies in Tourism 2002, pp. 275–284. Springer Vienna, Vienna (2002). https://doi.org/10.1007/978-3-7091-6132-6_28
11. Wen, L.: Research on reservation intention and market delivery of online tourism consumers in China. Nanchang University (2012)
12. Yu, W., Wang, T., Ge, J.: Research on the design of smart tourism information service. Design **05**, 127–129 (2018)
13. Yu, Z.: A brief discussion on information network security in government institutions. Netw. Secur. Technol. Appl. **4**, 110–111 (2023)
14. Zhao, K.: Research on the Impact of Tourism App Characteristics on Users' Use Intention. Yantian University, Jilin (2019)
15. Zhang, D.: Analysis of personal information protection in smart city construction. Netw. Secur. Technol. Appl. **4**, 111–113 (2023)
16. 2021 National Travel Agency Statistical Survey Report: Market Management Department of the Ministry of Culture and Tourism, 10 May 2022. Index Number: 357A11-04-2022-0001
17. Author, F. Contribution title. In: 9th International Proceedings on Proceedings, pp. 1–2. Publisher, Location (2010)
18. LNCS Homepage. http://www.springer.com/lncs. Accessed 21 Nov 2016

# A Study Based on Logistic Regression Algorithm to Teaching Indicators

Yufang He[1], Kaiyue Shen[1], Haiyang Zhang[2], Wenjing Duan[1], Zhen Gong[1], Ruoyao Jia[1], and He Wang[1(✉)]

[1] School of Health Management, Changchun University of Chinese Medicine, Changchun 130117, China
781562619@qq.com
[2] School of Medical Information, Changchun University of Chinese Medicine, Changchun 130117, China

**Abstract.** Objective: This study aims to examine the factors that influence teachers' choice regarding the importance of instructional indicators. Methods: Based on a logistic regression algorithm to survey university faculty on the importance of teaching indicators, and the resultant data were processed by binary logistic regression analysis. Results: 731 questionnaires were collected in total. We constructed 3 dimensions of basic construction development, practice cultivation construction, and innovation education development, concluding that the difference of age, education, education work time, and title category are the factors that influence the choice of teachers' teaching goal emphasis level under each dimension, education constitutes an important influencing factor with the most profound level impact. Conclusion: Accelerate faculty gradient structure, strengthen teachers' teaching awareness and enhance teachers' sense of belonging in education career may become vital links to promote high quality construction and development of universities.

**Keywords:** Logistic regression algorithm · Teachers · Teaching indicators

## 1 Introduction

"Building a high-quality education system" is the policy direction and key requirement for the development of China in the 14th Five-Year Plan period [1]. Higher education construction is essential for high-quality human resources development, as the most important strategic resources for the development and core competitiveness of universities, university teachers are not only the object of internal governance services in universities, but also one of the main subjects to promote the innovative development of internal governance system and modernization of governance ability in universities [2], they bear the mission of nurturing talents for the country, and are also an important factor influencing the construction of China's "Double First Class" [3]. As the core force of university teaching, teachers' attitude towards teaching and their behavior directly affect the quality and effect of talent cultivation [4]. Therefore, this study takes teachers as the

H. Jin et al. (Eds.): IAIC 2023, CCIS 2059, pp. 219–227, 2024.
https://doi.org/10.1007/978-981-97-1280-9_17

research object, based on logistic regression algorithm, examining multidimensional factors that influence the importance teachers place on teaching metrics, with a view to providing corresponding development ideas for building a high-quality education and teaching system.

## 2  Objects and Methods

### 2.1  Questionnaire Content and Target

In order to explore the importance of each teaching indicators and the factors influencing it among teachers in general higher education institutions, this study was conducted with teachers in general higher education institutions as the respondents, this study was conducted with general higher education teachers as the respondents, combining the questionnaire design with the relevant indicators in the <Index system for the audit and evaluation of undergraduate education in general higher education institutions (Type II audit and evaluation)>. Survey content includes: (1) the basic information of the survey object; (2) evaluation status of the importance of each indicator of college construction, classifying the evaluation degree of teachers into very important, relatively important, generally important, not too important and unimportant, assigning values of 5, 4, 3, 2 and 1 in order; (3) expectations for the construction and development of universities. 731 questionnaires were distributed and collected in total.

### 2.2  Questionnaire Content and Target

Reduced factor analysis was performed on the importance rating scale using SPSS 27.0 software with a view to classifying multiple dimensions. Based on the logistic regression algorithm test [5], and construct a binary logistic regression model (1), $Y$ in the above equation takes the value of 0 or 1, $X$ is the influence factor in this study, the test is $\alpha = 0.05$, and the difference is considered statistically significant when $P < 0.05$. Taking the dominance ratio as a measure 2), in logistic regression, when OR $> 1$, $X$ will likely promote the emergence of $Y$ and thus the influence relationship [6].

$$Y_i = \beta_0 + \beta_1 X_{1i} + \beta_2 X_{2i} + \mu_i \ (i = 1, 2 \cdots n) \tag{1}$$

$$OR = exp(beta = exp[ln(Odds_1) - ln(Odds_0)] \tag{2}$$

## 3  Results

### 3.1  Basic Demographic Information

731 teachers of general colleges and universities were surveyed in total. The sample covers a total of 28 universities, Changchun University of Traditional Chinese Medicine, Northeastern Electric Power University, Changchun University of Science and Technology ranked in the top 3, accounting for 15.3%, 12.6% and 11.1% respectively. Table 1 shows the percentage of each sample characteristic.

**Table 1.** Description of data sample characteristics

| Sample characteristics | | Frequency (times) | Proportion (%) |
|---|---|---|---|
| Gender | Male | 326 | 44.6 |
| | Female | 405 | 55.4 |
| Age | Under 30 | 96 | 13.1 |
| | 31-35 | 141 | 19.3 |
| | 36-40 | 200 | 27.4 |
| | 41-45 | 149 | 20.4 |
| | 46-50 | 61 | 8.3 |
| | Over 50 | 84 | 11.5 |
| Education | Ph. | 152 | 20.8 |
| | Master | 268 | 36.7 |
| | Bachelor | 242 | 33.1 |
| | Specialist and below | 69 | 9.4 |
| Education Work Hours | 5 years or less | 241 | 33 |
| | 6-10 years | 241 | 33 |
| | 11-15 years | 75 | 10.3 |
| | 16-20 years | 57 | 7.8 |
| | Over 20 years | 117 | 16 |
| Title | Professor/Researcher, etc. (senior teaching title) | 87 | 11.9 |
| | Associate Professor/Associate Researcher, etc. (senior title) | 184 | 25.2 |
| | Lecturer/Assistant Researcher etc. (Intermediate level) | 252 | 34.5 |
| | Teaching Assistant (junior level) | 127 | 17.4 |
| | None | 81 | 11.1 |
| Position | Instructional | 281 | 38.4 |
| | Teaching and research | 271 | 37.1 |
| | Scientific | 92 | 12.6 |
| | Tutors | 87 | 11.9 |

## 3.2 Dimensional Construction

Reduced factor analysis, KMO and Bartlett's sphericity tests were conducted on the content of the scale for evaluating the importance of teaching indicators in universities within the questionnaire, the KMO value of 0.961 tends to be close to 1, with a more significant level of significance. Table 2 shows the content of the 3 dimensions after the comparison of weight mean values. Therefore, this study will explore the various factors that influence the selection of teachers' teaching indicators' importance in three dimensions: capital development, practice development construction, and innovation education development.

## 3.3 Data Assignment

Calculate the mean value of the survey respondents' importance scores under the three dimensions. Dividing the respondents into two categories, selecting very important to be recorded as 1 and not very important to be recorded as 0, and treating them as dependent variables, the influencing factors of different dimensions were used as independent variables. Table 3 shows the descriptive statistics and assignments.

**Table 2.** Dimensional breakdown

| Dimensionality | Factor | Contents |
|---|---|---|
| | Cultivating Excellence | Cultivation of talents through scientific and educational synergy and integration of |
| | Resource building | Construction of applied teaching materials, teaching facilities and conditions, teaching resources |
| Capital development | Professional Development | Cultivation of innovative, applied and complex talents; construction of a demand-based management system |
| | Teacher Development | Building thinking and party affairs; building a team of classified teachers; building teaching capacity in practice, research and other areas |
| | Undergraduate status | Educational climate, training of students, construction of teaching mechanisms and annual assessment |
| | Training programmes | Development and implementation of comprehensive student development |
| Practice development | Civic Education | Thinking courses and the building of moral character |
| | Hands-on teaching | Construction of the training base and the construction of the thesis (design) |
| | Teacher Quality and Teaching | The construction of teacher competence, ethics and teaching commitment |
| Innovative Education | Innovation and Entrepreneurship Education | Student motivation for innovation and the construction of an innovation platform |
| | Classroom teaching | Construction of information resources and translation of results |

**Table 3.** Descriptive statistics and assignment of independent variables

| Variable name | Meaning of the values |
|---|---|
| Age | 1: under 30; 2: 31-35 ; 3: 36-40; 4: 41-45 ; 5: 46-50 ; 6: over 50 |
| Education | 1: Ph.; 2: Master's ; 3: Bachelor and below |
| Education work hours | 1: Over 20 years; 2: 16-20 years; 3: 11-15 years; 4: 6-10 years; 5: 5 years or less |
| Title | 1: Professor/Researcher, etc. (senior teaching title); 2: Associate Professor/Associate Researcher, etc. (senior title); 3:Lecturer/Associate Researcher, etc. (intermediate title); 4: Assistant Professor (junior title); 5: None |
| Position | 1: Instructional; 2: Teaching and research; 3: Scientific; 4: Tutor |

## 3.4 Single Factor Analysis Results

The results of univariate analysis showed, characteristics such as age, education, duration of educational work, title and type of position showed significant and statistically significant differences ($p < 0.05$) under three dimensions: development of basic construction, construction of practice training, development of innovative education (Table 4).

## 3.5 Binary Logistics Regression Analysis

Binary logistic regression was used to develop judgments on each influencing factor. Hosmer-Lemeshow test was conducted (Table 5). The results of both the capital developments and practice development building passed the Hosmer-Lemeshow test ($p > 0.05$), indicating that the overall fit of the regression model was good and the independent variables could effectively explain (and predict) the dependent variables.

**Table 4.** Single factor analysis results

| Feature | | Capital Development (n) No:447 | Yes:254 | $x^2$ | P | Practice Development (n) No:456 | Yes:275 | $x^2$ | P | Innovative Education (n) No:433 | Yes:276 | $x^2$ | P |
|---|---|---|---|---|---|---|---|---|---|---|---|---|---|
| Age | Under 30 | 50 | 46 | 19.456 | 0.002 | 45 | 51 | 29.73 | <0.001 | 49 | 47 | 28.47 | <0.001 |
| | 31-35 | 87 | 54 | | | 81 | 60 | | | 86 | 55 | | |
| | 36-40 | 146 | 54 | | | 147 | 53 | | | 146 | 54 | | |
| | 41-45 | 107 | 42 | | | 102 | 47 | | | 100 | 49 | | |
| | 46-50 | 40 | 21 | | | 39 | 22 | | | 37 | 24 | | |
| | Over 50 | 47 | 37 | | | 42 | 42 | | | 37 | 47 | | |
| Education | Ph. | 62 | 90 | 104.931 | <0.001 | 68 | 84 | 95.35 | <0.001 | 57 | 95 | 113.145 | <0.001 |
| | Master | 150 | 118 | | | 131 | 137 | | | 138 | 130 | | |
| | Bachelor or below | 265 | 46 | | | 257 | 54 | | | 260 | 51 | | |
| Education Work Hours | Over 20 years | 61 | 56 | 37.576 | <0.001 | 54 | 63 | 51.98 | <0.001 | 48 | 69 | 61.751 | <0.001 |
| | 16-20 years | 36 | 21 | | | 31 | 26 | | | 33 | 24 | | |
| | 11-15 years | 38 | 37 | | | 34 | 41 | | | 33 | 42 | | |
| | 6-10 years | 191 | 50 | | | 190 | 51 | | | 190 | 51 | | |
| | 5 years or less | 151 | 90 | | | 147 | 94 | | | 151 | 90 | | |
| Title | Professor / Researcher | 33 | 54 | 45.625 | <0.001 | 39 | 48 | 26.61 | <0.001 | 28 | 59 | 58.805 | <0.001 |
| | Associate Professor / Associate Researcher | 119 | 65 | | | 106 | 78 | | | 107 | 77 | | |
| | Lecturer / Research Associate | 162 | 90 | | | 157 | 95 | | | 157 | 95 | | |
| | Assistant Professor | 103 | 24 | | | 98 | 29 | | | 103 | 24 | | |
| | None | 60 | 21 | | | 56 | 25 | | | 60 | 21 | | |
| Position | Instructional | 174 | 107 | 9.663 | 0.022 | 155 | 126 | 11.01 | 0.012 | 160 | 121 | 10.835 | 0.013 |
| | Teaching and Research | 169 | 102 | | | 177 | 94 | | | 166 | 105 | | |
| | Scientific | 69 | 23 | | | 65 | 27 | | | 66 | 26 | | |
| | Tutor | 65 | 22 | | | 59 | 28 | | | 63 | 24 | | |

**Table 5.** Hosmer-Lemeshow test

| | Card Parties | Freedom | Significance |
|---|---|---|---|
| Capital developments | 7.234 | 8 | 0.512 |
| Practice development building | 12.695 | 8 | 0.123 |
| Innovative Educational Development | 18.562 | 8 | 0.017 |

**Binary Logistic Regression Analysis of Capital Development Dimensions.** Table 6 shows the results of the binary logistic regression analysis at the capital development level. Age, education, education work hours, and title were influential factors in the choice of importance as teachers ($p < 0.01$). There was a positive multiplicative effect of the importance of each instructional goal choice for teachers under 45 years of age on teachers over 50 years of age. Teachers with 11 years or more in education had a positive multiplicative effect on the importance of each instructional goal choice for teachers with 5 years of experience, with teachers with 20 years or more teaching experience having a 2.318 times higher level of importance tendency than teachers with 5 years of teaching experience. The degree of education affects the degree of important choice most significantly, with the level of importance propensity for teachers with doctoral degrees being 6.148 multiple of that for teachers with bachelor's degrees or less. Senior title faculty tendency level of importance is 2.676 times higher than that of non-title faculty.

**Table 6.** Binary logistic regression results for the capital development dimension

| Features | B | S.E. | Wald | P | OR | 95% CI Low | Up |
|---|---|---|---|---|---|---|---|
| Age (reference category: over 50) | | | 15.547 | 0.008 | | | |
| Under 30 | 1.398 | 0.543 | 6.624 | 0.01 | 4.046 | 1.396 | 11.732 |
| 31-35 | 1.023 | 0.514 | 3.966 | 0.046 | 2.782 | 1.016 | 7.614 |
| 36-40 | 0.615 | 0.5 | 1.51 | 0.219 | 1.85 | 0.694 | 4.932 |
| 41-45 | 0.103 | 0.47 | 0.048 | 0.826 | 1.109 | 0.441 | 2.785 |
| 46-50 | -0.05 | 0.426 | 0.016 | 0.901 | 0.948 | 0.411 | 2.187 |
| Education (reference category: bachelor or below) | | | 58.339 | <0.001 | | | |
| Ph. | 1.816 | 0.262 | 48.072 | <0.001 | 6.148 | 3.679 | 10.272 |
| Master | 1.395 | 0.218 | 40.968 | <0.001 | 4.035 | 2.632 | 6.185 |
| Education work hours(reference category: 5 years or less) | | | 23.66 | <0.001 | | | |
| Over 20 years | 0.84 | 0.47 | 3.204 | 0.073 | 2.318 | 0.923 | 5.818 |
| 16-20 years | 0.435 | 0.39 | 1.243 | 0.265 | 1.545 | 0.719 | 3.32 |
| 11-15 years | 0.765 | 0.325 | 5.548 | 0.018 | 2.149 | 1.137 | 4.06 |
| 6-10 years | -0.62 | 0.246 | 6.428 | 0.011 | 0.536 | 0.331 | 0.868 |
| Title (reference category: none) | | | 17.496 | 0.002 | | | |
| Professor/Researcher, etc. (senior teaching title) | 0.984 | 0.424 | 5.382 | 0.02 | 2.676 | 1.165 | 6.149 |
| Associate Professor/Associate Researcher, etc. (senior title) | -0.01 | 0.376 | 0.001 | 0.982 | 0.991 | 0.474 | 2.074 |
| Lecturer/Assistant Researcher etc. (Intermediate level) | 0.028 | 0.35 | 0.006 | 0.937 | 1.028 | 0.518 | 2.04 |
| Teaching Assistant (junior level) | -0.6 | 0.383 | 2.461 | 0.117 | 0.548 | 0.259 | 1.162 |
| Position (reference category: tutor) | | | 2.809 | 0.422 | | | |
| Instructional | -0.08 | 0.314 | 0.058 | 0.81 | 0.927 | 0.501 | 1.716 |
| Teaching and Research | -0.3 | 0.323 | 0.829 | 0.362 | 0.745 | 0.395 | 1.404 |
| Scientific | -0.5 | 0.393 | 1.623 | 0.203 | 0.606 | 0.281 | 1.309 |
| Constants | -2.17 | 0.616 | 12.355 | <0.001 | 0.115 | | |

**Binary Logistic Regression Analysis of Practice Training Construction Dimensions.** Table 7 demonstrates the results of the binary logistic regression analysis at the level of practice development construction. Age, education, and working time in education were influential factors in the choice of importance as teachers ($p < 0.01$). Teachers under 45 years of age had a positive multiplicative effect on the importance of each instructional goal choice for all teachers over 50 years of age. Teachers who have been working in education for 20 years or more are 3.948 multiple more important in choosing each instructional goal than those who have been working for 5 years. The degree of education affects the degree of important choice most significantly, with the level of importance propensity of teachers with doctoral degrees being 5.312 multiple higher than that of teachers with bachelor's degrees or less. Forest maps were drawn compared to the capital development dimension (Fig. 1).

**Table 7.** Binary logistic regression results for the practice development construction dimension

| Features | B | S.E. | Wald | P | OR | 95% CI Low | 95% CI Up |
|---|---|---|---|---|---|---|---|
| Age (reference category: over 50) | | | 27.11 | <0.001 | | | |
| Under 30 | 1.625 | 0.538 | 9.136 | 0.003 | 5.081 | 1.771 | 14.578 |
| 31-35 | 1.387 | 0.508 | 7.445 | 0.006 | 4.004 | 1.478 | 10.845 |
| 36-40 | 0.575 | 0.496 | 1.345 | 0.246 | 1.777 | 0.672 | 4.698 |
| 41-45 | 0.119 | 0.463 | 0.066 | 0.798 | 1.126 | 0.455 | 2.788 |
| 46-50 | -0.242 | 0.417 | 0.337 | 0.561 | 0.785 | 0.347 | 1.777 |
| Education (reference category: bachelor or below) | | | 59.41 | <0.001 | | | |
| Ph. | 1.67 | 0.263 | 40.317 | <0.001 | 5.312 | 3.172 | 8.894 |
| Master | 1.503 | 0.214 | 49.353 | <0.001 | 4.495 | 2.955 | 6.836 |
| Education work hours(reference category: 5 years or less) | | | 38.119 | <0.001 | | | |
| Over 20 years | 1.373 | 0.467 | 8.639 | 0.003 | 3.948 | 1.58 | 9.865 |
| 16-20 years | 0.984 | 0.385 | 6.538 | 0.011 | 2.675 | 1.258 | 5.688 |
| 11-15 years | 1.098 | 0.328 | 11.227 | 0.001 | 2.997 | 1.577 | 5.696 |
| 6-10 years | -0.61 | 0.244 | 6.239 | 0.012 | 0.543 | 0.336 | 0.877 |
| Title (reference category: none) | | | 7.357 | 0.118 | | | |
| Professor/Researcher, etc. (senior teaching title) | 0.483 | 0.42 | 1.319 | 0.251 | 1.62 | 0.711 | 3.693 |
| Associate Professor/Associate Researcher, etc. (senior title) | 0.056 | 0.368 | 0.023 | 0.879 | 1.057 | 0.514 | 2.174 |
| Lecturer/Assistant Researcher etc. (Intermediate level) | -0.07 | 0.344 | 0.042 | 0.838 | 0.932 | 0.475 | 1.829 |
| Teaching Assistant (junior level) | -0.565 | 0.375 | 2.269 | 0.132 | 0.568 | 0.273 | 1.185 |
| Position (reference category: tutor) | | | 12.048 | 0.007 | | | |
| Instructional | -0.196 | 0.304 | 0.415 | 0.519 | 0.822 | 0.453 | 1.492 |
| Teaching and Research | -0.824 | 0.319 | 6.681 | 0.01 | 0.438 | 0.235 | 0.819 |
| Scientific | -0.661 | 0.379 | 3.032 | 0.082 | 0.516 | 0.246 | 1.087 |
| Constants | -1.962 | 0.603 | 10.584 | 0.001 | 0.141 | | |

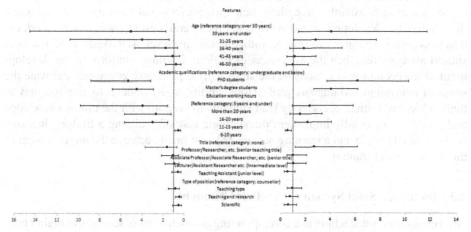

**Fig. 1.** Forest map

# 4 Discussion

## 4.1 The Construction of Faculty Gradient Structure Still Requires Optimization

Results of the survey showed that age was the main influencing factor constituting the importance of teachers' choice of teaching indicators, and the importance of teachers' choice of each teaching objective was positively multiplied for teachers under the age of 45 to those who were relatively older (over 50). Young and middle-aged teachers are more active and sensitive to new things, and they are able to perceive social trends and combine them with teaching indicators in a timely and organic manner. It is necessary to gradually optimize the construction of the gradient structure among teachers, to form an educational mechanism that promotes each other among teachers of all ages, and to strengthen the teachers' sense of belonging in the teaching system. At the same time, teachers should clarify their responsibilities and missions to actively participate in the overall educational program. In order to promote high quality development of the whole goal in education and teaching.

## 4.2 Create High-Level Teacher Qualification Team

Results of the survey show that teachers with higher levels of education are better able to grasp relatively accurately the balanced development and construction of various teaching goals. Teachers' own qualification levels should be promoted to accelerate the building of an inherently solid teaching force. With the dual role of educator and researcher [7], it is crucial for university teachers to properly lead young learners in the new era to further their studies. As the main body of the higher education career system, teachers should constantly strengthen their own professional level and fully integrate their individual development into the ranks of the overall development of school education to avoid stagnation or even discontinuity in teaching level. At the same time, teachers should always strengthen the awareness of lifelong learning, conform to the development of the economic era, focus on the study of real problems, constantly cultivate the sense of innovation and improve professional ability, assist students to give full play to their subjective initiative, correctly lead students to keep up with the times, knowledge and practice, profoundly implement the strategic task of "building a lifelong learning society for all people and a learning nation" [8], and finally achieve the improvement of the level of talent training.

## 4.3 Building a Solid System of Teachers' Thoughts

Awareness of moral teaching and corresponding competencies of teachers in higher education is the key orientation to carry out educational career [9]. Thought is the precursor of action. As a group that trains the country's young talents, university teachers should always have the cutting-edge thinking of keeping up with the times, that is, the overall awareness of being based on the current situation and looking at future trends, a state of constant attention and continuous thinking [10]. During the teaching process, teachers should keep their enthusiasm and vitality for education, standardize their teaching methods and philosophy of governance, establish positive teaching concepts and teaching attitudes, and always maintain learning interactivity with students. To achieve the promotion and integration of the whole teaching indicators under the multi-dimension of basic construction development, practice cultivation construction and innovation education development. In turn, the construction of a high-quality teaching pattern model is realized.

# References

1. Rong, H., Xiaochang, D.: A study on measuring the level of high quality development of higher education in China. J. East China Normal Univ. (Educ. Sci. Ed.) **40**(7), 100–113 (2022)
2. Xiaofei, C.: Realistic dilemmas of university teachers' participation in university governance in the new era and strategies to cope with them. J. Hengyang Normal Coll. **44**(1), 46–51 (2023)
3. Guowei, X., Tingting, L.: Research on the construction of university teachers in the new era. J. Zhengzhou Light Ind. Univ. (Soc. Sci. Ed.) **23**(5), 103–108 (2022)
4. Xinning, W., Huiyan, L., Qian, Z.: Research on the problems and countermeasures of teaching incentive mechanism for teachers in higher education. Contemp. Educ. Theory Pract. **15**(02), 142–148 (2023)
5. Zou, X., Hu, Y., Tian, Z., Shen, K.: Logistic regression model optimization and case analysis. In: 2019 IEEE 7th International Conference on Computer Science and Network Technology (ICCSNT), Dalian, China, pp. 135–139(2019)
6. Grant, R.L.: Converting an odds ratio to a range of plausible relative risks for better communication of research findings. Bmj **348** (2014)
7. Qihui, X., Surui, L., Yushan, H.: How to enhance university teachers' willingness to popularize science: environment, resources, costs and satisfaction. Sci. Popularization Res. **18**(1), 60–69+108 (2023)
8. Lingli, Z., Yichuan, C., Changsheng, Y., Jiacheng, L.: Optimizing the construction of university teachers with silver-age teachers: mission of the times, key tasks and development strategies. China Higher Educ. Res. **2**, 24–30 (2023)
9. Jingjing, S.: An analysis of the path to improve the capacity of university teachers in the context of Curriculum Civics. Natl. Gen. Lang. Teach. Res. **157**(10), 47–49 (2022)
10. Meijing, L., Shiyin, Z.: Three dimensions of frontier thinking cultivation for teachers of college Civics inthe context of intelligent education. Ideological and Political Science

# A Spellcheck Technique for Tibetan Syllables Based on Grammatical Vector Multiplication

AnJian-CaiRang[1,2](✉) and LaMaoCuo[1]

[1] The Computer College, Qinghai Minzu University, Qinghai 810007, China
ajcr@163.com
[2] College of Intelligence and Computing, Tianjin University, Tianjin 300350, China

**Abstract.** Automated spellcheck is a highly challenging research topic in the field of natural language processing, with broad application potential for corpus construction, text editing, speech captioning, and text recognition. Tibetan, a type of Pinyin script, is composed of 1–7 basic elements spliced horizontally and vertically. As a result, the frequency of spelling errors in Tibetan texts is very high. This article analyzes the rules of character formation in Tibetan grammar and converts complex Tibetan syllables into their corresponding structures. Prior studies have shown the positions of components in Tibetan syllables are consistent for characters with the same structure. In this study, we propose a structure-based component recognition model for Tibetan syllables (SBCRATS). In addition, four new spelling rules are developed by investigating 28 existing rules. Four Tibetan grammatical specifications are then vectorized and corresponding functions are introduced. These vectors provide a Tibetan script spell check methodology based on a vector multiplication model (MOSCTCBOGVM), which is simple and easy to implement. Algorithm performance was assessed using five real-world corpora and one corpus generated using Tibetan grammar. Five undergraduates majoring in Tibetan language marked the syllables identified by the model as being either correct or incorrect. A series of validation experiments showed the average spell check accuracy to be 99.86% at a rate of 51,050 words per second. This accuracy is comparable to the highest values reported in previous studies, but with unprecedented calculation speed.

**Keywords:** Tibetan syllable component · grammatical vector · vector multiplication · spell check

## 1 Introduction

Automated spell check is a challenging research topic with broad application potential in the fields of natural language processing, corpus construction, text editing, and speech recognition. Tibetan natural language processing technology has developed rapidly in recent years, as the language is prone to a high frequency of spelling errors. The introduction of statistical and deep learning technology has also facilitated the use of big data for improving information accuracy [1]. Tibetan letters are spliced into characters composed of syllables, some of which are not Tibetan scripts. However, Tibetan syllables can contain characters and may be composed of multiple scripts. As such, texts use syllables to represent characters and words in forming collocation sentences. As a result, Tibetan spell check requires character, syllable, word, and grammar level verification steps. Syllable-level checks identify spelling mistakes in letters and determine whether syllables conform to word formation principles in the grammar, regardless of the context. This study analyzes the spelling grammar of Tibetan syllables and proposes a spell check methodology based on vector multiplication.

English text proofreading has achieved impressive results [2, 3], some of which have been commercialized and widely used in writing, education, and publishing. English spellcheck is based on dictionary, rule-based, statistical, and automated language knowledge [3, 4], which requires a training corpus and significant resources. Domestic research in text proofreading began in the early 1980s and has accelerated in recent years [5–10]. The study of spellcheck technology specifically for Tibetan scripts originated in the 1990s and primarily involves two techniques: dictionary matching and spelling rules for Tibetan syllable grammar. For example, Guan et al. [11] used a dictionary to identify syllable errors but did not provide specific test data. Similar studies have investigated the characteristics of modern Tibetan text, through syllable preprocessing, word list matching, confusion set matching, bi-neighborship, and minimum editing distance [12]. Duojie et al. [13] constructed a large-scale ternary model for each component in sparse Tibetan syllables. Letter combinations have also been processed in multiple sections, to establish a library for spell checking. Experiments applying sectional libraries to modern Tibetan syllables have achieved error detection rates of 100% [14]. However, applying this model to our experimental corpus produced an accuracy of only 84%. Zhu et al. divided seven syllable components into three parts: the prefix, vowel or phonetic node, and suffix, establishing a corresponding table for each element [15]. Spellcheck was then performed by looking up the table, without considering the transliteration from Sanskrit, achieving a spelling accuracy of 99.8%. Cai et al. established a Tibetan script vector model based on rule constraints and numbers representing letters, by analyzing the rules of constituent characters in Tibetan grammar [16]. Spellcheck experiments produced an average error detection rate of 99.995% at 1,060 syllables per second. However, this approach replaces Tibetan letters with numbers and checks syllables using a Tibetan grammar, without utilizing the digital advantage of the vector itself.

Tibetan spellcheck requires identifying components that comprise individual syllables. Previous studies have developed modern component recognition algorithms by studying Tibetan script structures, writing styles, and grammatical rules, achieving accuracies of over 99% [17]. However, the underlying premise o2f most existing techniques is that Tibetan syllables must be correct, which is unnecessary for proofreading. In addition, some models are not easy to operate or implement, since they require knowing which letter is the root character (often unknown component lengths greater than 3 or involve as few as 809 experimental Tibetan syllables. As such, many of the previously reported results are insufficient for the entire Tibetan script. Prior methodologies have included Tibetan script shapes, character collocation rules, and Tibetan script length characteristics, combined with Tibetan grammar rules used to construct recognition algorithms and determine the position of other characters in a syllable [18]. However, the rules of Tibetan script collocation are highly complicated and strict, as algorithms must be closely integrated with grammatical requirements. This article proposes a grammar-free structure-based recognition algorithm for Tibetan syllable components, to address the limitations and deficiencies of existing techniques. The algorithm is also significantly faster than some existing techniques, checking 34,057 syllables per second. The primary contributions of this work are as follows:

1) A dataset is established for Tibetan proofreading.
2) A structure-based syllable component recognition algorithm is developed, which does not rely on Tibetan grammar.
3) Four new spelling specifications are proposed for Tibetan syllables and their grammatical specifications are vectorized.
4) A spellcheck technique is introduced for Tibetan syllables, based on a grammar vector multiplication model.

The remainder of this paper is structured as follows. The first section analyzes the current status of spell check research. The second section proposes a structure-based recognition model for Tibetan syllable components based on a grammatical analysis. The third section introduces the proposed syllable spellcheck technique utilizing vector multiplication, which was verified experimentally. The fourth section provides conclusions and prospects for future research.

## 2  Vectorization of Tibetan Syllable Grammar

Tibetan syllables are composed of seven components, including root, prefix, superscript, subscript, vowel, suffix, and farther-suffix, each with its own restriction rules for script formation [19]. Obtaining Tibetan syllable components using a structure based recognition algorithm for Tibetan syllable components [20]. There are also mutual restriction rules between each component, including 28 grammatical rules for the prefix, superscript, root, and subscript, as shown in Table 1.

Vowels must satisfy the single constraint rule shown in Table 2 and five constraint rules exist between the suffix and farther-suffix, as seen in Table 3.

Our research has suggested that constraint rules between the prefix, subscript, root, vowel, suffix, and farther-suffix can be summarized into four types of specifications: the constraint specification (R1) between the prefix and root; the constraint specification (R2) between the superscript, root, and subscript; the vowel specification (R3); and the constraint specification (R4) between the suffix and farther-suffix. These four specifications are satisfied simultaneously, as shown in Fig. 1.

**Table 1.** Constraint specifications for the prefix, superscript, root, and subscript.

| No. | Combination rules | Optional root |
|-----|-------------------|---------------|
| 0 | Prefix ε + Superscript ε + Root + Subscript ε | ཀ, ཁ, ག, ང, ཅ, ཆ, ཇ, ཉ, ཏ, ཐ, ད, ན, པ, ཕ, བ, མ, ཙ, ཚ, ཛ, ཝ, ཞ, ཟ, འ, ཡ, ར, ལ, ཤ, ས, ཧ, ཨ |
| 1 | Prefix ε + Superscript ε + Root + Superscript ཱ | ཀ, ཁ, ག, པ, ཕ, བ, མ |
| 2 | Prefix ε + Superscript ε + Root + Superscript ི | ཀ, ཁ, ག, ཏ, ད, བ, ན, པ, བ, མ, ཏ |
| 3 | Prefix ε + Superscript ε + Root + Superscript ཱ | ཀ, ག, ན, ཟ, ར, ས |
| 4 | Prefix ε + Superscript ε + Root + Superscript ུ | ཀ, ཁ, ག, ཉ, ད, ཚ, འ, ཟ, ར, ཤ, ཧ, ཏ |
| 5 | Prefix ε + Superscript ར + Root + Superscript ε | རྐ, རྒ, རྔ, རྗ, རྙ, རྟ, རྡ, རྣ, རྦ, རྨ, རྩ, རྫ |
| 6 | Prefix ε + Superscript ལ + Root + Superscript ε | ལྐ, ལྒ, ལྔ, ལྗ, ལྕ, ལྟ, ལྡ, ལྤ, ལྦ |
| 7 | Prefix ε + Superscript ས + Root + Superscript ε | སྐ, སྒ, སྔ, སྙ, སྟ, སྡ, སྣ, སྤ, སྦ, སྨ, སྩ |
| 8 | Prefix ε + Superscript ར + Root + Superscript ཱ | རྐ, རྒ, རྨ |
| 9 | Prefix ε + Superscript ར + Root + Superscript ུ | རྨ |
| 10 | Prefix ε + Superscript ས + Root + Superscript ཱ | སྐ, སྒ, སྤ, སྦ, སྨ |
| 11 | Prefix ε + Superscript ས + Root + Superscript ི | སྐ, སྒ, སྤ, སྦ, སྨ, སྨ |
| 12 | Prefixཀ + Superscript ε + Root + Superscript ε | ཅ, ཉ, ཏ, ད, ན, ཙ, འ, ཟ, ཡ, ཤ, ས |
| 13 | Prefixག + Superscript ε + Root + Superscript ε | ཀ, ག, ང, པ, བ, མ |
| 14 | Prefixད + Superscript ε + Root + Superscript ε | ཀ, ག, ཅ, ཏ, ད, ཙ, འ, ཟ, ཤ, ས |
| 15 | Prefixས + Superscript ε + Root + Superscript ε | ཁ, ག, ང, ཆ, ཇ, ཉ, ཟ, ཏ, ན, ཙ, ཛ |
| 16 | Prefixའ + Superscript ε + Root + Superscript ε | ཁ, ག, ཆ, ཇ, ཟ, ཏ, པ, བ, ཙ, ཛ |
| 17 | Prefixག + Superscript ε + Root + Superscript ཱ | ཀ, ག, པ, བ, མ |
| 18 | Prefixག + Superscript ε + Root + Superscript ི | ཀ, ག, པ, བ |
| 19 | Prefixའ + Superscript ལ + Root + Superscript ε | ལྗ, ལྕ |

*(continued)*

**Table 1.** (*continued*)

| 20 | Prefix་ + Superscript ར + Root + Superscript ε | རྐུ, རྒུ, རྒྱུ, རྔུ, རྙུ, རྗུ, རྟུ, རྡུ, རྣུ, རྦུ |
| 21 | Prefix་ + Superscript ལ + + Root + Superscript ε | རྐུ, རྒུ, རྒྱུ, རྔུ, རྙུ, རྗུ, རྟུ |
| 22 | Prefix་ + Superscript ε + Root + Superscript ◌ | ཀྭ, ཁྭ, ལ |
| 23 | Prefix་ + Superscript ε + Root + Superscript ◌ | ཀྱ, ཟ, ར, ལ |
| 24 | Prefix་ + Superscript ε + Root + Superscript ◌ | ཀྲ, ཁྲ |
| 25 | Prefix་ + Superscript ར + Root + Superscript ◌ <br> Prefix་ + Superscript ལ + Root + Superscript ◌ <br> Prefix་ + Superscript ལ + Root + Superscript ◌ | རྐྭ, རྒྭ |
| 26 | Prefixམ + Superscript ε + Root + Superscript ◌ <br> Prefixམ + Superscript ε + Root + Superscript ◌ | ཁྭ, གྭ |
| 27 | Prefix་ + Superscript ε + Root + Superscript ◌ | ཁྭ, གྭ, ཤྭ, བྭ |
| 28 | Prefix་ + Superscript ε + Root + Superscript ◌ | ཁྭ, གྭ, དྭ, ཤྭ, བྭ |

**Table 2.** Vowel constraint rules.

| No. | Rule | Optional vowel |
|-----|------|----------------|
| 0 | Vowel | ི, ུ, ེ, ོ |

**Table 3.** Restriction rules between the suffix and farther-suffix.

| No. | Rule | Optional farther-suffix |
|-----|------|-------------------------|
| 0 | Suffix is empty ε | ε |
| 1 | If Suffix is one of ན, ར, ལ | ད, ε |
| 2 | If Suffix is one of ག, ང, བ, མ | ས, ε |
| 3 | If Suffix is one of འ | ི, ུ, ོ, ང, མ, ε |
| 4 | If Suffix is one of ད, ས | ε |

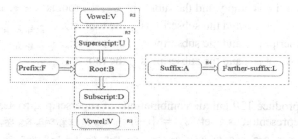

**Fig. 1.** Four types of specification rules for Tibetan syllables.

## The Constraint Specification R1 Between the Prefix and Root

Prefixes have unique corresponding roots and can be represented as elements of the set:
$f = \{$'ག','ད','བ','མ','འ'$\}$.

Root characters for individual prefixes are given by:

1) Prefix: 'ག'. Roots: ཅ, ཉ, ཏ, ད, ན, ཙ, ཞ, ཡ, ཤ, ས.
2) Prefix: 'ད'. Roots: ཀ, ག, ང, པ, བ, མ.
3) Prefix: 'བ'. Roots: ཀ, ཅ, ཏ, ད, ག, ང, ཟ, ཙ, ཞ, ཛ, ཤ, ཟ, ར, ས, ཧ.
4) Prefix: 'མ'. Roots: ཁ, ཆ, ཐ, ཚ, ག, ཇ, ད, ཛ, ང, ཉ, ན.
5) Prefix: ' འ '. Roots: ག, ཇ, ད, བ, ཛ, ཁ, ཆ, ཐ, ཕ, ཚ.
6) Prefix: 'ε'. Roots: ཀ, ཁ, ག, ང, ཅ, ཆ, ཇ, ཉ, ཏ, ཐ, ད, ན, པ, ཕ, བ, མ, ཙ, ཚ, ཛ, ཝ, ཞ, ཟ, འ, ཡ, ར, ལ, ཤ,
ས, ཧ, ཨ.

These associations produce 109 unique combinations of prefixes and roots, which
can be represented as a set:$S_{R1} = \{$εཀ, εཁ, εག, εང, εཅ, εཆ, εཇ, εཉ, εཏ, εཐ, εད, εན, εཔ, εཕ,
εབ, εམ, εཙ, εཚ, εཛ, εཝ, εཞ, εཟ, εའ, εཡ, εར, εལ, εཤ, εས, εཧ, εང, εྐ, εྒ, εྔ, εྗ, εྙ, εྟ, εྡ, εྣ, εྦ, εྨ,
εྩ, εྫ, εྮ, εྯ, εྱ, εྲ, གཅ, གཉ, གཏ, གད, གན, གཙ, གཞ, གཡ, གཤ, གས, དཀ, དག, དང, དད, དན, དཉ, དཀ, དག,
དཅ, དཉ, དད, དཌ, དབ, དཟ, དར, དལ, དས, ྐ, ྒ, ྔ, ྗ, ྙ, ྟ, ྡ, ྣ, མཀ, མཁ, མང, མཆ, མཇ, མཉ, མད, མཛ,
མཆ, འག, འཀ, འཆ, འཇ, འད, འན, འཕ, འབ, འཚ, འཛ, ྐ, ྒ, ྔ ལྐ$\}$.

## The Constraint Specification R2 Between The Superscript, Root, and Subscript

1) If the superscript is ར and the subscript is empty, the root is:ཀ, ག, ཉ, ཏ, ད, ན, བ, མ, ཙ, ཛ, ལ, ཤ

2) If the superscript is ལ and the subscript is empty, the root is:ཀ, ག, ང, ཅ, ཇ, ཏ, ད, པ, བ, ཧ, ཨ

3) If the superscript is ས and the subscript is empty, the root is:ཀ, ག, ང, ཉ, ཏ, ད, ན, པ, བ, མ, ཙ

4) If the superscript is empty and the subscript is ྱ, the root is:ཀ, ཁ, ག, ཅ, ཏ, ད, པ, ཕ, བ, མ, ཙ, ཞ, ཟ.

5) If the superscript is empty and the subscript is ྲ, the root is:ཀ, ཁ, ག, ཏ, ཐ, ད, ན.

6) If the superscript is empty and the subscript is ླ, the root is:ཀ, ཁ, ག, ན, བ, ར, ས.

7) If the superscript is ས and the subscript is ྱ, the root is:ཀ, ཁ, པ, ཕ, བ, མ.

8) If the superscript is ས and the subscript is ྲ, the root is:ཀ, ག, ན, པ, མ.

9) If the superscript is ར and the subscript is ྱ, the root is:ཀ, ཁ, བ.

These rules produce 120 unique combinations of superscripts, roots, and subscript, which can be represented as a set: $S_{R2} = \{$εηε, εη, εη, εη, εη, εηε, εη, εη, εη, εηε, εη, εη, εη, εη, εεε, εεε, εεε, εεε, εεε, εηε, εη, εηε, εη, εεε, εη, εηε, εη, εη, εηε, εη, εη, εεε, εη, εη, εη, εηε, εη, εεε, εεε, εη, εεε, εηε, εη, εεε, εη, εη, εηε, εωε, εχε, εχ, εη, εηε, εη, εηε, εη, εη, εη, εη, εη, εηε, εηε, εη, εη, εηε, εη, εε, εε, εε, εε, εε, εε, εε, εε, εε, εε, εε, εε, εε, εε, εε, εε, εε, εε, εε, εε, εε, εε, εε, εε, εε, εε, εε, εε, εε, εε, εε, εε, εε, εε, εε, οη$\}$.

## The vowel specification R3

Characters used as vowels in Tibetan grammar are given by the set: $S_{R1} = \{$ི, ུ, ེ, ོ$\}$

## The constraint specification R4 between the suffix and farther-suffix

1) If the suffix is empty, the farther-suffix is:ε.
2) If the suffix is ག, ར, or ལ, the farther-suffix is:ད, ε.
3) If the suffix is ང, ད, ན, or མ, the farther-suffix is:ས, ε.
4) If the suffix is འ, the farther-suffix is:ི, ུ, ོ, ར, ས, ε.
5) If the suffix is ད or ས, the farther-suffix is:ε.

These specifications produce 120 unique combinations of suffixes and farther-suffixes, which can be represented as a set: $S_{R4} = \{$εε, ནད, རད, ལད, ནε, རε, ལε, གས, ངས, ནས, མས, ηε, ངε, ནε, སε, ི, ུ, ོ, འང, འས, འε, དε, སε$\}$

## Vectorization and Functions of Tibetan Syllable Grammar

Tibetan syllables use $c_0, c_1, c_2, c_3, c_4, c_5, c_6$ to represent components such as prefixes, superscripts, roots, subscripts, vowels, suffixes, and farther-suffixes, respectively. The individual values of these terms can be represented collectively using the vector $M_0 = \{1, 1, 1, 1, 1, 1, 1\}$, which can be described by the following series of steps.

(1) A vector $M_1$ can be established for the constraint specification $R_1$, with initial values of $c_0$-$c_6$ all set to 1:$M_1 = \{1, 1, 1, 1, 1, 1, 1\}$.

If the combination $c_2c_0$ is included in the set $S_{R1}$, then $M_1 = \{1, 1, 1, 1, 1, 1, 1\}$. Otherwise, the prefix and root are set to 0 and $M_1 = \{0, 1, 0, 1, 1, 1, 1\}$. Thus:

$$M_1 = f_1(T)$$
$$= \begin{cases} \{1, 1, 1, 1, 1, 1, 1\}, c_2c_0 \in S_{R1} \\ \{0, 1, 0, 1, 1, 1, 1\}, c_2c_0 \notin S_{R1} \end{cases} \tag{1}$$

2) A vector $M_2$ can be established for $R_2$, with initial values all set to 1:$M_2 = \{1, 1, 1, 1, 1, 1, 1\}$.

If the combination $c_1c_2c_3$ is contained in the set $S_{R2}$, then $M_2 = \{1, 1, 1, 1, 1, 1, 1\}$. Otherwise, the superscript, root, and subscript are set to 0 and $M_2 = \{1, 0, 0, 0, 1, 1, 1\}$. Thus:

$$M_2 = f_2(T)$$
$$= \begin{cases} \{1, 1, 1, 1, 1, 1, 1\}, c_1c_2c_3 \in S_{R2} \\ \{1, 0, 0, 0, 1, 1, 1\}, c_1c_2c_3 \notin S_{R2} \end{cases} \tag{2}$$

3) A vector $M_3$ can be established for $R_3$, with initial values all set to 1:$M_3 = \{1, 1, 1, 1, 1, 1, 1\}$.

If a vowel is included in the set $S_{R3}$, $M_3 = \{1, 1, 1, 1, 1, 1, 1\}$. Otherwise, the vowel is set to 0 and $M_3 = \{1, 1, 1, 1, 0, 1, 1\}$. Thus:

$$M_3 = f_3(T)$$
$$= \begin{cases} \{1, 1, 1, 1, 1, 1, 1\}, c_4 \in S_{R3} \\ \{1, 1, 1, 1, 0, 1, 1\}, c_4 \notin S_{R3} \end{cases} \tag{3}$$

4) A vector $M_4$ can be established for $R_4$, with initial values all set to 1: $M_4 = \{1, 1, 1, 1, 1, 1, 1\}$.

If the combination $c_5c_6$ is contained in the set $S_{R4}$, $M_4 = \{1, 1, 1, 1, 1, 1, 1\}$. Otherwise, the suffix and farther-suffix are set to 0 and $M_4 = \{1, 1, 1, 1, 1, 0, 0\}$. Thus:

$$M_4 = f_4(T)$$
$$= \begin{cases} \{1, 1, 1, 1, 1, 1, 1\}, c_5c_6 \in S_{R4} \\ \{1, 1, 1, 1, 1, 0, 0\}, c_5c_6 \notin S_{R4} \end{cases} \tag{4}$$

## 3  A Spellcheck Technique for Tibetan Syllables Based on a Grammatical Vector Multiplication Model (MOSCTCBOGVM)

A Tibetan syllable is valid only if the four vectors $M_1$, $M_2$, $M_3$, and $M_4$ are satisfied simultaneously. This condition can be implemented using the following steps.

Step 0: Apply the SBCRATFS method discussed in Sect. 1.3 to a Tibetan syllable T and perform component recognition. If the recognition result t is equal to $(t_6, t_5, t_4, t_3, t_2, t_1, t_0)$, proceed to step 1. Otherwise, if t is False, a Tibetan syllable T is misspelled and the check ends.

Step 1: Establish grammatical specification vectors $M_1$, $M_2$, $M_3$, and $M_4$ for the Tibetan syllable T using Eqs. (1), (2), (3), and (4), as discussed in Sect. 2:

$M_1 = f_1(T)$

$M_2 = f_2(T)$

$M_3 = f_3(T)$

$M_4 = f_4(T)$

Step 2: Calculate the function $\delta$:

$$M = \delta(M_1, M_2, M_3, M_4)$$

$$= \prod_{i=1}^{4} M_i \tag{5}$$

By multiplying the four vectors $M_1, M_2, M_3$, and $M_4$ using a scalar product of the corresponding elements. Here, M is a 7-dimensional vector given by $M = \{m_1, m_2, m_3, m_4, m_5, m_6, m_7\}$.

Step 3: Calculate the function $g$ by summing the components of M:

$$g = g(M)$$

$$= \sum_{i=1}^{7} m_i. \tag{6}$$

If $g$ is not equal to 7, some of the values of $M_1$, $M_2$, $M_3$, or $M_4$, must be 0, which implies some components of T do not conform to at least one of the specifications imposed by $R_1, R_2, R_3$, and $R_4$. In other words, T does not conform to grammatical rules. Otherwise, $M_1, M_2, M_3$, and $M_4$ do not include a value of 0 and the components of Tibetan scripts conform to the constraint specifications imposed by $R_1, R_2, R_3$, and $R_4$, indicating T satisfies Tibetan script grammar. In summary:

$$h(g) = \begin{cases} 1, g = 7 \\ 0, g \neq 7 \end{cases}. \tag{7}$$

A value of $h = 1$ implies T is legal and $h = 0$ indicates T is illegal. The functions presented above can be combined to obtain the Tibetan script proofreading function $h$ used in the grammatical vector multiplication model:

$$r = h(g(\delta(f_1(T), f_2(T), f_3(T), f_4(T)))). \tag{8}$$

Step 4: The positions of inconsistent elements are displayed between M and $M_0$, denoting irregular locations corresponding to components of the Tibetan script T.

# 4 Experiment

The effectiveness of the proposed automated Tibetan spellcheck system, based on a grammatical vector multiplication model, was verified using three sets of experiments. Experiment 1 investigated the effectiveness of spellcheck for typos in a real corpus; experiment 2 implemented spellcheck for typos in all legal Tibetan syllables; and experiment 3 examined spellcheck effectiveness for texts containing both Tibetan and Sanskrit characters. The experimental hardware environment included a processor with six cores and an intel(R) Core (TM) i5-9400F CPU@ 2.90 GHz, 16 GB of memory, an 11 GB GPU, and the Ubuntu20.10 operating system. All software was developed in Anaconda3 using Python3.9.

The corpus used in Experiment 1 contained 156,000 Tibetan syllables. It was divided into four parts to assess recognition rates and processing times for non-genuine characters. Each corpus included modern Tibetan scripts, while special Tibetan and Sanskrit characters, numbers, and other symbols were deleted. Corpus 1 consisted of a folklore essay with 6,569 syllables; corpus 2 was a business management textbook with 142,040 syllables; corpus 3 included a political network article with 3,018 syllables; corpus 4 was an economic network article with 4,653 syllables; corpus 5 included all of the scripts that satisfied the requirements of Tibetan grammar (20,292 syllables); and corpus 6 contained a section of the poetry of Geysar, including 61,688 syllables and some Sanskrit scripts. The corpora were all acquired from the Qinghai Minzu University Corpus and are detailed in Table 4.

The proposed Tibetan spellcheck method, based on a grammatical vector multiplication model, was applied to the corpus data, the results of which are provided in Tables 5 and 6. Measured runtime included standardization, component recognition, vectorization, grammar specification, and error checking. Method A shown in the table refers to spellchecking based on the model proposed in this study and method B refers to an existing technique [16].

The results of experiment 1 indicate the proposed model can correctly detect spelling errors in Tibetan syllables with an accuracy of 99.86% at a rate of 51,050 syllables/s. The average number of syllables detected per second was 4,993 syllables/s higher than a comparable vector model-based Tibetan spellcheck method [16]. Experiment 2 showed the proposed method can recognize all valid Tibetan syllables with an accuracy of 100%. Experiment 3 showed the model can not only correctly recognize valid Tibetan syllables, but also Sanskrit characters with an accuracy of 100% for common characters.

**Table 4.** The corpus used in experiments 1, 2 and 3.

| Corpus | Corpus type | Corpus size (number of syllables) | Number of invalid syllables in the corpus | Number of Sanskrit syllables in the corpus |
|---|---|---|---|---|
| Experiment 1 | Corpus 1: a folklore essay | 6,559 | 35 | 0 |
| | Corpus 2: a business management textbook | 142,040 | 316 | 0 |
| | Corpus 3: a political network article | 3,018 | 2 | 0 |
| | Corpus 4: an economic network article | 4,653 | 1 | 0 |
| Experiment 2 | Corpus 5: all Tibetan script corpus | 20,292 | 1,511 | 0 |
| Experiment 3 | Corpus 6: a part of the poetry of Geysar, containing a certain number of Sanskrit characters | 61,688 | 151 | 33 |
| **Sum** | | **238,250** | | |

However, there were 27, 36, and 2 valid syllables identified as invalid in experiments 1, 2, and 4, respectively. This occurred primarily across 25 Tibetan syllables, as shown in Table 8. Societal integration and the transliteration of names or special scripts from other ethnic groups, translated into Tibetan, led to the splicing of individual characters and the creation of common words that do not conform to the grammatical rules of the Tibetan script. However, there were no Sanskrit, foreign scripts, or foreign words in corpus 3 or 5, so the accuracy of the spellcheck was 100%.

**Table 5.** Experimental spellcheck results for Tibetan syllables based on a grammatical vector multiplication model.

| Experiment | Corpus | Number of invalid syllables | | Accuracy (%) | | Runtime (syllables/s) | | Average number of syllables detected per second | |
|---|---|---|---|---|---|---|---|---|---|
| | | Method | | Method | | Method | | Method | |
| | | A | B | A | B | A | B | A | B |
| 1 | 1 | 62 | 62 | 99.5 | 99.5 | 0.12937 | 0.15116 | 50,699 | 43,391 |
| | 2 | 352 | 352 | 99.9 | 99.9 | 2.6888 | 2.9297 | 52,826 | 48,482 |
| | 3 | 2 | 2 | 100 | 100 | 0.05589 | 0.05859 | 53,998 | 51,510 |
| | 4 | 3 | 3 | 99.9 | 99.9 | 0.08568 | 0.0893 | 54,306 | 52,105 |
| 2 | 5 | 1,511 | 1,511 | 100 | 100 | 0.4673 | 0.5831 | 43,423 | 34,800 |
| Average | | | | 99.86 | 99.86 | | | 51,050 | 46,057 |

**Table 6.** Experimental results for Sanskrit spellcheck based on a grammatical vector multiplication model.

| Experiment | Corpus | Number of recognized Sanskrit characters | | Accuracy (%) | | Runtime (syllables/s) | | Average number of syllables detected per second | |
|---|---|---|---|---|---|---|---|---|---|
| | | Method | | Method | | Method | | Method | |
| | | A | B | A | B | A | B | A | B |
| 3 | 6 | 33 | 0 | 100 | 0 | 0.4673 | 0.5831 | 54,509 | 43,274 |

**Table 8.** Syllables identified as invalid in corpus 1, 2, and 4.

ཨགེ, གྲིན, གྲིང, ངགེ, འགེ, ཧྲུག, གྲ, གྲིགྲ, གིགྲ, གེ, གྲ, ཡྲ, གྲ, ཧྲ, གི, གུ, གྲ, ཝགེ, ཧྲུག, ངགེ, ཡགེ, ཧྲུགེ, ཟགེ, ཝགེ, ཧྲུགེ

# 5   Conclusion

A structure-based algorithm for identifying Tibetan syllable components (SBCRAFTS), by analyzing the rules of syllable formation and component splicing, was proposed in this study for automated spellcheck. A study of Tibetan grammar was summarized by constraint specifications between the prefix and root (R1), the superscript, root, and subscript (R2), the suffix and farther-suffix (R4), and by vowel constraint specifications (R3). These rules were formalized by a proposed vectorization function. Finally, a method for spellchecking Tibetan syllables was developed, based on the grammatical vector multiplication model. The algorithm is simple and easy to implement. Tests on a Tibetan corpus containing 238,250 syllables produced an average recognition accuracy of 99.86% (true scripts and non-true scripts) at a rate of 51,050 syllables. This average check rate is 4,993 syllables faster than a comparable vector-based Tibetan spellcheck model [16]. The proposed algorithm can recognize both Tibetan syllables and Sanskrit characters, to satisfy the needs of automated spellchecking of texts. In the future, a neural network based on this model will be developed by analyzing the digitization of Tibetan grammar rules.

**Acknowledgments.** We thank LetPub (www.letpub.com) for linguistic assistance and pre-submission expert review.

This work was supported by the by the NSFC projects (61741314), The State Key Laboratory of Tibetan Intelligent Information Processing and Application, Tibetan Information Processing And Machine Translation Key Laboratory Of Qinghai Province project (2021-Z-001).

# References

1. Blaivas, M., et al.: DIY AI, deep learning network development for automated im-age classification in a point-of-care ultrasound quality assurance program. J. Am. Coll Emerg. Phys. Open **1**, 124–131 (2020)
2. Kukich, K.: Techniques for automatically correcting words intext. ACM Comput. Surv. **24**(40), 377–438 (1992)
3. Peterson, J.L.: Computing practices computer programs for detecting and correcting spelling errors. Commun. ACM **12**, 676–687 (1980)
4. Pollock, J.J., Zamora, A.: Automatic spelling correction in scientific and scholarly text. Commun. ACM **27**(4), 358–368 (1984)
5. Chang, C.H.: A pilot study on automatic Chinese spelling error correction. Commun. Colips **4**(2), 143–149 (1994)
6. Wu, Y., Li, X., Liu, T., et al.: Research on and implementation of Chinese text proof-reading system. J. Harbin Inst. Technol. **33**(1), 60–64 (2001)
7. Zhang, Y.S., Yu, S.W.: Summary of text automatic proofreading technology. Appl. Res. Comput. **23**(6), 8–12 (2006)
8. Luo, W.H., Luo, Z., Gong, X.J.: Study of techniques of automatic proof reading for Chinese texts. J. Comput. Res. Dev. **41**(1), 244–248 (2004)
9. Maihefureti, A.W., et al.: Spelling check method of uyghur languages based on dictionary and statistics. J. Chin. Inf. Process. **28**(2), 66–71 (2014)
10. HaoLi, Aodengbala, Gong, Z., et al.: A research on automatic proof reading for Mongolian text based on Bayes algorithm. J. Inner Mongolia Univ. **41**(4), 440–442 (2010)
11. Guan, B., Cai, K.: Research on modern Tibetan syllables word automatically proof-reading. Comput. Eng. Appl. **48**(29), 151–156 (2012)
12. Ciren, Z.: Design of a Tibetan Spell-Checking System. In: International Conference on Chinese Information Processing (1998)
13. Duojie, Z.M.: Research on the application of the n-gram model in Tibetan text error correction partially by n-gram. Comput. Eng. Sci. **31**(4), 117–119 (2009)
14. An, J.: Tibetan word proofreading algorithm based on segmentation. J. Chin. Inf. Process. **27**(2), 58–64 (2013)
15. Zhu, J., Li, T., Liu, S.: The algorithm of spelling check base on TSRM. J. Chin. Inf. Process. **28**(3), 92–98 (2014)

16. Cai, Z., Sun, M.S., Cai, R.: Vector based spelling check for Tibetan characters. J. Chin. Inf. Process. **32**(9), 47–55 (2018)
17. Wangdui, B., Zhuoga, Chen, Y., Wu, Q.: Study on recognition algorithms for Tibetan construction elements. J. Chin. Inf. Process. **28**(3), 104–111 (2014)
18. Wang, W.L., Wang, S.L.: Implementation method and process of Tibetan basic characters positioning. China Tibetology **32**(4), 215–221 (2019)
19. Gesang, J., Gesang, Y.: Practical Tibetan Grammar Tutorial. Ethnic Publishing House, Chengdu China (2004)
20. AnJian-CaiRang.: Tibetan sorting method based on hash function. J. Artif. Intell. **4**(2), 85–98 (2022)

# Fine-Grained Image Classification Network Based on Complementary Learning

Xiaohou Shi[1], Jiahao Liu[2]([⊠]), and Yaqi Song[1]

[1] China Telecom Corporation Limited Research Institute, Beijing, China
{shixh6,songyq11}@chinatelecom.cn
[2] Johns-Hopkins University, Baltimore, MD 21218, USA
jliu306@jh.com

**Abstract.** The objects of fine-grained image categories (e.g., bird species) are various subclass under different categories. Because the differences between subclass are very subtle and most of them are concentrated in multiple local areas, the task of fine-grained image recognition is very challenging. At the same time, some fine-grained networks tend to focus on a certain region when judging the target category, resulting in the lack of other auxiliary regional features. To this end, Inception V3 is used as the backbone network, and an enhanced and complementary fine-grained image classification network is designed. While adopting the method of reinforcement learning to obtain more detailed fine grain image features, the complementary network can obtain the complementary discriminant area of the target through the method of attention erasure to increase the network's perception of the overall target. Finally, experiments are conducted on CUB-200–2011, FGVC Aircraft and Stanford dogs three open datasets. The experimental results show that the proposed model has better performance.

**Keywords:** Fine-grain · image recognition · Inception-V3 · Reinforcement complementary learning · Complementary learning · Interclass gap

## 1 Introduction

Fine-grained image classification, that is, identifying the subclass of different kinds of objects, is a hot research topic in the fields of image recognition, computer vision and other fields in recent years. Compared with traditional image classification tasks, its research content is mainly to identify different subclass under a certain category. For example, the detailed classification of different types of vehicles in urban management can be used as the basis for traffic detection and tracking reference, and the identification of different types of goods can help businesses analyze consumer buying habits and adjust sales strategies. Similar to other computer vision tasks, fine-grained image classification methods have many common problems, such as uneven illumination, large scene differences, and variable scales and perspectives. At the same time, the biggest challenge of fine-grained image classification comes from its characteristics of small inter-class differences and large intra-class differences. Therefore, how to find highly recognizable object components from these fine-grained images is a difficult problem to be solved in the current fine-grained recognition field. Some methods use additional manual labeling information [1–4] (such as bird's head, tail and other areas) to help the convolution neural network locate local areas with high distinguishability. Although these methods have achieved good results, they require a lot of manpower. Another kind of method uses weak supervision to locate local areas, and needs image label information in the experiment process. Fu et al. [5] proposed the RACNN model to identify by region detection and fine grain feature mutual reinforcement. This method can well locate the most discriminative local areas, but the increasing scale will lead to the loss of secondary features. Zheng et al. Therefore, in this paper, Inception-V3 is used as the feature extraction network, and a fine-grained image classification network based on reinforcement complementary learning is designed. The network can obtain more detailed fine-grained features of the target through reinforcement learning strategy. In order to deal with the multi-pose and multi-angle problems commonly existing in the target, the model is forced to learn other complementary discriminative regions through complementary learning strategy, Finally, the obtained features are spliced to improve the overall recognition effect of the network for the target.

## 2  Related Work

The method of strong supervision requires a lot of manual annotation to obtain important local information. However, with the increasing amount of data, manual annotation is obviously inappropriate. At present, for fine-grained recognition tasks, the common approach is to use weak supervision to make the model automatically focus on the salient regions and extract features. To this end, we will focus on the way of weak supervision:

(1) **The method based on local location**. In local localization, the method of constructing local localization subnetwork is relatively common. Literature [6–8] uses semantic component localization subnetwork to locate the key area of fine-grained image and then learn. In addition, it is also possible to learn fine-grained features through segmentation of semantic components. Huang et al. [9] established an interpretable model with high accuracy by combining prior knowledge and regional parts. The interpretation of the model is carried out through segmentation of semantic components and their contribution to classification. Different from the above methods, Ge et al. [10] established a complementary part model, and extracted the semantic parts of fine-grained objects by using the segmentation of Mask R-CNN and CRF, so that the model can focus on the most discriminating secondary parts.

(2) **End-to-end feature coding method**. Second-order bilinear features have good feature representation ability. Lin et al. proposed BCNN [11], which extracts features through two parallel convolution neural networks and then multiplies them by outer product. However, the feature dimension generated by this method is very large, which is not conducive to model training. Taking ResNet50 as an example, the resulting dimension is as high as 2048 * 2048. In order to reduce the dimension, the original bilinear feature is approximated by compressing bilinear feature [12], bilinear pooling [13] and Hadamard product [14], and the parameter quantity is

compressed by more than 90%; Dubey A et al. [15] introduced confusion in activation to reduce over-configuration, and used Pairwise Fusion regularization to reduce over-fitting (Fig. 1).

**Fig. 1.** Strengthen complementary learning strategies

(3) **Methods based on attention mechanism**. Zheng et al. [16] TASN method, which uses the trilinear attention module to model the relationship between channels to generate an attention graph, and uses the content represented by the graph to learn features; Liu [17] proposed that Full Convolutional Attention Localization Networks. The structure of FCANs is mainly composed of three parts: feature extraction, full convolution local area attention network, and classification network. Among them, the full convolution network locates multiple key areas of the image, and uses convolution features to generate fractional mapping for each part, and finally obtains the classification results; Zheng et al. [18] proposed MACNN, which is composed of convolution, channel grouping and local classification. The convolution feature based on the region is extracted from the input image through convolution layer. The peak response region feature of the feature map is used to cluster the channels with similar response regions to obtain local regions with discrimination. At the same time, the channel grouping loss function is used to increase the inter-class differentiation and reduce the intra-class differentiation; Zhang et al. [19] control the contribution of different regions to recognition through the gating mechanism.

## 3   RACL-NET (Reinforcement and Complementary Learning Network)

For a recognition network, the features it pays attention to tend to focus on a certain area of the target, which becomes the most important feature for identifying the target. However, we hope that the designed model can identify the target in a larger range, which can be achieved by relying on secondary features as well as no longer relying on a certain salient feature. So this paper designs a complementary learning reinforcement network model, which drives the other two subnetworks to carry out reinforcement learning and complementary learning (This is a kind of adversarial learning. Two parallel classifiers are forced to use complementary target regions for classification, and finally generate complete target localization together) respectively through the feature extraction of the backbone network, so as to realize the detailed and comprehensive recognition of the recognition target. Its network structure is shown below (Fig. 2):

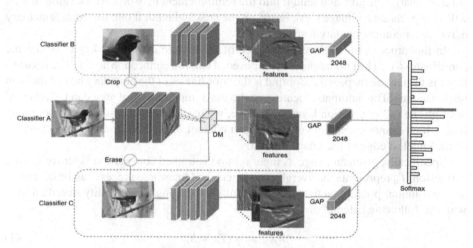

**Fig. 2.** The structure of RC-Net

The network structure is composed of Inception V3 [20], which are the basic network, the reinforcement network and the complementary network. By building a three-way classification network to aggregate the overall and local features of the object, we can obtain both the overall semantic information of the object and the local semantic information of the object. Then we can pool the global average of the features output from each network, three 2048 dimensional eigenvectors can be obtained and then splice the pooled features to form a 6144-dimensional eigenvector, add a 200-dimensional classification layer to the vector for end-to-end training, and finally get the classification results through SoftMax.

## 3.1 Drive Model

Traditional neural networks do not take advantage of the advantages of deep neural networks for location and recognition learning. Inspired by the attention region recommendation network APN, this paper proposes a module DM (Drive Model) that can drive complementary learning and reinforcement learning. DM is a very important structure in this model. It can help the backbone network find the rectangular region that has the greatest impact on the results during the training process. Specifically, It has two functions: on the one hand, it cuts and enlarges the area that has the greatest impact on the result and sends it into the reinforcement network; on the other hand, it erases the rectangular area in the original image and sends it into the complementary network. At the same time, the calculation cost of DM module is very small, and it can help the model to conduct end-to-end training.

DM receives the characteristic map of the basic network after training, and then it will generate a square area with $(x, y)$ as the center and half of $[l]$ as the side length, and cut and enlarge the area and send it into the reinforcement network. At the same time, it will also generate an image mask based on the area and input it into the complementary network for complementary learning.

In this process, the high response area of the feature map is the key to obtain the coordinate $(x, y)$. The DM module is composed of two full connection layers. The network input is the feature map, and the output is the coordinates of the boundary box of the high response area. The automatic location of the most important local area can be achieved through the full connection layer. Therefore, we have limited the size of the bounding box, which cannot exceed 2/3 of the longest edge of the overall image at most and 1/3 of the smallest edge of the image at least.

Specifically, given an image $X$, input it into the trained convolution layer for feature extraction, $T_n$ represents the overall parameters, and the whole process can be described as convolution, pooling, activation, and finally generating a probability distribution $p$, with the following calculation formula:

$$p(X) = f(T_n * X) \tag{1}$$

In this formula, $f(\cdot)$ represents the full connection layer, which converts the features extracted by the convolution neural network into feature vectors, and uses softmax to convert this vector into probability values. The next step is to generate the position and length parameter information of the square bounding box, specifically:

$$[x, y, l] = g(T_n * X) \tag{2}$$

Where $x, y, l$ is half of the horizontal and vertical coordinates and side length of the bounding box in $X$, $g(\cdot)$ represents the DM module, and its structure is composed of two fully connected layers. The weight parameters of network initialization have a great impact on the model, so the output characteristic graph of the last layer of the basic network is added. This is because the later the number of layers of the neural network is, the richer the semantic information of the characteristic graph is, and the more accurate the generated bounding box is. The region with the largest value can be obtained by adding the feature map, which is the most critical area in the image, and the parameter information of the region is the initialization parameter of the DM module. The specific formula is:

$$F = \sum_{n=1}^{d} fn \tag{3}$$

Where, $f$ represents the feature map output at the last layer of the convolutional neural network, $n$ represents the feature maps number, $d$ represents the total number of feature maps, $F$ represents the total feature map after adding each feature map. And then compares the $F$ and $I$. If $F$ is greater than $I$, then $F$ is 1, otherwise $F$ is 0, as shown in formula 5. Select the side length of the largest area as the side length of the bounding box, and the implementation formula is:

$$\bar{I} = \frac{1}{h \times w} \sum_{i=0}^{h} \sum_{j=0}^{w} F^{i \cdot j} \tag{4}$$

$$F^{i \cdot j} F^{i \cdot j} = \begin{cases} 0 F^{i \cdot j} < \bar{I} \\ 1 F^{i j j} > \bar{I} \end{cases} \tag{5}$$

In (4), h and w represent the width and height of the feature map, and $I$ represents the mean value of the feature map. By comparing the size of $I$ and $F^{i,j}$, the initialization coordinates of the bounding box center are generated.

After obtaining the initial coordinates, the model can automatically optimize the coordinates according to the training process, and then the region needs to be trimmed and enlarged to obtain a more detailed local region and then sent to the reinforcement network for learning. The coordinates of the upper left corner and the lower right corner of the local area are obtained according to the center coordinate and side length. The coordinates of the upper left corner are recorded as $(t_{lx}, t_{rx})$, and the coordinates of the lower right corner are recorded as $(t_{ly}, t_{ry})$. The calculation process is as follows:

$$\begin{aligned} t_{lx} &= x - l \quad t_{rx} = y - l \\ t_{ly} &= x + l \quad t_{ry} = y + l \end{aligned} \tag{6}$$

After obtaining the coordinate information, the clipping operation can be seen as the multiplication between the original image I and the template, expressed as:

$$X^{crop} = X \otimes M(x, y, l) \tag{7}$$

In this formula, $X^{crop}$ is the clipped area, $\otimes$ is the clipping operation between the original image and the template, $[M(\cdot)$ is the attention mask, and its expression is:

$$M(\cdot) = [\mu(i - t_{lx}) - \mu(i - t_{ly})] \times [\mu(j - t_{rx}) - \mu(j - t_{ry})] \tag{8}$$

In this formula, i and j are at any point in the feature map. If i and j are located inside the feature map, the value of $M(\cdot)$ is 1, otherwise the value is 0. At the same time, $\mu(\cdot)$ is a continuous differentiable function, whose expression is:

$$\mu(x) = \frac{1}{1 + \exp(-kx)} \tag{9}$$

In addition, in order to cut and enlarge the image, the bilinear interpolation method is used to expand the size of the extracted local area. According to the ratio of the original image and the local area, the enlarged local area can be obtained. The formula is as follows:

$$\partial = \frac{X_a^{crop}}{X_a} \tag{10}$$

$$X_{local} = X^{crop} \times \partial \tag{11}$$

$X_a^{crop}$ and $X_a$ represent the area of the local area and the overall area, $\partial$ is the area ratio, $X_{local}$ is the enlarged local area. Strengthen training on key areas of the image according to $X_{local}$ (Fig. 3).

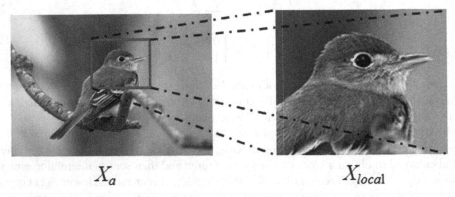

$$X_a \qquad\qquad\qquad X_{local}$$

**Fig. 3.** Local area amplification

In order to train the complementary network, the generated local area is changed into a mask image. The mask pixels are uniformly the mean value of the original image pixels, and the rest are replaced by white pixels, as shown in the following formula:

$$Mask_{pix} = \frac{\sum_{o=1}^{n} \sum_{p=1}^{s} X_{pix}^{crop}}{s \times n} \tag{12}$$

After that, the mask is erased from the original image according to the previously obtained position information, and the obtained mask image is sent to the complementary model training. The values of each position in the pixel matrix formed by the original image represent different pixels, 1 in the mask image represents black pixels, and the pixel values in the RGB channel are (0, 0, 0). Through the position calculation of the original image and the mask, the black pixel part is directly filled with the original pixel, and the mask pixel will replace the original pixel, which can also obtain the image after erasing the key area. As shown in Fig. 4:

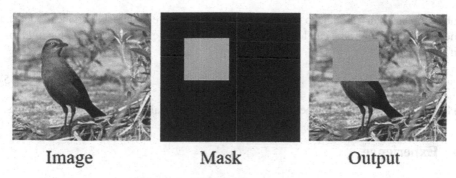

Image                    Mask                    Output

**Fig. 4.** Mask generation

## 3.2 Loss Function

The loss function has a great impact on the model, and the appropriate loss function has a positive impact on the model training. The commonly used loss function for fine-grained image recognition is the softmax function, and the formula is shown in Eq. 14:

$$L_{soft\,max} = -\sum_{i=1}^{m} \log \frac{e^{W_{yi}^T x_i + b_{yi}}}{\sum_{j=1}^{s} e^{W_j^T x_i + b_j}} \tag{13}$$

where, $m$ represents the size of a batch, $W$ represents the output result of the full connection layer, $yi$ represents the category of the $i$th image, $x_i$ represents the feature vector of the $i$th image before the full connection layer, $b$ represents the network offset, and $s$ represents the number of target categories.

The SoftMax is used to optimize the classification network. In order to enable the DM model to locate the key areas, the loss function of DM is designed. It is used to continuously optimize and strengthen the location information of the network, and at the same time, it provides more accurate mask location to enable the complementary network to learn secondary features.

$$L_{DM} = \max\{0, p^k - p^{k+1} + \Delta\} \tag{14}$$

In this formula, $p^k$ represents the probability value of the output sample of the backbone model, while $p^{k+1}$ represents the probability value generated by the reinforcement model. Here, the value of $p$ is obtained according to formula 1, the $\Delta$ represents the difference between the two models, which is 0.05. When $p^k > p^{k+1}$, there is no loss; when $p^k < p^{k+1}$, there is loss. Therefore, the loss function can help the reinforcement network to find more accurate features, and after extracting accurate features, it can help the backbone network to locate more accurately. The two strengthen each other.

At the same time, in the complementary model, because the features extracted from the backbone feature will be erased, the features extracted from the backbone model have no connection with the complementary model at all. However, the precise local area provided by the backbone model is conducive to the secondary feature learning of the complementary model, so it is only necessary to ensure that the backbone model and the reinforcement model can locate the key parts. From the above example, the total loss of the model is:

$$L_{total} = L_{softmax} + \partial L_{DM} \tag{15}$$

where $\partial$ is the modulation coefficient, which is used to balance the two loss functions.

## 4  Experience

### 4.1  Datasets

Datasets: In order to verify the performance of this model, three challenging fine-grained public data sets are compared, including CUB-200–2011, Stanford Cars, and FGVC-Aircraft.

(1) CUB-200–2011: The data set includes 200 different types of birds, including 5994 images in the training set and 5794 images in the test set, a total of 11788 images.
(2) Stanford Cars: The data set includes 196 types of vehicles of different brands and years, including 8144 images in the training set and 8041 images in the test set, a total of 16185 images.
(3) FGVC-Aircraft: The data set includes 100 different types of aircraft, including 6667 images in the training set and 3333 images in the test set, totaling 10000 images.

### 4.2  Erase Experiment

The purpose of designing the complementary network is to improve the ability of the model to pay attention to secondary features, which is very important for strengthening the complementary learning network. However, we also note that different data must have different feature distributions, which will affect the experimental results. A small number of data may need to be erased in key areas several times to obtain all features. In the AE-PSL [21] model, the experimenter erased the original image twice to obtain the overall characteristics of the target. Therefore, in this paper, it is necessary to conduct erasure experiments on the data set used to find the appropriate erasure times.

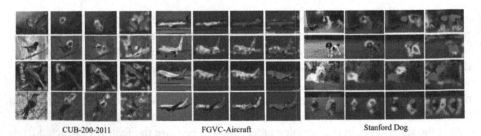

CUB-200-2011            FGVC-Aircraft            Stanford Dog

**Fig. 5.** Erase times experiment

From the above Fig. 5, after using CAM [22] to visualize the original image, the feature extraction network only focuses on the most important local area. Erasing this area, we can find that the focus of the feature extraction network is shifted to the secondary part. If the secondary part is erased again, the network cannot find the effective local area, so the best erasing number is 1, which is the same as the erasing number of the complementary model in this paper.

### 4.3  Experimental Steps

The method proposed in this paper mainly consists of three steps, which are to train the backbone model using the transfer learning method, and then train the reinforcement network and the complementary network in turn according to the training results of the feature extraction network until the three networks converge. 1) Migration learning is a very common way of training neural networks at present. It uses the training weight of Perception V3 on Image Net to train the feature extraction network. The parameters of pooling layer, input layer and convolution layer are reserved. The existing full connection layer and SoftMax layer are removed to fine-tune the network and train the data used in this paper. 2) The training of the reinforcement network is carried out according to the results of the feature extraction network. Through the calculation of the key area by the feature extraction network, the coordinate information of the most critical area is found to be cut and amplified to produce more detailed training results. The training principle of the complementary network is similar to that of the reinforcement model. The coordinate information generated by the feature extraction network is erased, and the image with only secondary local area is generated to strengthen the ability of the model to pay attention to secondary features.

### 4.4  Parameter Setting

The experimental environment is carried out under the version of pythoch1.71. The GPU is Nvidia Genforce 3060Ti, and the CPU is i7-10700K. The optimizer selects SGD, the initial learning rate is set to 0.0001, the momentum superparameter is 0.9, batch Size is 32, and epoch is set to 200.

### 4.5  Visualization of Experimental Results

In order to prove that the proposed reinforcement complementary learning network can better capture the characteristics of other auxiliary discriminant regions, this paper uses CAM algorithm to activate the class diagram of a single reinforcement network and reinforcement complementary network. The experimental results are shown in the following Fig. 6:

In Fig. 6, (a) represents the attention heat map after using only feature extraction network Inception V3. It can be found that the network focuses on the most important area of the target, and the range of the heat map is small. (b) It indicates the target area that the network pays attention to after using reinforcement complementary learning. It can be clearly found that the range of red heat map becomes larger and more local areas are concerned.

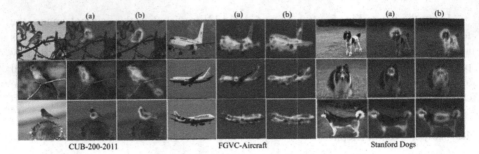

Fig. 6. Visualization of RACL-Net results

## 4.6 Experimental Verification and Analysis

**Ablation Experiment.** In order to further verify that the various network structures proposed in this paper can effectively improve the network performance, three sets of comparative tests have been conducted on CUB-200–2011. The experimental settings and results are shown in Table 1 (Tables 2 and 3):

**Table 1.** Ablation experiments on the CUB-200–2011 Datasets

| CUB-200–2011 | Experiment 1 | Experiment 2 | Experiment 3 |
|---|---|---|---|
| Inception-V3 | √ | √ | √ |
| Strengthen network | | √ | √ |
| Complementary network | | | √ |
| Top-1 Acc (%) | 83.5 | 85.6 | 89.5 |

**Comparison Test.** In order to verify the superiority of the algorithm in this paper, experiments were carried out on three open fine-grained image data sets CUB-200–2011, FGVC-Aircraft, and Standard-dogs, and the accuracy reached 89.5%, 93.6%, and 94.8%, and some of the latest models were selected for comparison, as shown in the following table:

**Table 2.** Accuracy of related methods in CUB_200_2011

| Method | Top-1 Acc (%) |
|---|---|
| B-CNN [11] | 84.1 |
| MA-CNN [6] | 86.5 |
| DFL-CNN [23] | 87.4 |
| DCL [24] | 87.8 |
| DB [25] | 88.6 |
| FDL [26] | 89.0 |
| **RC-Net** | **89.5** |

**Table 3.** Accuracy of related methods in FGVC_Aircraft

| Method | Top-1 Acc (%) |
|---|---|
| BCNN [12] | 84.1 |
| RA-CNN [5] | 88.4 |
| MA-CNN [6] | 89.9 |
| DFL-CNN [23] | 92.0 |
| **RC-Net** | **93.6** |

## 5  Conclusion

We propose a reinforcement complementary learning network to classify fine-grained images. The work done in this paper shows that the reinforcement model can help the network to obtain more detailed local features. At the same time, for the multi-pose and multi-angle problems commonly existing in fine-grained images, obtaining other complementary discriminant regions through the reinforcement model can also improve the effect of fine-grained image recognition. Finally, our reinforcement complementary learning network is weakly supervised, and it can be widely used in other classification tasks. In the future, we will explore more efficient fine-grained image classification methods, which will be carried out from the following two aspects: first, how to fuse more local regions to judge the fine-grained image classification to improve the model recognition effect; Secondly, how to build interpretable models of complementary regions to continuously improve the model recognition effect on a more detailed scale.

## References

1. Zhang, N., Donahue, J., Girshick, R., et al.: Part-based R-CNNs for fine grained category detection. In: ECCV European Conference on Computer Vision (ECCV), pp. 834–849 (2014)
2. On, S., Van Horn, G., Belongie, S., et al.: Bird species categorization using pose normalized deep convolutional nets[EB/OL]. (2014–06–11) [2021–09–15]. https://arxiv.org/ahttps://arxiv.org/abs/1406.2952
3. Lin, T.Y., Roychowdhurya, Maji, S.: Bilinear CNN models for fine-grained visual recognition. In: ICCV Proceedings of the 15th IEEE International Conference on Computer Vision (IEEE), Santiago, Chile, pp. 1449–1457 (2015)
4. Donahue, J., Jia, Y.Q., Vinyals, O., et al.: DeCAF A deep convolutional activation feature for generic visual recognition. In: Proceedings of the 31st International Conference on Machine Learning. New York JMLR. org2014647–655
5. Fu, J., Zheng, H., Tao, M.: Look closer to see better: recurrent attention convolutional neural network for fine-grained image recognition. In: IEEE Conference on Computer Vision and Pattern Recognition (CVPR), pp. 4438–4446 (2017)
6. Zheng, H., Fu, J., Tao, M., et al.: Learning multi-attention convolutional neural network for fine-grained image recognition. In: ICCV International Conference on Computer Vision (ICCV), pp. 5209–5217 (2017)

7. Sun, M., Yuan, Y., Zhou, F., et al.: Multi attention multi-class constraint for fine grained image recognition. In: ECCV Proceedings of the European Conference on Computer Vision (ECCV), pp. 805–821 (2018)

8. Wang, Y., Morariu, V.I., Davis, L.S.: Learning a discriminative filter bank within a CNN for fine-grained recognition. In: IEEE Conference on Computer Vision and Pattern Recognition (CVPR), pp. 4148–4157 (2018 )

9. Huang, Z., Xu, D., Tao, D., Zhang, Y.: Part-stacked CNN for fine-grained visual categorization. In: IEEE Conference on Computer Vision and Pattern Recognition (CVPR), pp. 1173–1182 (2016)

10. Ge, W., Lin, X., Yu, Y.: Weakly supervised complementary parts models for fine grained image classification from the bottom up. In: IEEE/CVF Conference on Computer Vision and Pattern Recognition (CVPR), pp. 3029–3038 (2019)

11. Lin, T., Roychowdhury, A., Maji, S.: Bilinear S.CNN models for fine-grained visual recognition. In: IEEE International Conference on Computer Vision (ICCV), pp. 1449–1457 (2015)

12. Shu, K., Fowlkes, C.: Low-rank bilinear pooling for fine-grained classification. In: IEEE Conference on Computer Vision and Pattern Recognition (CVPR), pp. 365–374 (2017)

13. Gao, Y., Beijbom, O., Zhang, N., et al.: Compact bilinear pooling. In: IEEE Conference on Computer Vision and Pattern Recognition (CVPR), pp. 317–326 (2016)

14. Dubey, A., Gupta, O., Raskar, R., et al.: Maximum entropy fine grained classification. ArXiv Preprint ArXiv,2018:1809.05934

15. Gao, Y., Eijbom, O., Hang, N., et al.: Compact bilinear pooling. In: 2016 IEEE Computer Vision and Pattern Recognition(CVPR), pp. 317–326 (2016)

16. Zheng, H., Fu, J., Zha, Z., et al.: Looking for the devil in the details: learning trilinear attention sampling network for fine-grained image recognition. In: 2019 IEEE/CVF Conference on Computer Vision and Pattern Recognition (CVPR), pp. 5007–5016 (2019)

17. Liu, X., Xia, T., Wang, J., et al.: Fully convolutional attention networks for finegrainedrecognitionEB/OL. 2017–03–212021–11–11. https//arxiv. org/pdf/1603. 06765. pdf

18. Zheng, H.L., Fu, J.L., Mei, T., et al.: Learning multiattention convolutional neural network for fine grained image recognition. In: Proceedings of the 2017 IEEE International Conference on Computer Vision. Piscataway IEEE, pp. 5219–5227 (2017)

19. Zhang, L., Huang, S., Liu, W., et al.: Learning a mixture of granularity-specific experts for fine-grained categorization. In: 2019 IEEE/CVF International Conference on Computer Vision (ICCV), pp. 8330–8339 (2019)

20. Szegedy, C., Vanhoucke, V., Ioffe, S., et al.: Rethinking the inception architecture for computer vision. In: 2016 CVPR Conference on Computer Vision and Pattern Recognition (CVPR), pp. 2818–2826 (2016)

21. Wei, Y., Feng, J., Liang, X., et al.: Object region mining with adversarial erasing: a simple classification to semantic segmentation approach. In: Proceedings of the IEEE Conference on Computer Vision and Pattern Recognition, pp. 1568–1576 (2017)

22. Selvaraju, R.R., Cogswell, M., Das, A., et al.: Grad-cam: Visual explanations from deep networks via gradient-based localization. In: Proceedings of the IEEE International Conference on Computer Vision, pp. 618–626 (2017)

23. DFL-CNN: Yang, Z., Luo, T.G., Wang, D., et al.: Learning to navigate for fine-grained classification. In: Proceedings of the 15th European Conference on Computer Vision. Cham: Springer, p. 420435 (2018)

24. DCL Chen, Y., Bai, Y., Zhang, W., et al.: Destruction and construction learning for fine-grained image recognition. In: IEEE/ CVF Conference on Computer Vision and Pattern Recognition (CVPR), pp. 5152–5161 (2019)

25. Sung, D.B., Cholakkal, H., Khan, S., et al.: Fine-grained recognition: accounting for subtle differences between similar classes. In: Proceedings of the AAAI Conference on Artificial Intelligence, vol. 34, no. 1, pp. 12047–12054 (2020)
26. FDL, Liu, C., Xie, H., Zhaz, J., et al.: Filtration and distillation: enhancing region attention for fine-grained visual categorization. In: AAAI Conference on Artificial Intelligence, pp. 11555–11562 (2020)
27. Luo, W., Zhang, H., Li, J., et al.: Learning semantically enhanced feature for fine-grained image classification. In: 2020 IEEE Signal Processing Letters (IEEE), vol. 27, pp. 1545–1549 (2020)
28. Zhao, B., Wu, X., Feng, J., et al.: Diversified visual attention networks for fine grained object classification. IEEE Trans. Multimedia, **19**(6), 1245−1256 (2017)
29. Dubey, A., Gupta, O., Guo, P., et al.: Pairwise confusion for fine-grained visual classification. In: European Conference on Computer Vision(ECCV), pp. 71–88 (2018)
30. Chen, Y., Bai, Y., Zhang, W., et al.: Destruction and construction learning for fine grained image recognition. In: IEEE/CVF Conference on Computer Vision and Pattern Recognition(CVPR), pp. 5152–5161 (2019)

# An Overview of Graph Data Missing Value Imputation

Jiahua Wu[1,2], Xiangyan Tang[1,2(✉)], Guangxing Liu[3], and Bofan Wu[4]

[1] School of Computer Science and Technology, Hainan University, Haikou 570228,
China
[2] Hainan blockchain technology engineering research center, Haikou 570228, China
tangxy36@163.com
[3] Bureau of Human Resources and Social Security of Mudan District, Heze 274005,
China
[4] Warrior Logistics, 450 S Denton Tap Rd 2651, Coppell, TX 75019, United States

**Abstract.** Graph data holds a significant position in various fields, enjoying widespread applications. However, practical applications Missing data not only diminishes the capacity for analyzing and extracting insights from graph data but also impairs the accuracy and reliability of associated tasks. Consequently, imputing missing data in graph datasets has garnered substantial attention, spanning diverse application domains. This paper presents an overview of research advancements and methodologies in graph data missing value imputation. To begin, we review prevalent types of missing data in graph datasets, providing examples illustrating the impact of these missing data types on various applications. Subsequently, we introduce widely adopted methods for imputing missing data in graph datasets. Each method is meticulously described, followed by an exploration of a series of cutting-edge research and techniques.In conclusion, we discuss the challenges and future research directions within the realm of graph data missing value imputation. We underscore the significance of interdisciplinary collaboration and the imperative for practical applications to propel further development in this field.

**Keywords:** Missing Data · Graph Imputation · Machine Learning · Graph Neural Networks

## 1 Introduction

Graph neural networks have been a highly prominent research direction in the field of artificial intelligence in recent years. They focus on processing and analyzing graph data, such as user-item relationships in social networks [19], chemical molecule structures [43], and recommendation systems [37]. Graph neural networks demonstrate strong expressive power and application potential in various domains.

H. Jin et al. (Eds.): IAIC 2023, CCIS 2059, pp. 256–270, 2024.
https://doi.org/10.1007/978-981-97-1280-9_20

However, in real-world applications, graph data often contain missing or incomplete information due to various reasons. For instance, in the healthcare domain, clinical data, medical records, or test results of patients frequently have missing values. This can lead to inaccurate diagnoses and treatment decisions, delaying patient care and recovery, and in critical cases, even endangering lives [22]. In the financial sector, incomplete information provided by customers or data loss due to technical glitches can impact risk assessment, credit scoring, and investment decisions [36]. In social science research, respondents may choose not to answer certain questions, leading to samples that may no longer represent the overall population and resulting in biased and distorted analysis results [4]. In geographical data, missing geographical location data or information gaps in certain regions can affect map-making and geographical analysis [50]. Additionally, in fields such as the Internet of Things (IoT), user behavior analysis, natural disasters, environmental monitoring, and education research [29,46], various data missing scenarios exist with practical implications.

In fact, data missing is a pervasive issue in almost any context of data collection and recording, resulting in consequences such as inaccurate analysis and predictions, resource wastage, trust issues, incomplete or unfair/discriminatory decisions, erroneous analysis and predictions, incomplete research, and scientific discoveries. Therefore, efficiently inferring or estimating missing graph data using limited information for meaningful computations and predictions is crucial in the fields of data science, statistical analysis, and machine learning.

## 2    Different Types of Graph Data Missing

The term "graph data missing" refers to the absence of nodes, edges, or other relevant information within graph data [44], which can potentially impact the analysis, mining, and application of the graph. Graph data missing can be categorized into the following types, and Fig. 1 presents some of the categories:

### 2.1    Missing Node Features

In graph data, each node typically has a set of features or attributes that describe its properties. Missing node features mean that some nodes have incomplete or unknown feature information [33]. Depending on the actual collection and application of graph data, different scenarios of missing node attributes may occur: a partially incomplete graph where some nodes have missing attributes; a graph where specific nodes have no attributes at all; a mixed missing graph containing both incomplete and missing attribute samples [34]. For example, in a social network, some users might not provide enough personal information, resulting in missing node features [32]. This introduces missing values in graph neural networks and can affect tasks related to node features, such as node classification, clustering, embedding, or graph generation [20].

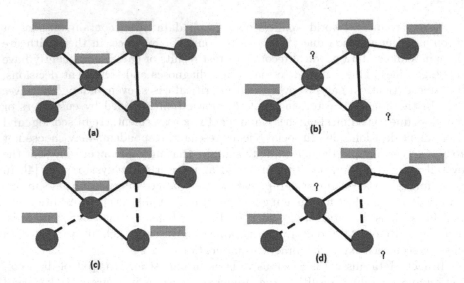

**Fig. 1.** (a) A graph without missing data. (b) A graph with missing node features. (c) A graph with missing edge existence. (d) A graph with missing graph structure.

### 2.2  Missing Edge Existence

Missing edge existence indicates that the connection between certain nodes in the graph is unknown or unrecorded. In such cases, it may be unclear which nodes are connected [26]. For example, in a transportation network, information about connections between certain roads may be missing, which can lead to problems in route planning and network analysis [51]. Missing edge data can result in an incomplete graph structure, affecting tasks related to graph connectivity, such as graph classification, link prediction, or information propagation models.

### 2.3  Missing Edge Weights or Attributes

Even when edge existence is known, information about edge weights or attributes may be missing. This can impact tasks related to edge weights, such as weighted graph analysis, recommendation systems, or social network analysis [17]. For example, in a recommendation system, user interactions may be known, but rating information may be missing. Edge weight or attribute information is often used to measure the similarity or distance between nodes and to predict the future existence or strength of edges. Missing this information can affect node classification and regression, reduce link prediction accuracy, weaken the performance of graph generation models, and impact graph clustering performance.

## 2.4  Missing Graph Structure

Missing graph structure indicates that part or all of the topological structure of the entire graph is unknown, and both the graph's topological structure and node feature information may be partially missing [47]. In such cases, it may not be possible to determine the complete set of nodes and edges, making traditional graph analysis tasks impossible. For example, in a social network, the relationship network of some users may be unknown, resulting in missing graph structure for the entire social network [15]. Missing graph structure data can have a global impact on the analysis and tasks of the entire graph because missing data can lead to incomplete graph structures and feature information.

## 2.5  Missing Time Steps in Time Series Graph Data

In time series graph data, missing time steps mean that some snapshots of the graph at certain time points are unknown or unrecorded. This can result in discontinuous time points in time series analysis, affecting time-related graph analysis tasks [21,24,27,41,49]. For example, in a sensor network, data from certain sensors at certain time points may be missing, leading to missing time steps in time series graph data. Missing time steps imply that information for some time points is missing, which may prevent the capture of important temporal patterns and evolution processes, especially in applications requiring continuous, complete time series graph data, negatively impacting downstream tasks [45].

# 3  Methods for Imputing Missing Graph Data

In dealing with missing graph data, various graph data imputation methods are commonly employed. Graph Data Imputation is a technique used to address missing information within graph data. Its goal is to fill in or estimate the missing information for nodes or edges, enabling more accurate utilization of this data in subsequent graph analysis, machine learning, or data mining tasks. The primary objective of graph data imputation is typically to preserve the integrity of the graph's topological structure and feature information while reducing biases and uncertainties introduced by data gaps [42]. Its development has gone through multiple stages and encompasses various methods and techniques.

## 3.1  Early Imputation Methods

Early imputation methods for graph data relied primarily on traditional statistical and mathematical techniques, often used to fill in missing attributes, features, or edge presence for nodes. While these methods are relatively simple, they provided a foundation for research into graph data imputation [28].

**Mean Imputation:** Mean imputation is a simple and common method for handling missing data, especially for node or edge feature data. The core idea is to estimate the values of missing data points using the mean of known data. It calculates the mean of the known data, and if there are multiple features, it can calculate the mean for each feature separately. Once the mean is computed, it is used to estimate the values of missing data points. This means that all missing data points are assigned the same value, which is the mean. Mean imputation is a straightforward method suitable for various applications of graph data. It is easy to implement and understand. However, it assumes that the data is uniformly distributed, which is often not the case, making it suitable only for cases where a quick handling of missing data is required [8,18].

**Linear Imputation:** Linear imputation is a common method for handling missing data, and it estimates the values of missing data points using the linear relationships between known data points. It establishes linear relationship models using the data of neighboring nodes or other related nodes. For example, linear regression models or linear weighted models can be used. Subsequently, this model is used to estimate the values of missing data points. Linear imputation methods utilize the linear relationships between known data points to better estimate missing data point values. This is particularly effective when there are clear linear relationships in graph data.

**Matrix Factorization:** Matrix factorization for graph data is a technique used to handle missing information within graph data. It focuses on representing graph data as matrices and uses matrix factorization methods to estimate missing data.

Matrix factorization decomposes the adjacency matrix or feature matrix of a graph into multiple low-dimensional matrices, often the product of two matrices. These low-dimensional matrices contain latent information that can be used to estimate missing data. Common matrix factorization methods include Singular Value Decomposition (SVD), Principal Component Analysis (PCA), and others. Once the matrix factorization is complete, the low-dimensional matrices resulting from the decomposition can be used to estimate missing data. For node feature imputation, low-dimensional feature matrices can be used to reconstruct node features. For edge presence imputation, low-dimensional adjacency matrices can be used to rebuild the connectivity relationships [23].

## 3.2   Imputation Methods Based on Graph Structure

Subsequently, with the rise of the field of graph data mining, more methods focused on the graph structure for imputing missing data, as opposed to general traditional methods. These methods typically emphasize estimating missing data by utilizing the adjacency matrix of the graph and the connection relationships between nodes.

**Nearest Neighbor Imputation:** The nearest neighbor imputation method uses the similarity between nodes to estimate missing data [16]. Specifically, for missing nodes or features, similarity scores between them and neighboring nodes are calculated. Then, the data from the most similar neighboring nodes is used to fill in the missing values. This can often be done by calculating distances between nodes, similarity metrics (such as cosine similarity or Jaccard similarity), or based on the features of neighboring nodes. Higher similarity scores indicate greater similarity between two nodes. This method is suitable for cases where there is high local similarity between nodes.

**Label Propagation:** Label propagation methods can be used to impute labels or attributes for nodes. They utilize known node label information to propagate labels to missing nodes through information transfer on the graph, thereby filling in missing labels [12,13]. This method does not rely on node feature information, making it suitable for cases with no feature information or incomplete feature information. It assumes that neighboring nodes are more likely to have similar labels, considering the graph's topological structure and connection relationships, thereby capturing the information propagation characteristics of graph data more effectively.

**Graph Partitioning:** Graph partitioning methods divide the graph into different subgraphs, each with a manageable size, making it easier to handle and estimate missing data [5]. Missing data is then estimated in each subgraph, and these methods can include linear imputation, label propagation imputation, generative model imputation, etc., depending on the task and data nature. By analyzing the connection relationships of subgraphs, missing values can be estimated, and the subgraphs can be merged to obtain the complete graph. Imputation methods based on graph partitioning are suitable for handling large-scale and complex graph data. Segmenting the data into subgraphs helps alleviate computational and storage burdens, and the imputation of subgraphs can be performed in parallel, accelerating the imputation process.

**Graph Signal Processing:** Imputation based on graph signal processing treats graph data as signals and utilizes graph signal processing theory and techniques to estimate the values of missing data points [25]. The process consists of signal propagation and filtering: in signal propagation, information from known data points is propagated to missing data points, taking into account the graph's topological structure and connection relationships. This can be achieved iteratively, where each node's information propagates to its neighboring nodes. In signal filtering, graph filters are used to smooth and estimate missing data point values. These filters are typically based on the graph Laplacian operator or its variants. Graph signal processing is particularly suitable for graph data with complex topological structures, as it fully considers the graph's topological structure and connection relationships, capturing node relationships more effectively. It does not rely on node feature information, making it suitable for cases with no feature information or incomplete feature information.

### 3.3   Graph Neural Network and Deep Learning Methods for Graph Data Imputation

The aforementioned early and traditional graph-structure-based methods have made significant contributions to the advancement of graph data imputation techniques. However, as research progresses, several inherent limitations have been identified. These methods often employ imputation strategies based on rules or simple mathematical models, making it challenging to capture the nonlinear, high-order relationships, and patterns in complex graph data. They typically require manual selection and design of imputation strategies, along with manual parameter tuning and model structuring, demanding domain expertise and substantial effort. When dealing with large-scale graph data, these methods may face limitations due to computational and storage resource constraints, making it difficult to scale efficiently using technologies like distributed computing and hardware acceleration. Their generalization capability to new data is often poor, as they are often modeled based on specific rules or assumptions.

Furthermore, with the continuous improvement in computational power and the development of deep learning frameworks, significant successes have been achieved in various fields such as computer vision and natural language processing, including methods for feature learning and representation learning. Researchers have been able to leverage Graph Neural Networks (GNNs) to handle large-scale graph data, and the success of these methods has inspired researchers to apply them to graph data imputation tasks to enhance performance.

When faced with missing information in graph data, GNNs and deep learning methods have become powerful tools. They estimate missing node or edge information by propagating information between nodes, aggregating neighbor information, and learning node representations. These methods combine the structural aspects of graph data with the strong representation learning capabilities of deep learning, offering innovative solutions to graph data imputation tasks [2,39,40,48]. The following section will introduce them in more detail:

**Graph Convolutional Neural Network (GCN):** GCN is a commonly used structure in graph neural networks. It was introduced in 2017 by Thomas Kipf and Max Welling [19] and is considered a significant breakthrough in the field of graph neural networks. The primary goal of GCN is to learn low-dimensional representations of nodes in a graph. In graph data imputation, GCN can be used to estimate the feature representations of missing nodes. Through GCN, the model can utilize information from known nodes to infer the features of missing nodes, facilitating data imputation.

The input data that GCN processes is a graph $G = (V, E)$, where $V$ represents a set of nodes and $E$ represents a set of edges. Each node $v_i$ is associated with a feature vector $X_i$ that represents the node's feature information.

In GCN, nodes exchange information with their neighboring nodes and aggregate this information to update their own representations. This message-passing process can be represented by the following formula:

$$H^{(l+1)} = \sigma \left( \widetilde{D}^{-\frac{1}{2}} \widetilde{A} \widetilde{D}^{-\frac{1}{2}} H^{(l)} W^{(l)} \right)$$

$H^{(l+1)}$ is the representation in the $l+1$-th layer. $\widetilde{A} = A + I_N$ is the adjacency matrix of the undirected graph $G$ with added self-connections. $I_N$ is the identity matrix. $\widetilde{D}_{ii} = \sum_j \widetilde{A}_{ij}$ and $W^{(l)}$ is a layer-specific trainable weight matrix. $\sigma$ is the activation function.

GCN is typically composed of multiple graph convolutional layers stacked together, allowing the model to learn node features at different abstraction levels. Stacking multiple layers helps the model capture more complex information and relationships within the graph. The node representations from the final layer can be connected to task-specific heads for performing specific tasks. In the context of graph data imputation, this can be connected to relevant components to decode and process the associated representations into the missing graph data that needs to be imputed.

**Graph Attention Network (GAT):** GAT is a deep learning model designed for handling graph data. It was introduced in 2018 by Petar Veličković and his team [35] and represents a significant innovation in the field of graph neural networks. The main idea behind GAT is to incorporate attention mechanisms, allowing nodes to dynamically focus on different parts of their neighboring nodes during the message-passing process, thereby enhancing the performance of representation learning on graph data.

For node $v_i$, GAT calculates the attention weight $a_{ij}$ between its adjacent nodes $v_j$. These weights are calculated using the following formula:

$$e_{ij} = a(W\overrightarrow{h}_i, W\overrightarrow{h}_j)$$

$$\alpha_{ij} = softmax_j(e_{ij}) = \frac{exp(e_{ij})}{\sum_{k \in N_i} exp(e_{ik})}$$

$h_i$ and $h_j$ represent the feature representations of nodes $v_i$ and $v_j$ respectively, $a$ performs self-attention on the nodes, and $W$ is the weight matrix used to map node features to the attention space. Using the calculated attention weights, GAT performs a weighted summation of the features of neighbor nodes to update the representation of the node. The formula representing the update is as follows:

$$\overrightarrow{h}_i' = \sigma \left( \sum_{j \in N_i} a_{ij} W \overrightarrow{h}_j \right)$$

Here, $\overrightarrow{h}_i'$ represents the updated feature representation of the node $v_i$. GAT usually uses multiple attention heads to learn different weights in parallel and capture different graph structure information from different subspaces. Multi-head attention helps improve the expressive power of the model.

**Graph Sample and Aggregation(GraphSAGE):** GraphSAGE can also be applied to graph data imputation tasks, estimating missing node features using information from neighboring nodes. It is another method for learning node representations in a graph, was proposed by Hamilton et al. in 2017 [11].

The core idea of GraphSAGE is to generate low-dimensional representations for each node by sampling and aggregating the features of neighboring nodes. At each layer, for each node $v_i$, it samples its neighboring nodes: randomly selecting a set of neighbor nodes (or sampling a fixed number of neighbors), represented as $\mathcal{N}(v_i)$. Using an aggregation function to summarize the features of neighbor nodes, it can be expressed as:

$$h_i^{(l+1)} = AGGREGATE_l \left( \left\{ W_l \cdot h_j^{(l)}, \forall j \in \mathcal{N}(v_i) \right\} \right)$$

Here, $AGGREGATE_l$ is the aggregation function for $l - th$ layer, $W_l$ is the weight matrix. Usually, GraphSAGE employs aggregation strategies such as mean pooling or max pooling or LSTM. The aggregated result represents information from neighboring nodes of node $v_i$ . GraphSAGE can be composed of multiple sampling and aggregation layers, with each layer updating the node's representation, repeating the process of sampling and aggregation until the desired number of layers is reached. Stacking multiple layers helps the model capture graph features at different abstraction levels, from local to global. The advantages of GraphSAGE include its computational efficiency, scalability to large-scale graphs, and effectiveness. It can be applied to various graph data tasks, with appropriate sampling and aggregation strategies chosen based on the specific task and dataset.

## 4   Introduce of Other Methods

In this section, we will introduce a series of cutting-edge research and methods aimed at leveraging Graph Neural Networks and deep learning techniques to address missing data in graph datasets. By delving into these latest studies, we will gain insights into how to combine graph neural networks with deep learning to tackle the common issue of missing data in practical graph datasets, thereby providing more powerful tools and methods for various application domains. These research endeavors not only offer a fresh perspective on handling graph data but also open up new possibilities for addressing missing data in graphs and imputation methods.

Rianne van den Berg and her team approached the problem of matrix completion in recommendation systems from a graph-based link prediction perspective [3]. They introduced Graph Convolutional Matrix Completion (GC-MC), which is an autoencoder framework for matrix completion based on graphs. They treat matrix completion as a link prediction problem on a graph, where the interaction data in collaborative filtering can be represented using a bipartite graph between users and items, and observed ratings or purchases are represented as links. Content information can naturally be incorporated into this framework in the form of node features.

Xu Chen and colleagues proposed the Shared Latent Space Hypothesis for graphs and developed a novel Distribution-Based Graph Neural Network called the Structural Attribute Transformer (SAT) for attribute-missing graphs. SAT decouples structure and attributes and achieves joint distribution modeling of structure and attributes through distribution matching techniques. It can not only perform link prediction tasks but also handle newly introduced node attribute completion tasks. On this basis, a quantitative method for assessing node attribute completion performance was introduced [6].

Jiaxuan You et al.proposed GRAPE, a graph-based framework for feature input and label prediction [44]. GRAPE uses a graph representation to address the issue of missing data, where observed values and features are treated as two types of nodes in a bipartite graph, with observed feature values serving as edges. In the GRAPE framework, feature input is treated as an edge-level prediction task, while label prediction is treated as a node-level prediction task. GRAPE solves both tasks using a graph neural network. Specifically, GRAPE adopts a GNN architecture inspired by the GraphSAGE model, and introduces edge embeddings during the message passing process, integrating discrete and continuous edge features in message computation.

Indro Spinelli and his team have proposed a more general Missing Data Imputation (MDI) framework by leveraging the latest developments in the field of Graph Neural Networks [31]. They formulated the MDI task based on a Graph Denoising Autoencoder, where each edge of the graph encodes the similarity between two patterns. One GNN encoder is employed to learn intermediate representations for each example by cross-classic projection layers and locally combining neighboring information. Another decoding GNN is utilized to learn to reconstruct the complete input dataset from these intermediate embeddings.

Zhiyong Cui and his team have introduced a novel neural network architecture called GMN for the prediction of spatiotemporal data [7]. They introduced two properties of the traffic state transition process and defined a graph Markov process. In contrast to other models based on Recurrent Neural Networks that treat traffic data as a multivariate time series, GMN processes the traffic state transition process as a graph Markov process. The proposed GMN can effectively consider the spatial relationships between neighboring links and the temporal dependencies between links at different time steps. Additionally, they combined spectral graph convolution operations and introduced the Spectral Graph Markov Network (SGMN).

Andrea Cini et al. proposed GRIN, a novel MTSI approach that utilizes modern graph neural networks [1]. It aims to reconstruct missing data in different channels of multi-dimensional time series by leveraging the relational information between the underlying sensor network and its functional dependencies through message passing and learning spatio-temporal representations, without relying on any assumptions about the distribution of missing values.

Yuankai Wu et al. proposed an Inductive Graph Neural Network Kriging (IGNNK) model [38]. It is used to recover data from unsampled sensors on both network and graph structures. By generating random subgraphs as samples and reconstructing all signals on each sample subgraph, IGNNK can effectively learn spatial message passing mechanisms. Furthermore, the learned model can successfully transfer to the same type of Kriging tasks on unknown datasets.

Emanuele Rossi and his team have proposed a general approach for handling missing features in graph machine learning applications [30]. This framework consists of an initial diffusion-based feature reconstruction step, followed by downstream Graph Neural Networks. The reconstruction step is based on minimizing Dirichlet energy, which leads to a diffusion-type differential equation on the graph. Discretizing this differential equation results in a very simple, fast, and scalable iterative algorithm known as Feature Propagation (FP). FP can withstand remarkably high rates of missing features, and it is both fast and requires significantly less memory.

Ivan Marisca and his team have proposed a graph-based architecture called SPIN [24], which learns representations of temporal and spatial points by leveraging a spatiotemporal propagation framework consistent with imputation tasks when provided with a set of highly sparse discrete observations.

Z Gao and his team have introduced a regularized graph autoencoder for graph attribute inputs, known as MEGAE, with the aim of mitigating the issue of spectral concentration by maximizing the graph spectrum entropy [9]. They first proposed a method for estimating graph spectrum entropy that does not require the eigenvalue decomposition of the Laplacian matrix and provided a theoretical upper bound on the error.

Soohwan Jeong and his team have proposed a graph-based interpolation method called Graph-Based Interpolation with Feature Information (GBIM) to enhance model performance by considering feature information and relationships between data points [14]. The proposed GBIM expresses the relationships between each data point by referencing a dependency model, using a GNN to estimate missing values. The proposed GBIM serves as an effective method for computing missing values and can be applied across various industries.

Dongliang Guo and his team have introduced the FairAC method, a fairness-aware attribute completion approach designed for information imputation in attribute-missing graphs and fair node embedding learning [10]. FairAC incorporates an attention mechanism to address attribute missing issues while alleviating two forms of unfairness: feature unfairness introduced by attributes and topological unfairness caused by attribute completion.

Wenxuan Tu and his team develop a novel graph imputation network termed Revisiting Initializing Then Refining (RITR) [34]. It can effectively utilize the close relationship between structure and attributes to guide the imputation of incomplete attributes and complete missing attributes using the most reliable visible information.

Certainly, it's important to note that there are other works related to missing data in graphs that are also valuable references.

# 5   Conclusion

In this paper, we have undertaken a comprehensive exploration of the field of missing data handling and imputation in graph data and provided an overview of research progress and methodologies. Our review of common types of missing data is intended to emphasize the far-reaching impact of these deficiencies across different application domains. The importance of addressing missing data becomes evident when we consider that it not only impairs the ability to analyze and extract insights from graph data but also compromises the accuracy and reliability of tasks dependent on such data.

We have elucidated various methods for imputing missing data in graph datasets, spanning early-stage techniques, those rooted in graph structure, and approaches leveraging graph neural networks and deep learning. Each method offers its own strengths and limitations.

Moreover, our exploration extended to the latest cutting-edge research and methodologies in the field, highlighting the ongoing efforts to enhance the accuracy and robustness of missing data imputation techniques.

In conclusion, we must acknowledge that the field of missing data handling and imputation in graph data is both dynamic and challenging. It is clear that interdisciplinary collaboration and real-world applications will play pivotal roles in advancing this field. By continually refining and innovating upon missing data imputation methods, we can fully harness the potential of graph data, ensuring its meaningful impact across a spectrum of applications. In an era where data is the lifeblood of decision-making, addressing the issue of missing data stands as an essential endeavor to unleash the true power of graph-based insights.

**Acknowledge.** This work was supported by National Natural Science Foundation of China (NSFC) (Grant No. 62162024, 62162022), the Major science and technology project of Hainan Province (Grant No. ZDKJ2020012), Hainan Provincial Natural Science Foundation of China (Grant No. 620MS021), Youth Foundation Project of Hainan Natural Science Foundation(621QN211).

# References

1. Andrea, C., Ivan, M., Alippi, C., et al.: Filling the g_ap_s: Multivariate time series imputation by graph neural networks. In: ICLR 2022, pp. 1–20 (2021)
2. Asif, N.A., et al.: Graph neural network: a comprehensive review on non-euclidean space. IEEE Access **9**, 60588–60606 (2021)
3. Berg, R.v.d., Kipf, T.N., Welling, M.: Graph convolutional matrix completion. arXiv preprint arXiv:1706.02263 (2017)
4. Brown, B.L., Hendrix, S.B., Hedges, D.W., Smith, T.B.: Multivariate analysis for the biobehavioral and social sciences: a graphical approach. John Wiley & Sons (2011)
5. Çatalyürek, Ü., et al.: More recent advances in (hyper) graph partitioning. ACM Comput. Surv. **55**(12), 1–38 (2023)
6. Chen, X., Chen, S., Yao, J., Zheng, H., Zhang, Y., Tsang, I.W.: Learning on attribute-missing graphs. IEEE Trans. Pattern Anal. Mach. Intell. **44**(2), 740–757 (2020)

7. Cui, Z., Lin, L., Pu, Z., Wang, Y.: Graph Markov network for traffic forecasting with missing data. Transp. Res. Part C: Emerg. Technol. **117**, 102671 (2020)
8. Enders, C.K., Baraldi, A.N.: Missing data handling methods. The Wiley handbook of psychometric testing: a multidisciplinary reference on survey, scale and test development, pp. 139–185 (2018)
9. Gao, Z., et al .: Handling missing data via max-entropy regularized graph autoencoder. In: Proceedings of the AAAI Conference on Artificial Intelligence. vol. 37, pp. 7651–7659 (2023)
10. Guo, D., Chu, Z., Li, S.: Fair attribute completion on graph with missing attributes. arXiv preprint arXiv:2302.12977 (2023)
11. Hamilton, W., Ying, Z., Leskovec, J.: Inductive representation learning on large graphs. In: Advances in Neural Information Processing Systems 30 (2017)
12. Huang, Q., He, H., Singh, A., Lim, S.N., Benson, A.R.: Combining label propagation and simple models out-performs graph neural networks. arXiv preprint arXiv:2010.13993 (2020)
13. Iscen, A., Tolias, G., Avrithis, Y., Chum, O.: Label propagation for deep semi-supervised learning. In: Proceedings of the IEEE/CVF Conference On Computer Vision and Pattern Recognition, pp. 5070–5079 (2019)
14. Jeong, S., Joo, C., Lim, J., Cho, H., Lim, S., Kim, J.: A novel graph-based missing values imputation method for industrial lubricant data. Comput. Ind. **150**, 103937 (2023)
15. Jiang, W., Wang, G., Bhuiyan, M.Z.A., Wu, J.: Understanding graph-based trust evaluation in online social networks: methodologies and challenges. Acm Comput. Surv. (Csur) **49**(1), 1–35 (2016)
16. Jiang, X., Tian, Z., Li, K.: A graph-based approach for missing sensor data imputation. IEEE Sens. J. **21**(20), 23133–23144 (2021)
17. Karasuyama, M., Mamitsuka, H.: Adaptive edge weighting for graph-based learning algorithms. Mach. Learn. **106**, 307–335 (2017)
18. Kim, J.K., Shao, J.: Statistical methods for handling incomplete data. CRC Press (2021)
19. Kipf, T.N., Welling, M.: Semi-supervised classification with graph convolutional networks. arXiv preprint arXiv:1609.02907 (2016)
20. Kossinets, G.: Effects of missing data in social networks. Social Netw. **28**(3), 247–268 (2006)
21. Kreindler, D.M., Lumsden, C.J.: The effects of the irregular sample and missing data in time series analysis. In: Nonlinear Dynamical Systems Analysis for the Behavioral Sciences Using Real Data, pp. 149–172. CRC Press (2016)
22. Li, M.M., Huang, K., Zitnik, M.: Graph representation learning in biomedicine and healthcare. Nature Biomed. Eng. **6**(12), 1353–1369 (2022)
23. Ma, X., Sun, P., Wang, Y.: Graph regularized nonnegative matrix factorization for temporal link prediction in dynamic networks. Phys. A **496**, 121–136 (2018)
24. Marisca, I., Cini, A., Alippi, C.: Learning to reconstruct missing data from spatiotemporal graphs with sparse observations. Adv. Neural. Inf. Process. Syst. **35**, 32069–32082 (2022)
25. Narang, S.K., Gadde, A., Ortega, A.: Signal processing techniques for interpolation in graph structured data. In: 2013 IEEE International Conference on Acoustics, Speech and Signal Processing, pp. 5445–5449. IEEE (2013)
26. Ouzienko, V., Obradovic, Z.: Imputation of missing links and attributes in longitudinal social surveys. Mach. Learn. **95**, 329–356 (2014)
27. Park, J., et al.: Long-term missing value imputation for time series data using deep neural networks. Neural Comput. Appl. **35**(12), 9071–9091 (2023)

28. Patrician, P.A.: Multiple imputation for missing data. Res. Nurs. Health **25**(1), 76–84 (2002)
29. Peugh, J.L., Enders, C.K.: Missing data in educational research: a review of reporting practices and suggestions for improvement. Rev. Educ. Rcs. **74**(4), 525–556 (2004)
30. Rossi, E., Kenlay, H., Gorinova, M.I., Chamberlain, B.P., Dong, X., Bronstein, M.M.: On the unreasonable effectiveness of feature propagation in learning on graphs with missing node features. In: Learning on Graphs Conference, pp. 11–1. PMLR (2022)
31. Spinelli, I., Scardapane, S., Uncini, A.: Missing data imputation with adversarially-trained graph convolutional networks. Neural Netw. **129**, 249–260 (2020)
32. Stomakhin, A., Short, M.B., Bertozzi, A.L.: Reconstruction of missing data in social networks based on temporal patterns of interactions. Inverse Prob. **27**(11), 115013 (2011)
33. Taguchi, H., Liu, X., Murata, T.: Graph convolutional networks for graphs containing missing features. Futur. Gener. Comput. Syst. **117**, 155–168 (2021)
34. Tu, W., Xiao, B., Liu, X., Zhou, S., Cai, Z., Cheng, J.: Revisiting initializing then refining: an incomplete and missing graph imputation network. arXiv preprint arXiv:2302.07524 (2023)
35. Velickovic, P., Cucurull, G., Casanova, A., Romero, A., Lio, P., Bengio, Y., et al.: Graph attention networks. Stat **1050**(20), 10–48550 (2017)
36. Wang, J., Zhang, S., Xiao, Y., Song, R.: A review on graph neural network methods in financial applications. arXiv preprint arXiv:2111.15367 (2021)
37. Wu, S., Sun, F., Zhang, W., Xie, X., Cui, B.: Graph neural networks in recommender systems: a survey. ACM Comput. Surv. **55**(5), 1–37 (2022)
38. Wu, Y., Zhuang, D., Labbe, A., Sun, L.: Inductive graph neural networks for spatiotemporal kriging. In: Proceedings of the AAAI Conference on Artificial Intelligence. vol. 35, pp. 4478–4485 (2021)
39. Wu, Z., Pan, S., Chen, F., Long, G., Zhang, C., Philip, S.Y.: A comprehensive survey on graph neural networks. IEEE Trans. Neural Netw. Learn. Syst. **32**(1), 4–24 (2020)
40. Xu, K., Hu, W., Leskovec, J., Jegelka, S.: How powerful are graph neural networks? arXiv preprint arXiv:1810.00826 (2018)
41. Yang, J., Yue, Z., Yuan, Y.: Deep probabilistic graphical modeling for robust multivariate time series anomaly detection with missing data. Reliability Engineering & System Safety, p. 109410 (2023)
42. Ye, Y., Zhang, S., Yu, J.J.Q.: Spatial-temporal traffic data imputation via graph attention convolutional network. In: Farkaš, I., Masulli, P., Otte, S., Wermter, S. (eds.) Artificial Neural Networks and Machine Learning – ICANN 2021: 30th International Conference on Artificial Neural Networks, Bratislava, Slovakia, September 14–17, 2021, Proceedings, Part I, pp. 241–252. Springer International Publishing, Cham (2021). https://doi.org/10.1007/978-3-030-86362-3_20
43. You, J., Liu, B., Ying, Z., Pande, V., Leskovec, J.: Graph convolutional policy network for goal-directed molecular graph generation. In: Advances in Neural Information Processing Systems 31 (2018)
44. You, J., Ma, X., Ding, Y., Kochenderfer, M.J., Leskovec, J.: Handling missing data with graph representation learning. Adv. Neural. Inf. Process. Syst. **33**, 19075–19087 (2020)
45. Yuan, H., Xu, G., Yao, Z., Jia, J., Zhang, Y.: Imputation of missing data in time series for air pollutants using long short-term memory recurrent neural networks.

In: Proceedings of the 2018 ACM International Joint Conference and 2018 International Symposium on Pervasive and Ubiquitous Computing and Wearable Computers, pp. 1293–1300 (2018)

46. Zhang, M., Wang, J.: Trend analysis of global disaster education research based on scientific knowledge graphs. Sustainability **14**(3), 1492 (2022)

47. Zheng, W., Huang, E.W., Rao, N., Katariya, S., Wang, Z., Subbian, K.: Cold brew: Distilling graph node representations with incomplete or missing neighborhoods. arXiv preprint arXiv:2111.04840 (2021)

48. Zhou, J., Zhou, J., et al.: Graph neural networks: a review of methods and applications. AI open **1**, 57–81 (2020)

49. Zhou, Y., et al.: For-backward lstm-based missing data reconstruction for time-series landsat images. GIScience Remote Sens. **59**(1), 410–430 (2022)

50. Zhu, D., et al.: Understanding place characteristics in geographic contexts through graph convolutional neural networks. Ann. Am. Assoc. Geogr. **110**(2), 408–420 (2020)

51. Żochowska, R., Soczówka, P.: Analysis of selected transportation network structures based on graph measures. Zeszyty Naukowe, Transport/Politechnika Śląska (2018)

# Speech Emotion Recognition Method Based on Cross-Layer Intersectant Fusion

Kaiqiao Wang[1,2] , Peng Liu[1,2](✉) , Songbin Li[1,2] , Jingang Wang[1,3] , and Cheng Zhang[4]

[1] Hainan Acoustics Laboratory, Institute of Acoustics, Chinese Academy of Sciences, Haikou 570105, China
[2] Lingshui, Marine Information, Hainan Observation and Research Station, Lingshui 572423, China
liup@dsp.ac.cn
[3] University of Chinese Academy of Sciences, Beijing 100049, China
[4] The University of Melbourne, Melbourne, VIC 3010, Australia

**Abstract.** Speech emotion recognition (SER) is a key technology in human-computer interaction (HCI) systems. Although the existing neural-based methods have achieved some satisfactory results in recognition accuracy, the failure of effective in-depth fusion of multi-scale features hinders the improvement of the accuracy of SER. In this paper, we address this issue from the two aspects of extracting exhaustive features and fusing features of multi-scale. In particular, we propose a recognition network based on **Cross-Layer Intersectant Fusion**, termed **CLIF**. It mainly consists of multi-scale feature extraction and cross-layer intersectant fusion. The former takes acoustic features as input and extracts feature maps with different receptive field ranges layer by layer through deepening convolution structures. Among these features, the lower level has more original information but also contains noise. The higher level has emotional semantics that is easier to classify but loses the perception of the details of the original acoustic features. Therefore, we use the cross-layer intersectant fusion module to achieve efficient utilization of low-level and high-level features. The experimental results demonstrate that the proposed CLIF is superior to the existing state-of-the-art speech emotion recognition algorithm. The overall recognition accuracies of CLIF can achieve 82.17% and 93.26% on IEMOCAP and CASIA datasets respectively.

**Keywords:** Speech emotion recognition · Human computer interaction · Deep learning · Attention mechanism

Supported in part by Youth Innovation Promotion Association, Chinese Academy of Sciences, in part by South China Sea Nova project of Hainan Province, in part by the Important Science and Technology Project of Hainan Province under Grant ZDKJ2020010, and in part by Frontier Exploration Project Independently Deployed by Institute of Acoustics, Chinese Academy of Sciences under Grant QYTS202015 and Grant QYTS202115.

H. Jin et al. (Eds.): IAIC 2023, CCIS 2059, pp. 271–285, 2024.
https://doi.org/10.1007/978-981-97-1280-9_21

# 1   Introduction

Emotion is a characteristic of human personality and plays an important role in complex daily social activities. With the development of intelligent technology such as machine learning, people hope that machines can also have rich human-like emotion expressions [4,22,28]. If computers can perceive emotions and give timely feedback like humans, this will greatly enhance the communication convenience of human-computer interaction and make the computer better serve humans. Therefore, speech emotion recognition has become a research hotspot in the field of speech signal processing.

The processing paradigm of speech emotion recognition mainly includes the front end and the back end. The front end is mainly acoustic feature extraction, and the commonly used ones are MFCC (Mel Frequency Cepstral Coefficient). The back end generally refers to the emotion prediction model. In the early research, machine learning models such as GMM [7] and SVM [19] are used more. In the past few years, neural network-based methods have become the mainstream emotion recognition method with the development of deep learning. The neural network can further extract the hidden high-dimensional features related to emotion categories by means of supervised learning on the basis of front-end acoustic features and then realize emotion prediction. In terms of specific implementation, CNN [27] and RNN [11] their combination are common strategies for constructing emotion recognition networks. On the basis of these backbone networks, some researchers have introduced the attention mechanism [5,8], which can further improve detection accuracy.

The above method based on deep learning makes it possible for machines to recognize human emotions automatically to a certain extent. However, this research is still in an incomplete stage and there are still some deficiencies [13,18,23]. On one hand, different layers in deep neural networks can extract multi-scale feature representations through different mapping functions. These features present the characteristics of the original speech sequence from different terms, which are all vital for the SER task. Nevertheless, previous SER studies have paid little attention to the multi-scale feature extraction, instead, they tend to focus on the deep-level features of neural networks [2,10,14,24]. On the other hand, an effective fusion of multi-scale features is of great significance. While compared with the traditional static fusion [18], interactive fusion allows the model to learn the potential relationship independently. Therefore, to get a sufficient model performance, it is essential to perceive multi-scale features and design an interactive fusion mechanism.

To address these issues, this paper proposes a network for SER with a cross-layer intersectant fusion mechanism based on multi-scale features, called CLIF. Our proposed model mainly contains two parts. One is multi-scale feature extraction, the other is the cross-layer intersectant fusion of multi-scale features. Specifically, a pre-trained convolution network is first employed to obtain multi-scale features with the local dependency of acoustic features. Secondly, for the deep convolution network, the receptive field of the shallow layer is relatively small so that more details of the original acoustic features are retained and the resolution of the feature representation is high. Yet the overall emotion representation ability of the shallow layer is weak. The receptive field of the deep layer is relatively large so that the detailed features are filtered out and the resolution of feature mapping is low. However, the overall emotion representation ability is strong. Therefore, if the shallow and deep features can be effectively fused, the detailed and overall emotion characteristics of acoustic features will be captured simultaneously, which will provide sufficient information for further emotion classification. To this end, this paper proposes a cross-layer intersectant fusion mechanism, which can dynamically fuse the multi-scale shallow features and deep features learned by the convolution network for SER.

We performed experiments on two different speech datasets, IEMOCAP [1] and CASIA [26], respectively. Experimental results indicate that our model performs better results than the existing state-of-the-art algorithm. The major contributions are summarized as follows:

(1) In this paper, a cross-layer intersectant fusion module is presented. By dynamically fusing the feature maps with different local correlations, this module can take full advantage of the rich information of the low-level features and the semantic information of the high-level features.
(2) Based on the cross-layer intersectant fusion module and other backbone structures, we propose a speech emotion recognition network named CLIF. The process paradigm provides a valuable reference for researchers. Experimental results indicate that our proposed method can achieve state-of-the-art on the IEMOCAP dataset with weighted accuracy.

## 2  Related Work

The processing paradigm of speech emotion recognition can be divided into two parts: acoustic feature extraction and emotion prediction. In acoustic feature extraction, MFCC [24,30,31] is a common solution since it can model human vocal characteristics to some extent. In addition, some researchers fed pitch [17], energy [17], LPCC (linear prediction Cepstral coefficients) [16], STFT(short term Fourier transform) [9] as front-end inputs into the emotion prediction model. Based on the above acoustic features, there are two types of the minimum prediction unit for speech emotion recognition, segment-level and utterance-level. In this paper, to make full use of the information in the original data samples, we use the fine-grained prediction strategy.

In terms of prediction models, early researchers used traditional machine learning methods such as GMM [7], SVM [19], and decision tree [3]. As an early exploration, these methods can hardly achieve satisfactory detection performance. With the rapid development of deep learning technology based on neural networks, more and more researchers began to study the solution of using neural networks for emotion recognition. Taking acoustic features as input, the deep neural network can further mine the speech sequence correlation information within the features, capture the latent emotion-related features between speech frames, and thus achieve better detection performance than traditional algorithms. For specific implementation, CNN and RNN are commonly used as backbone structures.

Recently, the application of attention mechanism has achieved further predominant performances [2,20,25,30]. For example, Nediyanchath et al. [20] employed a multi-head attention neural network with a gender recognition auxiliary task to attend to the dominant emotion-related features. Cao et al. [2] proposed a hierarchical network with a gated multi-features unit and attention mechanism to explore the contributions of static and dynamic features. Xu et al. [24] designed a multi-scale area attention in a deep CNN to attend emotion characteristics with varied granularities from the Log Mel spectrogram. Xu et al. [25] proposed an attention-based convolutional neural network(ACNN) model. Zhu et al. [30] proposed a novel GLobal-Aware Multi-scale neural network to learn multi-scale feature representation with a global-aware fusion module to attend emotion information.

Although attention mechanism based models have made great progress for SER, the failure of effective in-depth fusion of multi-scale features hinders the further improvement of SER system. Different from these works, the SER network with a cross-layer intersectant fusion mechanism based on multi-scale features proposed in this paper can simultaneously capture the shallow features and the deep features so that the SER system can achieve advanced accuracy.

## 3    Proposed Method

### 3.1    Model Overview

In this paper, our target is reaching an effective recognition for emotion category from a speech utterance. The difficulty mainly consists of two aspects, how to obtain sufficient and even redundant feature information from an utterance of speech, and how to construct an effective detection network to make predictions using the above feature representation. The overall framework of our proposed method is shown in Fig. 1. First, we pre-process the raw audio data and cut each speech utterance into several segments with a fixed interval. Secondly, we use the released speech processing tools (such as Librosa [12]) to extract MFCC acoustic features for each speech segment; Thirdly, we utilize a backbone network Resnet-50 to extract multi-scale features of acoustic features; Fourthly, we propose the Cross-layer intersectant fusion mechanism to restore and fuse the obtained multi-scale features; Finally, we get the emotion category with a full connection layer. The feature of an utterance can be defined as:

$$X = [x_1, x_2, ..., x_N] \tag{1}$$

where $N$ is the number of speech segments. We denote the MFCC feature which is obtained from a speech segment, as $x_i \in R^{D_m \times D_l}$, $D_n$ and $D_f$ represent MFCCs number and frame number respectively. Using our SER system $E$, we can get the final emotion classification result:

$$c = E(x_i) = FC(\psi(\phi(x_i))) \tag{2}$$

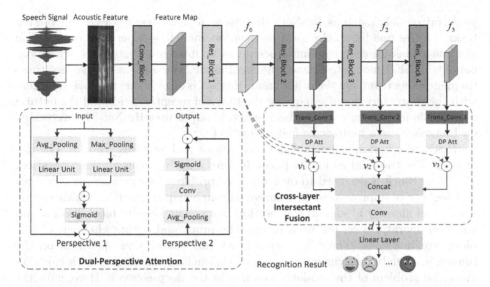

**Fig. 1.** Illustration of proposed speech emotion recognition method. In the figure, the output feature map of different Res_Blocks is denoted by $f_i$, $i \in (0, 1, 2, 3)$.

where $\phi$ and $\psi$ are the Multi-scale feature extraction and Cross-layer intersectant mechanisms which will be introduced in the following sections.

## 3.2 Preprocessing of Raw Audio

This paper applies the same data pre-processing strategies as Ref. [24]. Specifically, each utterance is divided into several audio segments of 2 s, with 1 s (in training) or 1.6 s (in testing) overlap between segments. Although being divided, the test is still based on the utterance, and the prediction results from the different segments in the same utterance are averaged as the final prediction result of the utterance. For the audio segments, the resampling rate is 16 kHz. Each sampling point is quantized with 16 bits. We divide each segment into frames of 128 ms window size with a window step of 32 ms, resulting in 63 frames. Using the Librosa audio processing library [12], we extract 26-dimensional MFCCs for each frame.

## 3.3 Multi-scale Feature Extraction

To conduct high-precision speech emotion recognition, powerful feature representation is essential, while the acoustic feature extracted from the speech signal is an original feature representation, which needs further feature extraction. A neural network structure with strong learning ability and multi-scale feature extraction is a feasible solution aimed at this question. Generally, increasing the depth of the network is conducive to improving performance, but the limited scale of emotion corpora greatly limits complexity in practice. Employing some powerful models (such as ResNet) that have achieved great success in related tasks, on the one hand, the local association modeling experience obtained in the image classification training process can be introduced into SER, which can benefit for SER model to extract the local association of acoustic feature time-frequency spectrum. On the other hand, ResNet is a backbone network structure with the characteristics of expanding the range of receptive fields and is a natural multi-scale feature extractor. Therefore, this paper uses ResNet-50 (pretrained on ImageNet) as a multi-scale feature extractor.

ResNet-50 mainly consists of a Conv_Block and four Res_Blocks. The former encodes the original inputs into a feature map by a convolution operation. The Res_Block is designed to overcome the problems of low learning efficiency and ineffective improvement of accuracy through skip-connection, which mainly consists of identity block and convolution block. The identity block has no convolution operations on the branch, and the input and output channels of this block are the same. The convolution block has convolution operation on the branch and enables to change the output channels of the block, which can alleviate the problem of the too-large channel in the deep network. Here, with the convolution operation of different kernel sizes in the convolution blocks, the different local dependencies of acoustic features in different layers will be mined. So that the detailed and overall emotion characteristics of acoustic features will be obtained in the shallow layer and the deep layer, respectively. Thus, after four Res_Blocks blocks, we can correspondingly obtain four high-level feature matrices with different scales: $f_i, i \in (0, 1, 2, 3)$.

### 3.4    Cross-Layer Intersectant Fusion

The main process of speech emotion recognition is to extract salient features and then classify them based on these features. Many existing mature neural networks have strong feature selection abilities. For example, the ResNet-50 pretraining model mentioned above can be used to extract multi-scale informative features. Therefore, the main problem is how to make full use of the obtained salient features for classification. Aiming at the problem, the existing methods can be divided into two categories: the first takes the highest-level features of the deep network as the input of the classification network. The second extracts multi-layer and multi-scale features for simple concatenation or fusion. These methods fail to effectively integrate the multi-scale features of different layers of the network. Thus, we propose a mechanism called cross-layer intersectant fusion to realize the effective fusion between different layer features.

**Feature Transposition.** To fuse the multi-scare features between different layers, it is necessary to restore these features to the same size. Compared with the intuitive upsampling strategy such as linear interpolation, transposed convolution is a learnable upsampling operation that can model the complex feature representation. Therefore, we utilize the transposed convolution to restore the feature maps $f_1$, $f_2$, $f_3$ to the same size of $f_0$. Mathematically, the mapping between the input and output size of the transposed convolution operation can be defined as:

$$o = (i - 1)s - 2p + k + u \tag{3}$$

where $i$ and $o$ represent the size of input and output. $s$, $p$, $k$, $u$ refer to stride, padding, kernel size, and output padding respectively. Here, transposed convolutions with different parameters are applied, as a result, we get three feature maps with the same size as $f_0$: $h_t$, $t \in (1, 2, 3)$.

**Dual-Perspective Attention.** The Dual-perspective attention mechanism is introduced to further obtain more salient feature maps. Specifically, perspective 1 is the channel perspective that uses for differentiating the channels of the feature map, and perspective 2 is the local perspective that focuses on learning the important local characteristics of the feature map. Firstly, perspective 1 performs channel attention on the feature map $h_t$ that is output by the previous layer:

$$u_t = Sigmoid(v_{1t} + v_{2t}) \cdot h_t \tag{4}$$

where $u_t$ is the channel attention output of $t$-th feature map, $\cdot$ is an element-wise product operation, $v_{1t}$ and $v_{2t}$ are the learned channel attention scores before normalization, which are calculated by:

$$v_{1t} = W_g(Relu(W_s(Avg\_Pooling(h_t)) + b_s)) + b_g \tag{5}$$

$$v_{2t} = W_c(Relu(W_z(Max\_Pooling(h_t)) + b_z)) + b_c \tag{6}$$

here $Avg\_Pooling$, $Max\_Pooling$, and $Relu$ are the average pooling operation, max-pooling operation, and nonlinear activation function respectively. $W.$ is the learnable parameter metrics, $b.$ is the bias terms. Next, perspective 2 calculates local attention on $u_t$:

$$z_t = Sigmoid(Conv(Avg\_Pooling(u_t))) \cdot u_t \tag{7}$$

where $Conv$ is convolution operation, $z_t$ is the final output of dual-perspective attention.

**Feature Fusion.** With a dual-perspective attention mechanism, we obtain a more powerful feature representation by increasing the direct influence of the important channels and local characteristics. Then, the fusion of $f_0$ and $z_t$ is realized by matrix multiplication:

$$v_t = f_0 * z_t \tag{8}$$

Finally, the three feature maps are concatenated and the final classification result $c$ is got through a convolution layer and a linear layer:

$$c = Fc(Conv([v_1, v_2, v_3])) \tag{9}$$

## 4    Experiments

### 4.1    Datasets

Two different emotion corpora with different languages, IEMOCAP [1] (for English) and CASIA [26] (for Chinese), are used to evaluate the speech emotion recognition methods. IEMOCAP is performed by 10 actors in 9 types of emotion states-angry, happy, excited, sad, frustrated, fear, surprise, neutral, and other. The performance is divided into two parts, improvised and scripted, according to whether the actors perform according to a fixed script. Due to the imbalances in the dataset, researchers usually choose the most common emotions, such as neutral, happy, sad, and angry states to conduct comparisons for different methods. Excited emotion and happy emotion have a similarity to some extent and there are too few happy utterances, thus researchers sometimes replace happy with excited or combine excited and happy to increase the amount of data [8,15,24,29]. In this study, following other published work, we use improvised data in the IEMOCAP with four types of emotion states-neutral, excited, sad, and angry. CASIA is a mandarin balanced speech emotion corpus and is performed by four subjects (two males and two females) in six different emotion states-angry, fear, happy, neutral, sad, and surprise. It contains a total of 9600 utterances, 7,200 utterances expressed in the same statements, and 2400 utterances expressed in different statements, respectively.

### 4.2    Baseline System

We take the latest speech emotion recognition models as the benchmark models. DT-SVM: Sun et al. [19] proposed a speech emotion recognition method based on the decision tree support vector machine model with Fisher feature selection. In this method, the decision tree SVM framework is first established by calculating the confusion degree of emotion, and then the features with higher distinguishability are selected for each SVM of the decision tree according to the Fisher criterion. Speech emotion recognition is finally realized.

MTL+Attention: Taking Log mel-Filter Bank Energies (LFBE) spectral features as the input, Nediyanchath et at. [14] presented a deep learning network. The introduced multi-head self-attention and position embedding help in attending to the dominant emotion features by identifying the positions of the features in the sequence. Besides, they utilize gender prediction to construct a multi-task learning framework to improve emotion prediction performance.

TLFMRE: Chen et al. [3] thought that one drawback of the existing SER methods is that emotion recognition doesn't consider the differences among different categories of people. In Ref. [3], they extracted and fused personalized and non-personalized speech features and divided the high-level features into different sub-classes by adopting the fuzzy C-means clustering algorithm. Multiple random forest is used to recognize different emotional states.

Head Fusion: Xu et al. [25] proposed a head fusion strategy based on multi-head self-attention to better capture the sequence correlation. Further, they presented an attention-enhanced CNN network and conducts empirical experiments by injecting speech data with different types of common noises to verify the robustness of the proposed model.

MAA: Xu et al. [24] believe that the characteristics of different emotions reflect in different local patterns in time-frequency acoustics features. According to this, they introduced a multi-scale area attention to attend emotional characteristics with varied granularities. For generalization prediction, they also utilized data augmentation strategy, such as vocal tract length perturbation.

GLAM: Similar to Ref. [24], Zhu et al. [30] demonstrated that the limitation of existing SER methods lies in that rich emotional features at different scales and important global information is not able to be well captured. Considering this, they proposed a GLobal-Aware Multi-scale (GLAM) neural network. For implementation, they utilized multiple convolution kernels with different local receptive fields and a global-aware fusion module.

## 4.3 Experimental Setup

Table 1. Model parameters

| Attribute | Value | Attribute | Value |
| --- | --- | --- | --- |
| MFCC frames | 63 | padding $p_1$ | $(1, 1)$ |
| MFCC features | 26 | kernel size $k_1$ | $(3, 3)$ |
| sampling rate | 16,000 | output padding $u_1$ | $(0, 1)$ |
| segments length | 2 s | stride $s_2$ | $(4, 4)$ |
| overlap length(training) | 1 s | padding $p_2$ | $(1, 1)$ |
| overlap length(testing) | 1.6 s | kernel size $k_2$ | $(5, 5)$ |
| batch size | 32 | output padding $u_2$ | $(0, 1)$ |
| learning rate | 0.001 | stride $s_3$ | $(8, 8)$ |
| optimizer | Adam | padding $p_3$ | $(1, 1)$ |
| Conv_Block | $(3, (1, 1))$ | kernel size $k_3$ | $(7, 7)$ |
| stride $s_1$ | $(2, 2)$ | output paddin $u_3$ | $(2, 3)$ |

We randomly divide the training set and testing set of two datasets into a ratio of 4:1. We carry out a 5-fold cross-validation in the experiment and the average prediction result is reported. In addition, we apply the released pre-trained ResNet-50 model weights[1] in our backbone network. In the network training phase, the set batch size is 32. Adam algorithm [6] is used to optimize the model with an initial learning rate of 0.001. Through these experimental strategies, most of the weight parameters of our network can be learned automatically, and

---

[1] https://pytorch.org/vision/stable/models.html.

only a small part needs to be defined in advance. For the model preproduction, these hyper-parameters are listed in Table 1. Among them, (3, (1, 1)) represents the number of channels and the kernel size of Conv_Block. $s_m$, $p_m$, $k_m$, and $u_m$ denote the $m$-th transposed convolution parameters, which are stride, padding, kernel size, and output padding. To achieve comprehensive evaluation, two mostly used metrics, WA (Weighted Accuracy) and UA (Unweighted Accuracy), are used. WA is the accuracy of all samples in the testing set, and UA is the average accuracy of all emotion categories.

### 4.4   Results and Analysis

**Experiment on IEMOCAP Dataset.** The experimental results on the IEMOCAP dataset are shown in Table 2. Compared with recent approaches, the proposed architecture improves performances on two metrics, which demonstrates the effectiveness of multi-scale feature extraction and cross-layer intersectant fusion technique. Specifically, our model achieves an accuracy of 82.17% on WA metric, and an accuracy of 80.83% on WA metric. Compared with GLAM, which is the best model on IEMOCAP dataset, our model is 0.99% and 1.58% higher in WA and UA metrics respectively.

**Table 2.** Performance comparison on IEMOCAP.

| Method | WA(%) | UA(%) |
|---|---|---|
| Head Fusion [25] | 76.18 | 76.36 |
| MTL+Attention [14] | 76.40 | 70.10 |
| MAA [24] | 79.34 | 77.53 |
| GLAM [30] | 81.18 | 79.25 |
| **CLIF(ours)** | **82.17** | **80.83** |

**Table 3.** Performance comparison on CASIA.

| Method | WA(%) | UA(%) |
|---|---|---|
| DT-SVM [19] | 85.08 | 83.75 |
| TLFMRE [3] | 85.83 | 85.61 |
| MAA [24] | 86.26 | 89.24 |
| GLAM [30] | 90.80 | 88.93 |
| **CLIF(ours)** | **93.26** | **91.66** |

**Experiment on CASIA Dataset.** To further investigate the universality of our proposed model, we also conducted experiments on the mandarin dataset

CASIA. As shown in Table 3, CLIF achieves the best performance of 93.26% and 91.66% in terms of WA and UA. In particular, CLIF is superior to GLAM not only in IEMOCAP dataset but also in CASIA dataset. It surpassed GLAM 2.46% and 2.73% in WA and UA metrics respectively, which is a significant improvement.

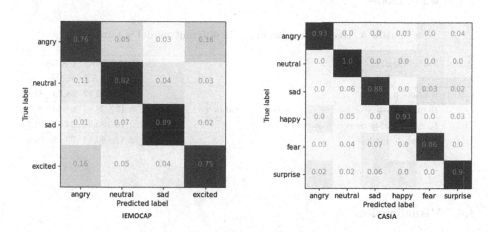

**Fig. 2.** Confusion matrix to show each class accuracy of two datasets

To further explore the specific performance of CLIF in each class of the two datasets, we present the confusion matrices of IEMOCAP and CASIA in Fig. 2. As mentioned earlier, IEMOCAP is an imbalanced dataset, the number of neutral is the largest, but it can be seen from the Fig. 2 that the neutral prediction accuracy is not the highest. The highest prediction accuracy is sad, whose number ranks second. The numbers of angry and excited are relatively small, and the performances are also poor. It can be seen that the recognition difficulty of each emotion category is not only related to their number but also related to their own emotion characteristics. Intuitively, sad has stronger emotion characteristics than neutral. Excited and angry are similar to some extent, which leads to the model easily confusing them. CASIA is a balanced dataset and its script has obvious emotion words, which makes its prediction accuracy of each category good. Relatively speaking, the accuracy of sad and fear is relatively poor since they have certain similarities.

## 4.5   Ablation Study

In this part, we investigate the effectiveness of each introduced part of the proposed method CLIF. The results are shown in Table 4. In #1, we evaluate the feature extraction capacity of standard ResNet-50, it can be observed that the pre-trained ResNet-50 has strong feature extraction ability since the pure ResNet-50 can obtain a not bad accuracy. In #2, we replace the proposed cross-layer intersectant fusion with simple feature concatenation. Compared with

#1, it can be demonstrated that the feature fusion operation can improve the recognition accuracy of the neural model. In #3, we discuss the effectiveness of the introduced dual-perspective attention. It can enable adaptive selection when the feature space remains redundant, thus making the model better capture emotion-related features. The recognition results in #4 prove the effectiveness of the whole model.

**Table 4.** Recognition accuracy in different model variants.

| Index | Model Variants | IEMOCAP | | CASIA | |
|---|---|---|---|---|---|
| | | WA% | UA% | WA% | UA% |
| #1 | ResNet-50 | 77.66 | 76.42 | 88.59 | 87.83 |
| #2 | Replace cross-layer intersectant fusion with feature concatenation | 79.17 | 78.45 | 91.40 | 88.93 |
| #3 | Remove dual-perspective attention | 81.14 | 80.28 | 94.57 | 91.27 |
| #4 | The whole model | 82.17 | 80.83 | 93.26 | 92.63 |

## 4.6    Visualization

To intuitively demonstrate how features from different layers contribute to the final emotion recognition system, we conducted feature visualization experiments on two datasets. The t-SNE algorithm [21] is utilized to reduce feature dimensions and visualize the output features of different layers.

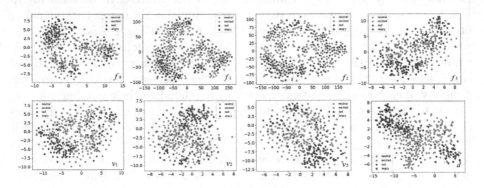

**Fig. 3.** The distribution difference of output features of each layer on the IEMOCAP dataset. The output features $f_0$, $f_1$, $f_2$, $f_3$, $v_1$, $v_2$, $v_3$, $d$ are visualized. The distance between points indicates the distribution difference between features. The closer the distance is, the smaller the distribution difference is. The different colors in the figure represent the true labels of features.

The experimental results of IEMOCAP dataset are shown in Fig. 3. From the figure we can see that The shallow feature $f_0$ contains more original acoustic information, and there is no obvious difference between the four categories.

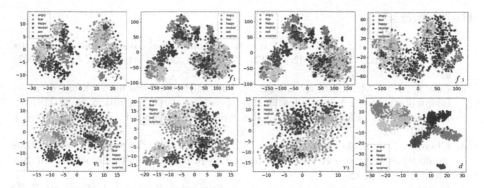

**Fig. 4.** The distribution difference of output features of each layer on the CASIA dataset. The feature distributions of different emotions are represented by different colors.

With the deepening of network processing, the distribution distance of different emotion categories gradually increases. The feature before full connection layer calculation $d$ shows the largest distribution difference between four emotions. We can conclude that our proposed speech emotion recognition method really can capture the differences between the four speech emotion categories for classification. In addition, we can also see that there is a large overlap between the features of excited and angry. We attribute to that on the one hand, there is a certain similarity between the two emotions in human senses. On the other hand, the sample category of angry is relatively small, accounting for only 10.2% of the total. The model learning of this emotion category is not sufficient. As shown in Fig. 4, the visualization results on CASIA dataset are similar to IEMOCAP, in the network structure we designed, with the deepening of the network layer, the six emotion features show a trend of increasing distribution differences, of which the feature difference of layer $d$ is the largest.

## 5   Conclusions

The existing deep learning-based SER methods fail to extract rich and effective feature representation from original acoustic features, thus the recognition performance is not satisfactory. In this paper, we propose an advanced speech emotion recognition network based on cross-layer intersectant fusion. Specifically, we first utilize a deep convolution network to conduct multi-scale feature extraction. These extracted features are then fed into the proposed cross-layer intersectant fusion mechanism to obtain salient features. Experimental results show that our proposed method can achieve advanced accuracies on two different language datasets. In the future, we will explore a lightweight network for speech emotion recognition.

# References

1. Busso, C., et al.: IEMOCAP: interactive emotional dyadic motion capture database. Lang. Resour. Eval. **42**, 335–359 (2008). https://doi.org/10.1007/s10579-008-9076-6
2. Cao, Q., Hou, M., Chen, B., Zhang, Z., Lu, G.: Hierarchical network based on the fusion of static and dynamic features for speech emotion recognition. In: 2021 IEEE International Conference on Acoustics, Speech and Signal Processing (ICASSP), ICASSP 2021, pp. 6334–6338. IEEE (2021)
3. Chen, L., Su, W., Feng, Y., Wu, M., She, J., Hirota, K.: Two-layer fuzzy multiple random forest for speech emotion recognition in human-robot interaction. Inf. Sci. **509**, 150–163 (2020)
4. Cummins, N., Scherer, S., Krajewski, J., Schnieder, S., Epps, J., Quatieri, T.F.: A review of depression and suicide risk assessment using speech analysis. Speech Commun. **71**, 10–49 (2015)
5. Kim, J., An, Y., Kim, J.: Improving speech emotion recognition through focus and calibration attention mechanisms. arXiv preprint arXiv:2208.10491 (2022)
6. Kingma, D.P., Ba, J.: Adam: a method for stochastic optimization. In: Bengio, Y., LeCun, Y. (eds.) 3rd International Conference on Learning Representations, ICLR 2015, Conference Track Proceedings, San Diego, CA, USA, 7–9 May 2015 (2015). http://arxiv.org/abs/1412.6980
7. Kockmann, M., Burget, L., et al.: Application of speaker-and language identification state-of-the-art techniques for emotion recognition. Speech Commun. **53**(9–10), 1172–1185 (2011)
8. Li, P., Song, Y., McLoughlin, I.V., Guo, W., Dai, L.R.: An attention pooling based representation learning method for speech emotion recognition. International Speech Communication Association (2018)
9. Li, Y., Zhao, T., Kawahara, T., et al.: Improved end-to-end speech emotion recognition using self attention mechanism and multitask learning. In: Interspeech, pp. 2803–2807 (2019)
10. Liu, J., Liu, Z., Wang, L., Guo, L., Dang, J.: Speech emotion recognition with local-global aware deep representation learning. In: 2020 IEEE International Conference on Acoustics, Speech and Signal Processing (ICASSP), ICASSP 2020, pp. 7174–7178. IEEE (2020)
11. Liu, Y., Sun, H., Guan, W., Xia, Y., Zhao, Z.: Discriminative feature representation based on cascaded attention network with adversarial joint loss for speech emotion recognition. In: Proceedings of the Interspeech, pp. 4750–4754 (2022)
12. McFee, B., et al.: librosa: audio and music signal analysis in Python. In: Proceedings of the 14th Python in Science Conference, vol. 8, pp. 18–25 (2015)
13. Muppidi, A., Radfar, M.: Speech emotion recognition using quaternion convolutional neural networks. In: 2021 IEEE International Conference on Acoustics, Speech and Signal Processing (ICASSP), ICASSP 2021, pp. 6309–6313. IEEE (2021)
14. Nediyanchath, A., Paramasivam, P., Yenigalla, P.: Multi-head attention for speech emotion recognition with auxiliary learning of gender recognition. In: 2020 IEEE International Conference on Acoustics, Speech and Signal Processing (ICASSP), ICASSP 2020, pp. 7179–7183. IEEE (2020)
15. Neumann, M., Vu, N.T.: Improving speech emotion recognition with unsupervised representation learning on unlabeled speech. In: 2019 IEEE International Conference on Acoustics, Speech and Signal Processing (ICASSP), ICASSP 2019, pp. 7390–7394. IEEE (2019)

16. Nwe, T.L., Foo, S.W., De Silva, L.C.: Speech emotion recognition using hidden Markov models. Speech Commun. **41**(4), 603–623 (2003)
17. Pao, T.L., Chen, Y.T., Yeh, J.H., Li, P.J.: Mandarin emotional speech recognition based on SVM and NN. In: 18th International Conference on Pattern Recognition, ICPR 2006, vol. 1, pp. 1096–1100. IEEE (2006)
18. Rajamani, S.T., Rajamani, K.T., Mallol-Ragolta, A., Liu, S., Schuller, B.: A novel attention-based gated recurrent unit and its efficacy in speech emotion recognition. In: 2021 IEEE International Conference on Acoustics, Speech and Signal Processing (ICASSP), ICASSP 2021, pp. 6294–6298. IEEE (2021)
19. Sun, L., Fu, S., Wang, F.: Decision tree SVM model with fisher feature selection for speech emotion recognition. EURASIP J. Audio Speech Music Proces. **2019**(1), 1–14 (2019)
20. Tarantino, L., Garner, P.N., Lazaridis, A., et al.: Self-attention for speech emotion recognition. In: Interspeech, pp. 2578–2582 (2019)
21. Van Der Maaten, L.: Accelerating t-SNE using tree-based algorithms. J. Mach. Learn. Res. **15**(1), 3221–3245 (2014)
22. Wang, S., et al.: Advances in data preprocessing for biomedical data fusion: an overview of the methods, challenges, and prospects. Inf. Fus. **76**, 376–421 (2021)
23. Wang, X., Wang, M., Qi, W., Su, W., Wang, X., Zhou, H.: A novel end-to-end speech emotion recognition network with stacked transformer layers. In: 2021 IEEE International Conference on Acoustics, Speech and Signal Processing (ICASSP), ICASSP 2021, pp. 6289–6293. IEEE (2021)
24. Xu, M., Zhang, F., Cui, X., Zhang, W.: Speech emotion recognition with multiscale area attention and data augmentation. In: 2021 IEEE International Conference on Acoustics, Speech and Signal Processing (ICASSP), ICASSP 2021, pp. 6319–6323. IEEE (2021)
25. Xu, M., Zhang, F., Zhang, W.: Head fusion: improving the accuracy and robustness of speech emotion recognition on the IEMOCAP and RAVDESS dataset. IEEE Access **9**, 74539–74549 (2021)
26. Zhang, J., Jia, H.: Design of speech corpus for mandarin text to speech. In: The Blizzard Challenge 2008 Workshop (2008)
27. Zhang, S., Zhang, S., Huang, T., Gao, W.: Speech emotion recognition using deep convolutional neural network and discriminant temporal pyramid matching. IEEE Trans. Multimedia **20**(6), 1576–1590 (2017)
28. Zhang, Y.D., et al.: Advances in multimodal data fusion in neuroimaging: overview, challenges, and novel orientation. Inf. Fus. **64**, 149–187 (2020)
29. Zhao, Z., Bao, Z., Zhang, Z., Cummins, N., Wang, H., Schuller, B.: Attention-enhanced connectionist temporal classification for discrete speech emotion recognition. ISCA (2019)
30. Zhu, W., Li, X.: Speech emotion recognition with global-aware fusion on multi-scale feature representation. In: 2022 IEEE International Conference on Acoustics, Speech and Signal Processing (ICASSP), ICASSP 2022, pp. 6437–6441. IEEE (2022)
31. Zou, H., Si, Y., Chen, C., Rajan, D., Chng, E.S.: Speech emotion recognition with co-attention based multi-level acoustic information. In: 2022 IEEE International Conference on Acoustics, Speech and Signal Processing (ICASSP), ICASSP 2022, pp. 7367–7371. IEEE (2022)

# Clothed Human Model Estimation from Unseen Partial Point Clouds with Meta-learning

Chenghao Fang[1], Kangkan Wang[1,2(✉)], Shihao Yin[1], and Shaoyuan Li[3]

[1] Key Lab of Intelligent Perception and Systems for High-Dimensional Information of Ministry of Education, Nanjing University of Science and Technology, Nanjing, China
[2] Jiangsu Key Lab of Image and Video Understanding for Social Security, School of Computer Science and Engineering, Nanjing University of Science and Technology, Nanjing, China
wangkangkan@njust.edu.cn
[3] College of Computer Science and Technology, Nanjing University of Aeronautics and Astronautics, Nanjing, China

**Abstract.** This work aims to address the problems of clothed human reconstruction from unseen partial point clouds. Existing methods focus on estimating vertex offsets on top of parametric models for clothing details but with the limitation of the fixed topology, or reconstructing non-parametric shapes using implicit functions which are of lack semantic information. Moreover, due to limited training data, large variations in clothing details, and domain gaps between training data and real data, these methods often have poor generalization ability on real data. In this paper, we propose a generalizable approach for estimating dressed human models from single-frame partial point clouds based on meta-learning. Specifically, we first learn a meta-model that can efficiently estimate the parameters of unclothed human models from unseen data by a fast fine-tuning. Based on the unclothed human models, we further meta-learn the point-based human models in clothing with local geometric features, which is topologically flexible and rich in human details. Our approach outperforms previous work in terms of reconstruction accuracy, as evidenced by qualitative and quantitative results obtained from a variety of datasets.

**Keywords:** Clothed human reconstruction · Meta-learning · Point clouds

## 1 Introduction

In recent years, 3D human model estimation has received significant attention in a wide range of real-world applications, including AR/VR, robotics, and environmental perception in autonomous driving. The goal of 3D human model estimation is to accurately reconstruct a complete 3D model of a clothed human. However, this task poses significant challenges due to the wide range of human poses, body shapes and intricate clothing details. The problem is further complicated by the presence of discrepancies between the training data and the real-world data. Effective and reliable estimation of clothed human models remains a fundamental problem in 3D computer vision.

With the rapid development of depth sensor technology, point clouds can now be acquired at low cost even with hand-held devices. Therefore, various 3D human estimation methods have been proposed from point clouds. Using deep learning techniques, some methods [10,13,28] attempt to directly predict the parameters of a statistical body shape model (such as SMPL [15]) from single frame partial point clouds. However, these methods mainly focus on recovering undressed human shapes without surface details such as clothes. In practice, many real-world applications require the recovery of detailed human shapes beyond the unclothed human model. To represent clothing details, some works [2,17,29] attempt to estimate vertex offsets on top of parametric models, which are efficient but have the limitation of the fixed topology. In addition, some research [5,30] has focused on restoring the non-parametric configurations of dressed individuals by using implicit functions [24] from point clouds. Despite the non-parametric representation can model diverse clothing topology, it is always lack of semantic information, which leads to problems such as misshapen or broken bodies. Point clouds are another 3D representation with flexible topology, and some recent works [16,18] have successfully predicted high-resolution point clouds to capture clothing details for human bodies, but they only deal with complete point clouds. In contrast to previous works, we aim to estimate detailed human models directly from single-frame partial point clouds, which face two main challenges. First, the point clouds are disordered, making it difficult to extract local geometric information directly from the point clouds. Second, most methods do not generalise well to real-world scenarios due to the domain gap between training and real data.

To address the above problems, in this work, we propose a novel approach to estimate dressed human models from single-frame partial point clouds with meta-learning. More specifically, we first use the meta-learning algorithm [20] to learn a meta-model that enables fast fine-tuning to estimate accurate unclothed parametric models based on global features extracted over the whole point clouds, given only a few monocular point clouds of unseen clothed humans as inputs. Then, the parametric models are mapped to UV space as the body template for point-based human models in clothing, and we structurally extract local geometric features of the input point clouds to meta-learn clothing deformations based on the UV position information. Our approach has been proven effective based on the qualitative and quantitative analysis of the experimental results obtained from different datasets.

In summary, our contributions are: (1) we propose a novel hierarchical framework to estimate clothed human models from partial point clouds based on meta-learning; (2) we design a novel local feature extraction method compatible with point clouds and meta-learning; (3) our method can estimate unclothed and clothed human models separately, with fine generalization towards real data.

## 2    Related Work

**Human Mesh Recovery from Point Clouds.** Human meshes can be recovered from point clouds using template-based methods, template-free methods, or a combination of both [3,9]. More recent work aims to derive three-dimensional body models from point clouds captured from a single image. Jiang et al. [10] proposes a skeleton-aware network to regress SMPL [15] parameters from point clouds. Wang et al. [28] uses a coarse-to-fine spatial-temporal mesh attention convolution module to predict human meshes. Liu et al. [14] proposes an occlusion-aware network for human parametric

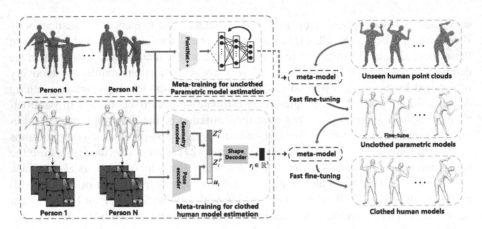

**Fig. 1. Overview of the proposed framework**. Our proposed method involves two cascading stages based on meta-learning, e.g. unclothed parametric model estimation and clothed human model estimation. A detailed description can be found in Sect. 3.

model estimation with Hough voting. However, these methods mainly focus on recovering the undressed human models, and due to the domain gaps between training data and real data, they usually fail to generalize well to real-world scenarios.

**Shape Representations for Clothed Human Modeling.** Due to their compatibility with pipelines and efficiency, surface meshes [2,4,17] have become a popular choice for modeling clothing in the 3D industry. Some recent works use graphical convolution [29], multi-layer perceptrons (MLPs) [7,21] to predict offsets on an unclothed human template for clothing details but the fixed topology limits the representation of details. Neural implicit surfaces, on the other hand, offer flexibility in surface topology and have emerged as a promising option for reconstructing the 3D shape of human bodies in recent years [5,25,30], but they lack of semantic information. As well, point clouds are a traditional method of representing 3D data that can handle diverse topological structures. SCALE [16] generates structured dense point clouds with a set of surface patches to simulate garment deformation. POP [18] learns a dense point cloud representation with fine-grained local point features and generates garment geometry based on auto-decoding. But they all deal only with complete point clouds and have the same problem of generalization ability.

**Meta-learning.** Meta-learning is a powerful approach to address the challenge of few-shot learning [19,26,31]. The aim is to train a learner that can quickly adapt to a new, unseen task with only a few training examples of that task. A recent class of meta-learning algorithms [6,20] proposes to learn a "meta-model" (e.g. an initialization for the parameters of a neural network) from different tasks. The meta-model can then be specialized to a new task via a few steps of gradient descent, known as fast fine-tuning. This class of algorithms has recently attracted considerable interest in the field of 3D vision. Sitzmann et al. [26] leverages gradient-based meta-learning algorithms to learn a universal initialization of implicit representations for static neural SDFs. Wang et al. [31] attempts to learn dynamic representations conditioned on human body poses from point clouds based on Reptile [20]. In this work, we propose to estimate clothed human models from partial point clouds based on meta-learning.

## 3   Method

Our goal is to learn a high-fidelity dressed human model from unseen partial point clouds. We achieve this goal with a hierarchical method based on meta-learning. An overview of the proposed method can be seen in Fig. 1. We first learn a meta-model for unclothed parametric model estimation from partial point clouds (Sect. 3.1). Based on the unclothed parametric models, we then meta-learn clothed human models with local geometric features (Sect. 3.2). After obtaining the meta-model through meta-training at each stage, our approach can efficiently estimate both unclothed parametric models and clothed human models on arbitrary unseen human point clouds via fast fine-tuning, without any additional parameters.

### 3.1   Unclothed Parametric Model Estimation with Meta-learning

**Meta-task Setting.** Meta-learning proposes to decompose the training data into different tasks and ensures task-specific convergence of the meta-model, so that the meta-model learns a prior knowledge from a series of tasks. Thus, the diversity and representativeness of the meta-task has a direct impact on the generalization ability of the learned model. In our approach, all training data is grouped into various meta-tasks according to subject/cloth-type/action, i.e. each meta-task represents a specific body type. The aim is to ensure that these tasks cover a wide range of dressed body variations, so that the learned meta-models can be adapted more effectively to unseen human bodies.

**Learning a Meta-model for Unclothed Parametric Model Estimation.** To estimate the unclothed human parametric models, we extract global deep features by capturing the 3D geometry information from the human body through PointNet++ [22]. With the global features as inputs, a fully connected network is employed to learn the SMPL parameters $(\beta, \theta, t)$ for the unclothed human mesh recovery. We use the loss function, denoted as $\mathcal{L}_{HPM}$, to encourage the predicted parameters and the corresponding parametric models to be close to the ground truth:

$$\mathcal{L}_{HPM} = \lambda_{para}(\|\beta - \hat{\beta}\|_2^2 + \|R(\theta) - R(\hat{\theta})\|_2^2 + \|t - \hat{t}\|_2^2)$$
$$+ \lambda_{3D} \sum_{i=1}^{N} \|M(\beta, \theta, t)_i - \hat{v}_i\|_2^2, \tag{1}$$

where the parameters $\hat{\beta}, \hat{\theta}$ and $\hat{t}$ represent the ground truth SMPL parameters. $R(\cdot)$ is the vectorized rotation matrix ($R(\theta) \in \mathbb{R}^{24 \times 9}$) corresponding to the pose parameters.

$M(\boldsymbol{\beta}, \boldsymbol{\theta}, \boldsymbol{t})_i$ and $\hat{\boldsymbol{v}}_i$ correspond to the $i$-th vertex of the predicted and actual undressed mesh, respectively. $N = 6890$. $\lambda$'s are weights that balance the loss terms.

We propose to leverage the meta-learning algorithm of Reptile [20] to train the above network as a meta-network $f_\phi(\cdot)$ with initial parameters $\phi$. In the forward pass, we sample a batch of training tasks and perform $m$ gradient descent steps according to the update rule (Line 7 in Algorithm 1) at inner-loop on each training task. After getting the final parameter $\phi_m^{(j)}$ updated on $j$-th meta-task, we treat $(\phi_m^{(j)} - \phi)$ as a gradient. In the last step, the mean of all calculated gradients is used as the final gradient direction to update the parameters $\phi$ at outer-loop (Line 10 in Algorithm 1). The full algorithm is formalized in Algorithm 1.

---

**Algorithm 1.** Meta-learning unclothed parametric model with Reptile [20]

---

**Input:** Meta-training set $\mathcal{D}_{meta}$ is all meta-tasks, inner learning rate $\alpha$, outer learning rate $\beta$, inner-loop iteration $m$, max training iteration $N$
**Output:** Predicted SMPL parameters from partial point clouds
1: Initialize: meta-network parameters $\phi$
2: **for** $i = 1, \dots, N$ **do**
3:     Sample a batch of $M$ training tasks $\{\hat{\mathbf{X}}^{(j)}\}_{j=1}^M$ from Meta-training set $\mathcal{D}_{meta}$
4:     **for** $j = 1, \dots, M$ **do**
5:         $\phi_0^{(j)} = \phi$
6:         **for** $k = 1, \dots, m$ **do**
7:             $\phi_k^{(j)} = \phi_{k-1}^{(j)} - \alpha \nabla_\phi \mathcal{L}_{HPM}(f_\phi(\hat{\mathbf{X}}^{(j)})|_{\phi=\phi_{k-1}^{(j)}})$
8:         **end for**
9:     **end for**
10:     $\phi \leftarrow \phi + \beta \frac{1}{M} \sum_{j=1}^M (\phi_m^{(j)} - \phi)$
11: **end for**

---

### 3.2 Clothed Human Model Estimation with Meta-learning

**Representing Humans with Dense Point Clouds.** For a high-fidelity representation of the clothed human, we first map the unclothed parametric models to the UV space to represent the human body in a dense point cloud inspired by [18]. The UV position map $I \in \mathbb{R}^{H \times W \times 3}$ is a 2D dimensional parameterization of the body manifold, where each valid pixel point $\boldsymbol{u}_i = (u, v)_i$ corresponds to a point $p_i \in \mathbb{R}^3$ on the surface of the posed body, $H$ and $W$ indicate UV map size. Specifically, we utilize the SMPL UV position map as the unclothed body template for the point-based human models in clothing, which is topologically flexible and rich in human details. The meta-task setting strategy follows Sect. 3.1 and each parametric model is converted to the UV position map.

**Meta-learning Clothing Deformation with Local Geometric Features.** With the new meta-tasks, we meta-learn clothing deformations based on the unclothed, point-based human model with a novel local geometric feature. Practically, we decompose the local geometric features into pose features $Z^P$ and clothing geometric features $Z^G$,

---

**Algorithm 2.** Meta-learning clothed human model with Reptile

---

**Input:** Meta-training set $\mathcal{D}_{meta}$, inner learning rate $\alpha$, outer learning rate $\beta$, inner-loop iteration $m$, max training iteration $N$

**Output:** Predicted point-based human models in clothing

1: Initialize: meta-network parameters $\psi$
2: **for** $i = 1, \ldots, N$ **do**
3:　　Sample a batch of $M$ training tasks $\{\hat{\mathbf{X}}^{(j)}\}_{j=1}^{M}$ from Meta-training set $\mathcal{D}_{meta}$
4:　　**for** $j = 1, \ldots, M$ **do**
5:　　　　$\psi_0^{(j)} = \psi$
6:　　　　$I_j \sim M(\beta, \theta, t)_j \in \hat{\mathbf{X}}^{(j)}$　　　　　　　▷ Obtain the corresponding UV position map
7:　　　　**for** $k = 1, \ldots, m$ **do**
8:　　　　　　$\psi_k^{(j)} = \psi_{k-1}^{(j)} - \alpha \nabla_\psi \mathcal{L}_{HCM}(g_\psi(\hat{\mathbf{X}}^{(j)}, I_j)|_{\psi=\psi_{k-1}^{(j)}})$
9:　　　　**end for**
10:　　**end for**
11:　　$\psi \leftarrow \psi + \beta \frac{1}{M} \sum_{j=1}^{M}(\psi_m^{(j)} - \psi)$
12: **end for**

---

as shown in Fig. 1. We first introduce the pose encoder, a UNet [23] that is used to encode the UV position map into a pose feature tensor $Z^P \in \mathbb{R}^{H \times W \times 64}$. $Z^P$ contains information about the pose and shape of the unclothed human and can provide common deformation properties for the cross-clothed human to make the meta-model pose-aware.

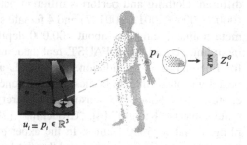

**Fig. 2.** An illustration of the clothing geometric feature extraction from the point clouds based on the UV map.

Due to the disorder of point clouds and the complexity of clothing details, we then extract the local features of the point clouds based on the UV map rather than using global features to learn the clothing geometric information. Specifically, we use the valid points in 3D space provided by the UV map of the unclothed human model as the sampling centers. Then, we use a geometry encoder to extract features from the point clouds sampled around each sampling center by radius $r$, resulting in a clothing geometric feature tensor $Z^G \in \mathbb{R}^{H \times W \times 64}$. Figure 2 illustrates the process of the clothing geometric feature extraction. The $Z^G$ and $Z^P$ correspond in dimension, allowing the geometric information of the point clouds to be ordered according to the structure of the UV space, thus providing clothing geometric features for each point on the unclothed body template.

Finally, $Z^G$, $Z^P$, and the UV position map are fed into an MLP as used in [18], denoted as Shape Decoder in Fig. 1, to predict the 3D offset on top of unclothed body template for clothing deformation. According to Algorithm 2, we train this meta-network which is denoted as $g_\psi(\cdot)$, by minimizing the loss function:

$$\mathcal{L}_{HCM} = \lambda_d \frac{1}{M} \sum_{i=1}^{M} \min_{j \in N} \|\mathbf{x}_i - \mathbf{y}_j\|_2^2 + \lambda_{rd} \sum_{j=1}^{N} \|\mathbf{r}_j\|_2^2, \qquad (2)$$

where the $\lambda$'s are weights that balance the loss terms, $M$ and $N$ are the point number of the generated point cloud and the ground truth surface, respectively. The term corresponding to $\lambda_d$ is the Chamfer Distance from the ground truth point cloud $X$ to the set of sampled points $Y$ on the surface of the generated point clouds. The other term is an $L2$ regularizer on the displacements to prevent unnormal predicted displacements.

## 4  Experiment

**Datasets.** Experiments were carried out on various datasets including SURREAL [27], DFAUST [3], CAPE [17] as well as real data. We use four different viewpoints to render 3D models into single-view depth images, which are then transformed into partial point clouds. The CAPE dataset contains multiple human subjects. Each subject wears different clothing and performs different actions. We mainly choose 4 male subjects (00032, 00096, 00122, 00127) and 4 female subjects (00134, 03223, 03331, 03383) for meta-training, sampling about 100,000 depth images. We use other unseen datasets, including SURREAL, DFAUST, real data, and two subjects (00215, 00159) in CAPE, for fine-tuning and validation. SURREAL and DFAUST consist of a wide range of posed unclothed human bodies, and we generate 10,000 frames each. The real data is captured by a Kinect V2 sensor with severer occlusion and random noise, including 2 male subjects "Kungfu" [8], "Crouching" [28] and 1 female subject "Girl" [28], sampling a total of 5000 frames. In the experiments, we sample every 5th frame on the unseen dataset for fine-tuning and the rest for validation.

**Architecture and Experimental Settings** During the estimation stage of the non-clothed parametric model, the raw point clouds are downsampled uniformly to a total of 4000 points. We use the standard PointNet++ [22] and adopt the fully connected network as in [29]. In the clothed human model estimation stage, we use the UV map of SMPL model with $128 \times 128 \times 3$ resolution and downsample the raw point clouds to 20,000 points. The pose encoder is a standard UNet [23]. In order to extract local geometric features, we conduct a search for neighbouring points situated within a specified radius of 0.3 m and use a 3-layer MLP [Conv2d, InstanceNorm, ReLU] to extract the clothing geometric features. The Adam optimizer [12] is used in the optimization, and the learning rate is $1e^{-4}$ for both inner-loop and outer-loop in meta-learning. Experiments are performed on a 2080 Ti GPU.

**Metrics.** We use MAVE [29] to quantitatively assess the discrepancy in the reconstruction of the unclothed parametric model across all vertices of the reconstructed 3D models (in mm). In addition, we quantitatively evaluate the reconstruction error of the detailed human model with the Chamfer Distance (in $10^{-4}\,\mathrm{m}^2$, mentioned in Eq. 2) from the input point clouds to the predicted points.

**Table 1.** Reconstruction errors (MAVE) in Meta-HPM and (Chamfer Distance) in Meta-HCM with different percent fine-tuning data.

| Percent | 100% | 50% | 20% | 10% | 5% |
|---|---|---|---|---|---|
| Meta-HPM | 13.98 | 19.58 | 21.52 | 32.94 | 46.67 |
| Meta-HCM | 0.100 | 0.112 | 0.125 | 0.163 | 0.204 |

**Table 2.** Reconstruction errors (MAVE) in Meta-HPM and (Chamfer Distance) in Meta-HCM with different fine-tuning batches.

| Batches | 0 | 25 | 50 | 75 | 100 |
|---|---|---|---|---|---|
| Meta-HPM | 116.03 | 33.23 | 28.16 | 25.20 | 21.50 |
| Meta-HCM | 0.320 | 0.154 | 0.133 | 0.128 | 0.125 |

error (in cm): 0 ▬▬▬▬ 10

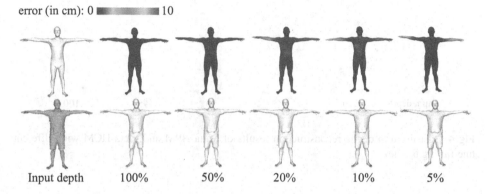

Input depth    100%    50%    20%    10%    5%

**Fig. 3.** Visualisation of the reconstruction results of Meta-HPM and Meta-HCM with different percent fine-tuning data.

## 4.1 Ablation Study

**Few-Shot Learning of the Meta-model.** We first evaluate the learning capabilities of our approach in terms of few-shot learning. As shown in Table 1, we report the performance of the meta-models fine-tuned on the reduced amount of data. The two meta-models are tested independently on CAPE data, denoted as Meta-HPM and Meta-HCM. Figure 3 shows an example of the recovered models with different percentages of fine-tuning data. We observe that both of the two meta-models achieve a very low reconstruction error with only 20%–50% of the unseen data. To balance the fine-tuning time and quality, we use 20% of the unseen data uniformly in the experiments for fine-tuning.

**Fine-Tuning Efficiency of the Meta-model.** Another highlight of meta-learning is its ability to quickly adapt to unseen data. With a fixed amount of fine-tuning data, we compare the reconstruction errors of the prediction results under different fine-tuning batches. The results are reported in Table 2. A significant reduction in the reconstruction error can be seen in the first 25 fine-tuning batches for both meta-models. As the number of fine-tuning batches increases, the meta-model gradually adapts to the unseen data, resulting in low reconstruction errors at 75–100 batches. With a small amount of data, this fine-tuning process can take just a few minutes. The visualisation results are shown in Fig. 4.

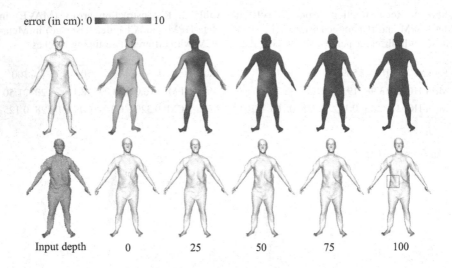

**Fig. 4.** Visualisation of the reconstruction results of Meta-HPM and Meta-HCM with different fine-tuning batches.

**Clothing Geometric Feature Extraction.** We further evaluate the effectiveness of our local features against the global features for clothing geometric feature extraction by comparing the reconstruction errors of the estimated clothed human models. The global features are extracted directly from the whole point clouds by PointNet++ [22].

| Methods | CAPE | Crouching | Girl |
|---|---|---|---|
| Global features | 0.155 | 0.406 | 0.413 |
| Local features | **0.127** | **0.373** | **0.381** |

**Table 3.** Reconstruction errors (Chamfer Distance) in geometric feature extraction using global features and local features.

As the point clouds are disordered, the global features cannot provide location-specific geometric information for the unclothed body templates, while our local features are extracted from the point clouds based on the UV position information. Table 3 reports the quantitative comparison. We can see that the approach based on local features has a significant improvement in the reconstruction accuracy compared to the global features, particularly on real data.

## 4.2 Comparison to State-of-the-Art Methods

**Evaluation of the Unclothed Parametric Model Estimation.** To validate our method for estimating unclothed human parametric models with meta-learning, we employ the approaches presented by Kanazawa et al. [11] and Wang et al. [29] as our baselines, both of which utilize deep learning techniques to predict SMPL parameters. We adopt the code provided by the authors and train the models on our datasets. Kanazawa et al. [11] demonstrates effective performance with the input of color image, so we substitute its feature encoder with PointNet++ to handle point cloud as input, referring to it as

**Table 4.** Reconstruction error (MAVE) with different methods for the estimation of unclothed parametric models from point clouds.

| Methods | Synthetic data | | | Real data | | |
|---|---|---|---|---|---|---|
| | CAPE | SURREAL | DFAUST | Kungfu | Crouching | Girl |
| Point-based HMR [11] | 44.18 | 48.98 | 47.01 | 50.10 | 49.12 | 50.35 |
| Wang et al. [29] | 25.51 | 29.33 | 28.35 | 30.93 | 28.33 | 31.49 |
| Our method | **21.50** | **22.85** | **24.13** | **27.86** | **23.59** | **26.44** |

error (in cm): 0 ▬▬▬ 10

(a)          (b)          (c)          (d)          (e)          (f)          (g)          (h)

**Fig. 5.** The visualization of the reconstruction accuracy (MAVE) of the different methods on CAPE, DFAUST and "Girl", "Kungfu". (a, e) Input point cloud. (b, f) Point-based HMR [11]. (c, g) Wang et al. [29]. (d, h) Our method.

**Point-based HMR.** Wang et al. [29] successfully predicts 3D human parametric models using a coarse-to-fine strategy, but necessitates self-supervised fine-tuning for the real data. All methods are tested on synthetic and real datasets, where the self-supervised fine-tuning method in [29] is used to obtain SMPL parameters as pseudo-labels for the real data. Qualitative comparisons are presented in Fig. 5, while quantitative results are outlined in Table 4. It is evident that the predicted parametric models from both Point-based HMR and Wang et al. [29] struggle to align well with the point clouds from unseen data due to domain gaps. Our results demonstrate a significant reduction in reconstruction error compared to the state-of-the-arts.

**Evaluation of the Clothed Human Model Estimation.** We further compare with [1, 30], and [16] on the reconstruction of dressed human models. These methods are also trained on our datasets. The reconstruction errors achieved by these different methods can be observed in Table 5, while Fig. 6 shows four reconstructed examples generated by utilizing these methods. Both Bhatnagar et al. [1] and Wang et al. [30] use implicit representations and SMPL-based parametric models to fit the surface, but they can only

**Table 5.** Reconstruction error (Chamfer Distance) obtained from various methods of detailed human reconstructions on synthetic and real data.

| Methods | CAPE | | Real data | | |
|---|---|---|---|---|---|
| | Subj 00215 | Subj 00159 | Kungfu | Crouching | Girl |
| Bhatnagar et al. [1] | 0.638 | 0.765 | 1.225 | 0.985 | 1.113 |
| Wang et al. [30] | 0.354 | 0.494 | 1.170 | 0.974 | 1.090 |
| Ma et al. [16] | 0.137 | 0.132 | 0.393 | **0.366** | 0.390 |
| Our method | **0.123** | **0.128** | **0.388** | 0.373 | **0.381** |

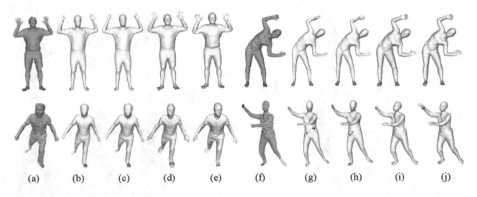

(a)    (b)    (c)    (d)    (e)    (f)    (g)    (h)    (i)    (j)

**Fig. 6.** The visual representation of the reconstruction results for the CAPE dataset (top row, "Subj 00215" and "Subj 00159") and the real dataset (bottom row, "Crouching" and "Girl"). (a, f) Input point cloud. (b, g) Bhatnagar et al. [1]. (c, h) Wang et al. [30]. (d, i) Ma et al. [16]. (e, j) Our method. Our method can recover the detailed human models that are consistent with the input point clouds in both human shape and pose.

represent rough clothing details and may produce impractical results, such as distorted arms. On the other hand, Ma et al. [16] produces a dense point cloud of the clothed body, where the points are grouped into local patches. It performs well on synthetic datasets but shows coarse details on real datasets due to domain gaps and random noise. In contrast, our approach captures fine and smooth local geometric features and efficiently reconstructs dressed human models with finer details, as demonstrated in Fig. 6. Furthermore, based on the prior knowledge learned through meta-learning, our approach can quickly adapt to the point clouds of unseen human bodies to achieve low reconstruction error and recover a more reasonable and complete detailed human model.

### 4.3   More Test on Real Data

To demonstrate the effectiveness of our approach on real data, we conduct experiments on real clothed body point clouds captured by a Kinect V2 sensor. The results obtained with our method are shown in Fig. 7. In the presence of significant occlusions and various deformations in the real dataset, our method shows robustness and accuracy in estimating both the human parametric models and the clothing details. The reconstruction

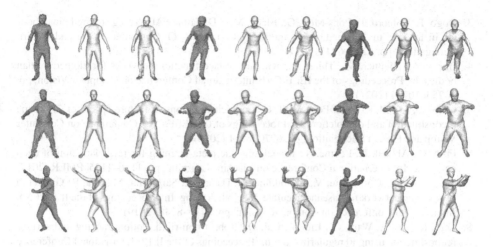

**Fig. 7.** Our method has been further validated on real data. Each result includes the raw depth scan, the predicted undressed model and the full detailed model. From top to bottom: "Crouching" [28], "Kungfu" [8] and "Girl" [28].

results demonstrate that our meta-learning based approach has strong generalization ability.

## 5    Conclusion

In this paper, we propose a novel hierarchical approach that utilizes meta-learning to accurately estimate dressed human models from single-frame partial point clouds. Our approach significantly improves the generalization ability and reconstruction accuracy on unseen data. First, we fine-tune a learned meta-model based on global features extracted over the entire point clouds to estimate unclothed parametric models from unseen human point clouds. Then, these unclothed parametric models are mapped to a dense point cloud representation with rich clothing details and flexible topology, on which we meta-learn clothing deformations with local geometric features. The proposed method has proven its effectiveness through the experimental results conducted on different datasets.

## References

1. Bhatnagar, B.L., Sminchisescu, C., Theobalt, C., Pons-Moll, G.: Combining implicit function learning and parametric models for 3D human reconstruction. In: Vedaldi, A., Bischof, H., Brox, T., Frahm, J.-M. (eds.) ECCV 2020. LNCS, vol. 12347, pp. 311–329. Springer, Cham (2020). https://doi.org/10.1007/978-3-030-58536-5_19
2. Bhatnagar, B.L., Tiwari, G., Theobalt, C., Pons-Moll, G.: Multi-Garment Net: learning to dress 3D people from images. In: Proceedings of the IEEE/CVF International Conference on Computer Vision, pp. 5420–5430 (2019)

3. Bogo, F., Romero, J., Pons-Moll, G., Black, M.J.: Dynamic FAUST: registering human bodies in motion. In: Proceedings of the IEEE Conference on Computer Vision and Pattern Recognition, pp. 6233–6242 (2017)

4. Burov, A., Nießner, M., Thies, J.: Dynamic surface function networks for clothed human bodies. In: Proceedings of the IEEE/CVF International Conference on Computer Vision, pp. 10754–10764 (2021)

5. Chibane, J., Alldieck, T., Pons-Moll, G.: Implicit functions in feature space for 3D shape reconstruction and completion. In: Proceedings of the IEEE/CVF Conference on Computer Vision and Pattern Recognition, pp. 6970–6981 (2020)

6. Finn, C., Abbeel, P., Levine, S.: Model-agnostic meta-learning for fast adaptation of deep networks. In: International Conference on Machine Learning, pp. 1126–1135. PMLR (2017)

7. Gundogdu, E., Constantin, V., Seifoddini, A., Dang, M., Salzmann, M., Fua, P.: GarNet: a two-stream network for fast and accurate 3D cloth draping. In: Proceedings of the IEEE/CVF International Conference on Computer Vision, pp. 8739–8748 (2019)

8. Guo, K., Xu, F., Wang, Y., Liu, Y., Dai, Q.: Robust non-rigid motion tracking and surface reconstruction using l0 regularization. In: Proceedings of the IEEE International Conference on Computer Vision, pp. 3083–3091 (2015)

9. Guo, K., Xu, F., Yu, T., Liu, X., Dai, Q., Liu, Y.: Real-time geometry, albedo, and motion reconstruction using a single RGB-D camera. ACM Trans. Graph. (ToG) 36(4), 32:1–32:9 (2017)

10. Jiang, H., Cai, J., Zheng, J.: Skeleton-aware 3D human shape reconstruction from point clouds. In: Proceedings of the IEEE/CVF International Conference on Computer Vision, pp. 5431–5441 (2019)

11. Kanazawa, A., Black, M.J., Jacobs, D.W., Malik, J.: End-to-end recovery of human shape and pose. In: Proceedings of the IEEE Conference on Computer Vision and Pattern Recognition, pp. 7122–7131 (2018)

12. Kingma, D.P., Ba, J.: Adam: a method for stochastic optimization. arXiv preprint arXiv:1412.6980 (2014)

13. Li, J., et al.: LiDARCap: long-range marker-less 3D human motion capture with lidar point clouds. In: Proceedings of the IEEE/CVF Conference on Computer Vision and Pattern Recognition, pp. 20502–20512 (2022)

14. Liu, G., Rong, Y., Sheng, L.: VoteHMR: occlusion-aware voting network for robust 3D human mesh recovery from partial point clouds. In: Proceedings of the 29th ACM International Conference on Multimedia, pp. 955–964 (2021)

15. Loper, M., Mahmood, N., Romero, J., Pons-Moll, G., Black, M.J.: SMP: a skinned multi-person linear model. ACM Trans. Graph. (TOG) 34(6), 1–16 (2015)

16. Ma, Q., Saito, S., Yang, J., Tang, S., Black, M.J.: SCALE: modeling clothed humans with a surface codec of articulated local elements. In: Proceedings of the IEEE/CVF Conference on Computer Vision and Pattern Recognition, pp. 16082–16093 (2021)

17. Ma, Q., et al.: Learning to dress 3D people in generative clothing. In: Proceedings of the IEEE/CVF Conference on Computer Vision and Pattern Recognition, pp. 6469–6478 (2020)

18. Ma, Q., Yang, J., Tang, S., Black, M.J.: The power of points for modeling humans in clothing. In: Proceedings of the IEEE/CVF International Conference on Computer Vision, pp. 10974–10984 (2021)

19. Min, C., Kim, T., Lim, J.: Meta-learning for adaptation of deep optical flow networks. In: Proceedings of the IEEE/CVF Winter Conference on Applications of Computer Vision, pp. 2145–2154 (2023)

20. Nichol, A., Achiam, J., Schulman, J.: On first-order meta-learning algorithms. arXiv preprint arXiv:1803.02999 (2018)

21. Patel, C., Liao, Z., Pons-Moll, G.: TailorNet: predicting clothing in 3D as a function of human pose, shape and garment style. In: Proceedings of the IEEE/CVF Conference on Computer Vision and Pattern Recognition, pp. 7365–7375 (2020)
22. Qi, C.R., Yi, L., Su, H., Guibas, L.J.: PointNet++: deep hierarchical feature learning on point sets in a metric space. In: Advances in Neural Information Processing Systems, vol. 30 (2017)
23. Ronneberger, O., Fischer, P., Brox, T.: U-Net: convolutional networks for biomedical image segmentation. In: Navab, N., Hornegger, J., Wells, W.M., Frangi, A.F. (eds.) MICCAI 2015. LNCS, vol. 9351, pp. 234–241. Springer, Cham (2015). https://doi.org/10.1007/978-3-319-24574-4_28
24. Saito, S., Huang, Z., Natsume, R., Morishima, S., Kanazawa, A., Li, H.: PIFu: pixel-aligned implicit function for high-resolution clothed human digitization. In: Proceedings of the IEEE/CVF International Conference on Computer Vision, pp. 2304–2314 (2019)
25. Shen, K., et al.: X-avatar: expressive human avatars. In: Proceedings of the IEEE/CVF Conference on Computer Vision and Pattern Recognition, pp. 16911–16921 (2023)
26. Sitzmann, V., Chan, E., Tucker, R., Snavely, N., Wetzstein, G.: MetaSDF: meta-learning signed distance functions. In: Advances in Neural Information Processing Systems, vol. 33, pp. 10136–10147 (2020)
27. Varol, G., et al.: Learning from synthetic humans. In: Proceedings of the IEEE Conference on Computer Vision and Pattern Recognition, pp. 109–117 (2017)
28. Wang, K., Xie, J., Zhang, G., Liu, L., Yang, J.: Sequential 3D human pose and shape estimation from point clouds. In: Proceedings of the IEEE/CVF Conference on Computer Vision and Pattern Recognition, pp. 7275–7284 (2020)
29. Wang, K., Zheng, H., Zhang, G., Yang, J.: Parametric model estimation for 3D clothed humans from point clouds. In: 2021 IEEE International Symposium on Mixed and Augmented Reality (ISMAR), pp. 156–165. IEEE (2021)
30. Wang, S., Geiger, A., Tang, S.: Locally aware piecewise transformation fields for 3D human mesh registration. In: Proceedings of the IEEE/CVF Conference on Computer Vision and Pattern Recognition, pp. 7639–7648 (2021)
31. Wang, S., Mihajlovic, M., Ma, Q., Geiger, A., Tang, S.: MetaAvatar: learning animatable clothed human models from few depth images. In: Advances in Neural Information Processing Systems, vol. 34, pp. 2810–2822 (2021)

# LCA-BERT: A Local and Context Fusion Sentiment Analysis Model Based on BERT

Xiaoyan Zhang[✉] and Zhuang Yan

School of Computer Science and Technology, Xi'an University of Science and Technology, Xi'an 710600, China
1366480684@qq.com

**Abstract.** Existing sentiment classification models often use static mask matrices for attention calculation when modeling local contexts, which may overlook or excessively focus on other neighboring tokens and fail to adequately represent contextual information in sentiment analysis models. In this paper, we propose a sentiment analysis model called LCA-BERT that enhances local information and integrates contextual information and attention mechanisms based on BERT. Specifically, we replace the static mask matrix in the self-attention network of the BERT model with a dynamic mask matrix, enabling the model to more effectively capture local information. Furthermore, we introduce the concepts of quasi-attention and deep-global context to mitigate the impact of textual noise while capturing contextual semantic information. Comparative experiments with baseline models demonstrate that our proposed model performs better in aspect-based sentiment analysis tasks.

**Keywords:** online comments · dynamic mask matrix · quasi-attention

## 1 Introduction

With the rapid advancement of the Internet era, the Internet has emerged as the primary channel for accessing information, gradually evolving into a vital platform for expressing viewpoints, venting emotions, and sharing life experiences. Particularly in the wake of the major outbreak of the COVID-19 pandemic, individuals have increasingly turned to online platforms such as Meituan, Twitter, and Weibo for commenting, enabling them to gain business insights and receive personalized recommendations based on the generated sentimental content [1]. Consequently, the analysis of sentiment-bearing comments on these web platforms, characterized by their immense volume, holds significant research value and practical applications. Therefore, the current focus of research is on determining how to automatically and efficiently analyze the underlying sentimental information conveyed in these texts and online comments.

However, the sentiments expressed in the texts and comments generated on these platforms are not always straightforward; more often than not, they tend to be complex and multifaceted. For example (as shown in Fig. 1), when it comes to a food delivery platform, a user might give a positive evaluation based on the delectable taste of the dishes while simultaneously complaining about the exorbitant prices. In such a scenario, the term 'delicious' represents a positive sentiment, whereas the mention of the price conveys a negative sentiment. Clearly, using a single sentiment polarity to classify the sentiment of this sentence would be highly inadequate. Therefore, scholars such as Jo [2] and Pontiki [3] have proposed a sentiment analysis task called Aspect-Based Sentiment Analysis (ABSA), which aims to identify the diverse sentiments associated with different aspects within the text.

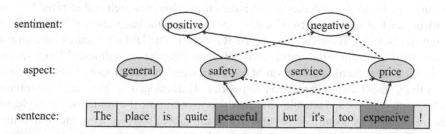

**Fig. 1.** The sentimental polarity of contextual information in different aspects.

In the early stages of research, deep neural networks were primarily used for aspect-based sentiment analysis. However, in recent years, the emergence of robust models such as Transformer and BERT, which leverage self-attention mechanisms as their core, has significantly enhanced the capacity for feature extraction. Scholars such as Sun, Huang, and Qiu [4], as well as Li [5], have applied pre-trained BERT models to aspect-based sentiment analysis, resulting in notable performance improvements. LCFS-BERT, introduced by Phan et al. [6], is a BERT model that incorporates local context-focused syntax. This model examines the syntactic relationships among words, enabling a better understanding of the specific context related to the target word. By identifying syntactic aspect words in sentences, utilizing self-attention mechanisms for semantic learning, and leveraging syntactic proximity to reduce the impact of loosely related words, the precision of the aspect word sentiment classifier is enhanced. Building on the widely used topic modeling technique called Latent Dirichlet Allocation (LDA), scholars have seamlessly integrated LDA with prior knowledge to enhance performance, as demonstrated by Xu et al. [7].

Beginning in 2019, researchers shifted their focus to co-attention mechanisms. Yang et al. [8] devised a method that alternates attention modeling between the target-level and the context-level, enabling a refined focus on pivotal aspects of the target and a more effective assimilation of contextual representations. In 2020, Xu et al. [9] discovered that viewpoints derived from BERT for aspect-based sentiment analysis tasks are not solely deduced from sentence context; instead, they are influenced by the domain or category to which the aspect pertains and the fine-grained semantics of the aspect itself.

The aforementioned research has significantly enhanced the performance of aspect-based sentiment analysis tasks and has paved the way for further advancements in aspect and sentiment analysis. However, despite these advancements, aspect-based sentiment analysis tasks still have several limitations. Pioneering research conducted by Shaw

[10], Yang [11], Guo [12], and others since 2018 has indicated that incorporating locality in experimental frameworks can lead to further improvements in model performance. Moreover, Fan et al.'s [13] study in 2019 highlighted the potential inadequacy of Transformer models in capturing local structures, possibly due to attention calculations in static masking matrices. Currently, the prevailing approach to context modeling involves the use of attention mechanisms. However, most of the semantic features derived after context modeling do not adequately capture contextual information, thereby hindering the overall effectiveness of sentiment classification by the model.

In response to the aforementioned issues in previous research and inspired by Fan's work on masked attention networks, this paper proposes the integration of Dynamic Mask Attention Network (DMAN) into the BERT model. The DMAN focuses on capturing features in local content that are semantically more relevant to sentiment polarity, thereby enabling a more flexible allocation of attention weights. The proposed model incorporates the Context-aware Perception Network [14] developed by Yang et al., combining context information and attention mechanisms to overcome the limitations of neglecting distant features by local features and to capture global features from multiple dimensions. This allows the model to fully consider contextual information from different linguistic backgrounds during feature extraction, thus obtaining more profound global semantic information.

Furthermore, the model employs effective methods to mitigate noise interference in sentiment classification. In the process of attention computation, it is observed that most existing research methods rely on the softmax function with a value range of [0, 1] to calculate attention weights for each position. Consequently, all hidden input vectors have a positive impact on the output vector, introducing certain noise interference. Inspired by Tay et al.'s "quasi" attention approach [15], this paper adopts the "quasi" attention for attention weight computation. By training the parameters of a linear layer, the model learns to dynamically allocate attention in different contextual contexts, with "+ 1", "0", and "−1" indicating "increase," "ignore," and "decrease" in attention, respectively. This effectively mitigates the influence of textual noise on the model.

Specifically, this paper proposes a sentiment analysis model called LCA-BERT, which enhances local modeling and integrates context information with attention mechanisms. The main contributions are as follows:

1. By introducing dynamic masking into the BERT model and replacing the static masking matrix, the issues of neglecting or excessively focusing on neighboring token information are mitigated. This allows the model to effectively capture local information.

2. By incorporating "quasi" attention into attention weight computation, the influence of hidden input vectors on output vectors is addressed. This enables attention weights to take negative values, transforming the value range of attention scores from [0, 1] to [−1, 1].

3. The paper employs a deep global context approach by combining context information with attention mechanisms. It integrates ordinary attention weights with "quasi" attention weights that incorporate fused context information. This approach fully utilizes the intrinsic information of the corpus to explore the connections between words in sentences.

## 2  Method

In order to enable the model to learn aspect-related information, this paper designs the aspect word sequence as aspect word phrases composed of either a single aspect word or multiple words. The objective of the proposed model is to predict the sentiment polarity expressed by different aspects in a sentence. The overall architecture of the model is illustrated in Fig. 2.

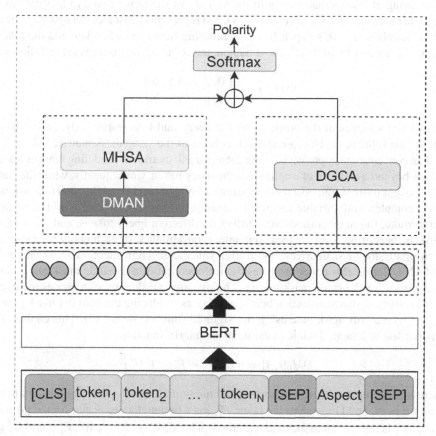

**Fig. 2.** Overall structure of the model

Suppose we are given a sentence s, which contains a word sequence {w1, w2,..., wn}, where some of the words are from the aspect list A, forming the aspect sequence sa. As shown in the example in Table 2, there exist two aspect terms, "price" and "safety," which are two different aspect sequences sa derived from the aspect list A within the entire sentence s. The objective of the proposed model in this paper is to predict the sentiment polarity expressed by sa within s. Firstly, this paper describes the application of a dynamic masking matrix to improve the local modeling of BERT. Secondly, the fusion of context information using a context-aware network is presented, which incorporates a deep global context attention computation method for capturing contextual information, as investigated by Yang et al. [14]. Finally, the utilization of "quasi" attention is described to reduce the interference of irrelevant words in sentiment classification, and relevant modifications are made to make it applicable in aspect-based sentiment analysis tasks.

## 2.1 Dynamic Mask Matrix

Based on the relevant content of Mask Attention Networks (MANs), this paper proposes the masking of tokens that are not in the vicinity of the target token to facilitate local semantic modeling. For instance, in Eq. (1), SM represents a distance-related static mask matrix established in this paper. If the relationship between each token and the tokens within a distance of b units is to be modeled, the static mask matrix is set as follows:

$$SM[t, s] = \begin{Bmatrix} 0, |t - s| > b \\ 1, |t - s| \leq b \end{Bmatrix} \tag{1}$$

where t and s represent the positions of the query and key, respectively, and SM[t, s] denotes the value at the t-th row and s-th column of the static mask matrix, SM.

The conventional approach of using a static mask matrix for modeling ignores tokens that are beyond a distance of b units from the query token. Consequently, while the static mask assigns more weight to a specific neighborhood, it lacks flexibility and fails to adapt to the complex and variable nature of semantic information in real-world scenarios. Furthermore, the neighborhood size differs for different query tokens, and the number of tokens benefiting from the local semantic representation of each query token varies. Additionally, the mask matrices should align with the different attention heads and layers in MANs. Hence, this paper proposes the introduction of a dynamic mask matrix as a replacement for the static mask matrix. The dynamic mask matrix integrates the query token, relative distance, attention heads, and layers, replacing the hard 0/1 mask gate in Eq. (1) with a soft mask gate using the sigmoid function in Eq. (2). This enables the construction of a more flexible dynamic mask matrix function.

$$DM_i^l[t, s] = \sigma(h_t^l W^l + P_{t-s}^l + U_i^l) \tag{2}$$

where s and t represent the positions of the query and key, respectively. i refers to the attention head, and l represents the layer. $P_{t-s}^l$ is a parameterized scalar for the position t and s, $U_i^l$ is the weight associated with the i-th head, $W^l \in \mathbb{R}^{d \times 1}$, $W^l$, $P_{t-s}^l$ and $U_i^l$ are trainable parameters.

## 2.2 Deep Global Context Attention Encoding

Unlike traditional attention encoding, this paper proposes Deep-Global Context Attention encoding (DGCA) based on the context fusion network introduced by Yang et al. [14] and the method of "quasi" attention for calculating attention weights to achieve context-aware attention encoding. This approach effectively explores contextual information, captures the relationships between words in the input sequence, improves the model's performance, and enhances the understanding of the relationships between different words in the sentence while mitigating the impact of textual noise on the model.

In the study by Yang et al. [14], different types of context vectors were utilized, including global context and deep context. These can be used separately or combined in the attention matrix. In traditional attention encoding, the focus is on the relationships between specific words in the input sequence to obtain their attention weights. Global context refers to a function that spans the entire input layer and represents the overall meaning of the sequence, sharing state across layers. Typically, deep learning models employ multiple hidden layers beneath the current input layer to capture various types and aspects of syntactic and semantic information, which is referred to as deep context. As mentioned earlier, in practical applications, deep context and global context can be combined to form the deep global context used in this paper. It leverages the advantages of both global and deep context to extract as much information as possible from vocabulary, syntax, and semantics. Peters et al. [16] demonstrated in their 2018 study that higher-level hidden layers are often utilized to capture contextually relevant aspects of word meaning, while lower-level hidden layers model various aspects of syntax. Therefore, capturing all these signals and incorporating the implied information into feature vectors is highly beneficial. Figures 3 and 4 illustrate the computational forms of traditional attention encoding and deep global context encoding, respectively.

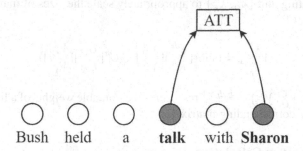

**Fig. 3.** Conventional Attention Mechanism

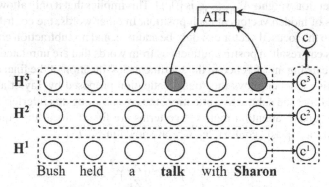

**Fig. 4.** Deep-Global Context Attention Mechanism

To enhance the contextual awareness of the self-attention model while maintaining the parallel computing flexibility of the self-attention network, the model incorporates contextual information into the original BERT model by modifying the query matrix Q and the key-value matrix K, as shown in Eq. (3).

$$[\begin{matrix} \hat{Q}^h \\ \hat{K}^h \end{matrix}] = (1 - [\begin{matrix} \lambda_Q^h \\ \lambda_K^h \end{matrix}])[\begin{matrix} Q^h \\ K^h \end{matrix}] + [\begin{matrix} \lambda_Q^h \\ \lambda_K^h \end{matrix}](C^h[\begin{matrix} U_Q \\ U_K \end{matrix}]) \tag{3}$$

where $C^h \in \mathbb{R}^{n \times d}$ represents the global context matrix for each head, $\{\lambda_Q^h, \lambda_K^h\} \in \mathbb{R}^{n \times l}$ is allocated to weight the feature vectors of the context, and $\{U_Q, U_K\} \in \mathbb{R}^{d_c \times d_h}$ denotes the weights of the linear layer used to transform the input context matrix $C^h$. In this paper, the tanh activation function is chosen instead of the sigmoid function used in the initial implementation by Yang et al.

This choice is made because tanh allows for both positive and negative contributions of context to Q and K, enriching the representation space of these two matrices and the resulting attention distribution. Previous research by Vaswani et al. [17] and Britz et al. [18] has noted that the tanh function can increase the magnitudes of $\widehat{Q}^h$ and $\widehat{K}^h$, and excessively large magnitudes of Q and K may push gradients to very small values, potentially hindering model learning. To mitigate this effect, this paper employs a symmetric gating unit $\{\lambda_Q^h, \lambda_K^h\}$ to appropriately scale the sizes of matrices Q and K, as shown in Eq. (4).

$$[\begin{matrix} \lambda_Q^h \\ \lambda_K^h \end{matrix}] = \tanh([\begin{matrix} Q^h \\ K^h \end{matrix}][\begin{matrix} V_Q^h \\ V_K^h \end{matrix}] + C^h[\begin{matrix} U_Q \\ U_K \end{matrix}][\begin{matrix} V_Q^C \\ V_K^C \end{matrix}]) \tag{4}$$

where $\{V_Q^h, V_Q^h, V_Q^C, V_K^C\} \in \mathbb{R}^{d_h \times 1}$ represents the trainable weights of a linear layer used to transform the corresponding matrix.

## 2.3 Quasi Attention Calculation

In the basic BERT model, attention calculation uses the softmax function, and the domain of the self-attention weights $A_{\text{Quasi-Attn}}^h$ is [0, 1]. This implies that it only allows for convex combinations of hidden vectors at each position. In other words, the contribution of the hidden state to the focused vector can only be additive, and no subtraction contribution is allowed. This can result in positive influences from words that are unrelated to the target aspect word, causing interference in the sentiment classification of the final aspect word. Therefore, this model employs the "quasi" attention proposed by Tay et al. in 2019 to learn both additive and subtractive attention, thus overcoming this interference. In order to derive the "quasi" attention matrix, two terms are first defined: the "quasi" context query $C_Q^h$ and the "quasi" context key $C_K^h$, as shown in Eq. (5).

$$[\begin{matrix} C_Q^h \\ C_K^h \end{matrix}] = C^h[\begin{matrix} Z_Q \\ Z_K \end{matrix}] \tag{5}$$

where $\{Z_Q, Z_K\} \in \mathbb{R}^{d_c \times d_h}$ represents the weight matrix obtained through linear transformation of the original context matrix, and $C^h$ represents the same context matrix as in Eq. (3). Therefore, the "quasi" attention matrix in this study can be derived and defined as shown in Eq. (6). Subsequently, the gating factor is formulated as bidirectional, leading to the derivation of the weight coefficient formula for the global context, denoted as Eq. (7).

$$A^h_{\text{Quasi - Attn}} = \alpha \cdot \text{sigmoid}(\frac{f_\psi(C^h_Q, C^h_K)}{\sqrt{d_h}}) \tag{6}$$

$$\lambda^h_A = 1 - (\beta \cdot \lambda^h_Q + \gamma \cdot \lambda^h_K) \tag{7}$$

In this equation, $\alpha$ is a scaling factor, and $f_\psi(\cdot)$ is a similarity measure used to capture the similarity between $C^h_Q$ and $C^h_K$. For simplicity, we parameterize $f_\psi$ using dot product. $\{\beta, \gamma\}$ are scalar values that control the weights of each component, regulating the weight coefficients of the "quasi" attention and adjusting the relative importance of global context information within the overall information. Finally, this study formulates the new attention matrix as a linear combination of the conventional softmax attention matrix and the "quasi" attention matrix:

$$\hat{A}^h = A^h_{\text{Self - Attn}} + \lambda^h_A A^h_{\text{Quasi - Attn}} \tag{8}$$

where $\lambda^h_A$ is a scalar that represents a compositional factor controlling the impact of context on attention computation. As evident from Eq. (4), the values of $\{\lambda^h_Q, \lambda^h_K\}$ range from 0 to 1, implying that the values of $\lambda^h_A$ fall within the range of -1 to 1. Therefore, by observing Eq. (8), it becomes apparent that there is a balanced contribution between conventional sentence information and global context information.

### 2.4 Classification

In this study, the first vector (i.e., [CLS]) of the output from the last layer of the BERT model is utilized as the input to the final classification layer, which is consistent with previous research [12]. For a given input sentence, this vector is referred to as $e_{CLS} \in \mathbb{R}^{1 \times d}$. Then, the probability of each emotion class y is determined using $y = softmacx(e_{CLS}A^h_{Self\text{-Attn}})$, where $e_{CLS} \in \mathbb{R}^{c \times d}$ represents the weight values of the classification layer and $y \in \mathbb{R}^{1 \times d}$. Finally, the label with the highest probability is chosen as the final prediction.

## 3 Experiment

### 3.1 DataSet

The Sentihood dataset, which comprises 5215 English sentences, was utilized in this study. It is a publicly available dataset derived from Yahoo Answers. Among the sentences, 3862 contain a single target, while 1353 contain multiple targets. The dataset encompasses sentiment polarity labels, namely none, positive, and negative, as well as aspect labels including general, price, safety, and transit-location. Following a similar approach as referenced in [3], this paper defines two subtasks for each dataset: (1) aspect classification and (2) aspect-based sentiment classification. In aspect classification, the objective is to determine whether an aspect "a" is mentioned in the input sentence with the target "t". For aspect-based sentiment classification, the aim is to identify the given model aspect from the input sequence and predict its polarity.

## 3.2 Experimental Setup

The code for this paper was implemented using Python programming language, specifically version 3.6.4. The PyTorch deep learning framework, version 1.10.2 + cu113, developed by Facebook, was employed. The experimental environment configuration details are provided in Table 1.

In the experiments conducted in this paper, a pre-trained BERT model was utilized for fine-tuning. Similar to the original BERT model, the model in this paper consists of 12 heads and 12 layers, with a hidden size of 768. For model training in this paper, Adam was used as the optimizer with an initial learning rate of $2e-5$. The batch size was set to 24, the dropout probability to 0.1, and the number of epochs to 20. Pre-trained weights from the BERT-uncased model were utilized.

**Table 1.** Experimental environmental parameters

| Parameter | Value |
|---|---|
| CPU | AMD Ryzen5 5600X |
| GPU | RTX3060 |
| Memory | 32 GB |

## 3.3 Experimental Results and Analysis

**Comparative Experiment**

Targeted aspect-based sentiment analysis aims to determine the sentiment expressed towards a specific aspect category concerning a given entity (target or aspect term) in a sentence. Consider the example sentence, "The environment of this hotel is excellent, but the prices are too high. However, the food in the restaurant is delicious and affordable." In this sentence, there is a negative sentiment expressed towards the aspect of "hotel prices" and positive sentiments towards aspects such as "hotel environment," "food prices," and "food taste." Such complex sentences require a higher level of proficiency from the model in recognizing the sentiment polarity associated with the target entity in the specified aspect.

To accurately and comprehensively evaluate the proposed model, this paper compares it with several representative models in the field of sentiment analysis. The selected models include classical logistic regression (LR) model with multi-class classification, H-LSTM model proposed by Ruder S, a hierarchical bidirectional long short-term memory network [19], Dmu-Entnet [20], BERT-single [21], LCFS-BERT [6], CG-BERT [22], and QACG-BERT [22]. The evaluation metrics chosen for comparison are accuracy, F1 score, and AUC score. In the conducted experiments, a detailed comparison of all the aforementioned models and the proposed model on the Sentihood dataset is presented in Table 2.

From Table 2, it can be observed that among the non-BERT-based baseline models, the logistic regression classifier with grammatical features performs surprisingly well

**Table 2.** Comparison of experimental results

| Model | Aspect | | | Sentiment | |
|---|---|---|---|---|---|
| | Acc | F1 | AUC | Acc | AUC |
| LR | – | 39.3 | 92.4 | 87.5 | 90.5 |
| LSTM | – | 68.9 | 89.8 | 82.0 | 85.4 |
| Dmu-Entnet | 73.5 | 78.5 | 94.4 | 91.0 | 94.8 |
| BERT-single | 73.7 | 81.0 | 96.4 | 85.5 | 84.2 |
| LCFS-BERT | 79.2 | 87.1 | 97.0 | 92.5 | 95.0 |
| CG-BERT | 79.7 | 87.1 | 97.5 | 93.7 | 97.2 |
| QACG-BERT | 79.9 | **88.6** | 97.3 | 93.8 | 97.8 |
| LCA-BERT | **80.3** | 88.4 | **97.6** | **94.5** | **97.8** |

**Table 3.** Results of ablation experiment

| Model | ablation | Aspect | | | Sentiment | |
|---|---|---|---|---|---|---|
| | | Acc | F1 | AUC | Acc | AUC |
| LCA-BERT | w/o DGCA | 74.7 | 83.5 | 96.7 | 89.4 | 87.2 |
| | w/o DMAN | 78.3 | 87.6 | 97.3 | 94.2 | 96.8 |
| | DGCA&DMAN | **80.3** | **88.4** | **97.6** | **94.5** | **97.8** |

in the aspect-based sentiment analysis task. Despite being an early method, it outperforms many deep learning models that were introduced later. However, considering the extensive feature engineering required, this approach is challenging to apply in current high-cost production scenarios. Therefore, researchers have made significant efforts to leverage long short-term memory networks (LSTM) and have achieved decent progress in aspect identification tasks. However, due to the limitations of the LSTM structure itself, they have not achieved significantly better performance than conventional machine learning methods in aspect-based sentiment classification.

The BERT model achieves excellent results in sentiment analysis tasks by predicting masked subwords, leading to a nearly 10% improvement in the accuracy metric. Subsequently, models such as LCFS-BERT, CG-BERT, and QACG-BERT leverage the strengths of the BERT model and incorporate network structures that prioritize aspect term recognition and attention weight allocation. These models also demonstrate outstanding performance. Finally, the proposed LCA-BERT model in this paper enhances the flexibility of recognizing local information while preserving the deep global contextual features. It introduces sub-modules specifically tailored to aspect sentiment analysis and achieves better results across all evaluation metrics for sentiment classification. The experimental results showcase the superior performance of the LCA-BERT model, which is designed based on the BERT model, in aspect-based sentiment analysis tasks.

**Analysis of Module Importance**

To further analyze the importance of each module structure in the LCA-BERT model, this paper conducted ablation experiments for validation. The corresponding experimental results are presented in Table 3. The results in Table 3 indicate that both the DMAN module, responsible for enhancing local modeling, and the DGCA module, responsible for acquiring contextual information, play important roles in aspect sentiment analysis tasks within the model. The ablation experiments, which involved removing each submodule mentioned above, demonstrated the effectiveness of the internal structure of the model for the tasks addressed in this paper.

## 4 Conclusion

This paper proposes a context-aware attention model that enhances local modeling and integrates contextual information for aspect sentiment analysis tasks. The model replaces the static mask matrix with a dynamic mask matrix, applies "quasi" attention weights, and introduces a context-aware network to fuse contextual feature information into the trained BERT model. Based on the experimental results and the analysis, it can be concluded that the proposed model significantly improves the performance of the BERT model in aspect sentiment analysis tasks and achieves good results. However, there is still room for improvement in aspect term recognition, potentially due to the interference between local and contextual information. Future research can focus on deeper investigation into feature fusion and noise reduction techniques to further enhance the model's performance.

## References

1. Kang, H., Yoo, S.J., Han, D.: Senti-lexicon and improved Nave Bayes algorithms for sentiment analysis of restaurant reviews. Expert Syst. Appl. **39**(5), 6000–6010 (2012)
2. Jo, Y., Oh, A.H.: Aspect and sentiment unification model for online review analysis. In: King, I., Nejdl, W., Li, H., WSDM 2011, pp. 815–824. ACM Press, Hong Kong (2011)
3. Pontiki, M., Galanis, D., Papageorgiou, H.: SemEval-2016 task 5: aspect based sentiment analysis. In: Bethard, S., Carpuat, M., Cer, D. SemEval-2016, pp.19–30. Association for Computational Linguistics, San Diego, California (2016)
4. Sun, C., Huang, L., Qiu, X.: Utilizing BERT for aspect-based sentiment analysis via constructing auxiliary sentence. In: Jill, B., Christy, D., Thamar, S., (eds.) NAACL HLT 2019, pp. 380–385. Association for Computational Linguistics, Minneapolis, Minnesota (2019)
5. Li, X., Bing, L., Zhang, W., Lam, W.: Exploiting BERT for end-to-end aspect-based sentiment analysis. In: Wei, X., Alan, Ritter., Tim, B., Afshin, R., (eds.). W-NUT 2019, pp. 34–41. Association for Computational Linguistics, Hong Kong, China (2019)
6. Phan, M.H., Ogunbona, P.O.: Modelling context and syntactical features for aspect-based sentiment analysis. In: Jurafsky, D., Chai, J., Schluter, N., Tetreault, J. (eds.) ACL 2020, pp. 3211–3220. Association for Computational Linguistics, Online (2020)
7. Hua, X., Fan, Z., Wei, W.: Implicit feature identification in Chinese reviews using explicit topic mining model. Knowl.-Based Syst. **76**(1), 166–175 (2015)

8. Yang, C., Zhang, H., Jiang, B., Li, K.: Aspect-based sentiment analysis with alternating coattention networks. Inf. Process. Manage. **55**(3), 463–478 (2019)

9. Xu, H., Shu, L., Philip, S.Y., Liu, B.: Understanding pre-trained BERT for aspect-based sentiment analysis. In: COLING 2020. International Committee for Computational Linguistics, Barcelona, Spain, pp. 244–250 (2020)

10. Shaw, P., Uszkoreit, J., Vaswani, A.: Self-attention with relative position representations. In: Walker, M., Ji, H., Stent, A., (eds.) NAACL HLT 2018, pp. 464–468. Association for Computational Linguistics, New Orleans, Louisiana (2018)

11. Yang, B., Tu, Z., Derek, F.W., Meng, F., Lidia, S.C., Zhang, T.: Modeling localness for self-attention networks. In: EMNLP 2018. Association for Computational Linguistics, Brussels, Belgium, pp. 4449–4458 (2018)

12. Guo, M., Zhang, Y., Liu, T.: Gaussian transformer: a lightweight approach for natural language inference. In: Pascal, V.H., Zhou, Z., (eds.) AAAI 2019, pp. 6489–6496. AAAI press, Honolulu, Hawaii, USA (2019)

13. Fan, Z., et al.: mask attention networks: rethinking and strengthen transformer. In: Toutanova, K., et al. (eds.). NAACL HLT 2021, pp. 1692–1701. Association for Computational Linguistics, Online (2021)

14. Yang, B., Li, J., Wong, D.F., Chao, L.S., Wang, X., Tu, Z.: Context-Aware Self-Attention Networks. In: AAAI 2019, vol. 33, no. 01, pp. 387–394. AAAI press, Honolulu, Hawaii, USA (2019)

15. Yi, T., Luu, A.T., Zhang, A., Wang, S., Hui, S.C.: Compositional de-attention networks. In: Proceedings of the 33rd International Conference on Neural Information Processing Systems, pp. 6135–6145. Curran Associates Inc, Red Hook, NY, USA (2019)

16. Matthew, E.P., et al.: Deep contextualized word representations. In: NAACL 2018, vol. 1, pp. 2227–2237. Association for Computational Linguistics, New Orleans, Louisiana (2018)

17. Ashish, V., et al.: Attention is all you need. In: 395 NIPS 2017, pp. 6000–6010. Curran Associates Inc, Red Hook, NY, USA (2017)

18. Britz, D., Goldie, A., Luong, M., Le, Q.: Massive exploration of neural machine translation architectures. In: EMNLP 2017, pp.1442–1451. Association for Computational Linguistics, Copenhagen, Denmark (2017)

19. Ruder, S., Ghaffari, P., Breslin, J.G.: A hierarchical model of reviews for aspect-based sentiment analysis. In: EMNLP 2016, pp. 999–1005. Association for Computational Linguistics, Austin, Texas (2016)

20. Liu, F., Cohn, T., Baldwin, T.: Recurrent entity networks with delayed memory update for targeted aspect-based sentiment analysis. In: NAACL 2018, vol. 2, pp. 278–283. Association for Computational Linguistics, New Orleans, Louisiana (2018)

21. Devlin, J., Chang, M.W., Lee, K., Toutanova, K.: BERT: pre-training of deep bidirectional transformers for language understanding. In: NAACL 2019, vol. 1, pp. 4171–4186. Association for Computational Linguistics, Minneapolis, Minnesota (2019)

22. Wu, Z., Ong, D.C.: Context-guided bert for targeted aspect-based sentiment analysis. In: AAAI 2021, vol. 35, no. 16, pp. 14094–14102. AAAI press, Online (2021)

# Author Index

Printed in the United States
by Baker & Taylor Publisher Services